AN INTRODUCTION TO SPECIAL RELATIVITY FOR RADIATION AND PLASMA PHYSICS

This textbook introduces the topic of special relativity, with a particular emphasis upon light–matter interaction and the production of light in plasma. The physics of special relativity is intuitively developed and related to the radiative processes of light. The book reviews the underlying theory of special relativity, before extending the discussion to applications frequently encountered by postgraduates and researchers in astrophysics, high-power laser interactions, and the users of specialized light sources, such as synchrotrons and free electron lasers. A highly pedagogical approach is adopted throughout, and numerous exercises are included within each chapter to reinforce the presentation of key concepts and applications of the material.

GREG TALLENTS is Professor in Physics at the York Plasma Institute at the University of York. He completed a PhD in Engineering Physics at the Australian National University in the late 1970s and has since researched the physics and applications of laser-produced plasmas and short wavelength plasma lasers. He has written a textbook *An Introduction to the Atomic and Radiation Physics of Plasmas* published in 2018 by Cambridge University Press.

AN INTRODUCTION TO SPECIAL RELATIVITY FOR RADIATION AND PLASMA PHYSICS

GREG TALLENTS

University of York

CAMBRIDGE UNIVERSITY PRESS

CAMBRIDGE
UNIVERSITY PRESS

Shaftesbury Road, Cambridge CB2 8EA, United Kingdom

One Liberty Plaza, 20th Floor, New York, NY 10006, USA

477 Williamstown Road, Port Melbourne, VIC 3207, Australia

314–321, 3rd Floor, Plot 3, Splendor Forum, Jasola District Centre,
New Delhi – 110025, India

103 Penang Road, #05-06/07, Visioncrest Commercial, Singapore 238467

Cambridge University Press is part of Cambridge University Press & Assessment,
a department of the University of Cambridge.

We share the University's mission to contribute to society through the pursuit of
education, learning, and research at the highest international levels of excellence.

www.cambridge.org
Information on this title: www.cambridge.org/9781009236065

DOI: 10.1017/9781009236072

First published 2023

A catalogue record for this publication is available from the British Library.

ISBN 978-1-009-23606-5 Hardback

Contents

Preface

This book introduces special relativity with an emphasis on the ways relativity impacts the interaction of light with matter, particularly the interaction of light with plasma material. Often called the fourth state of matter after solids, liquids, and gases, plasma is created when a gas is heated, so that free electrons and ions are produced. Free electrons, in particular, interact strongly with light and other electromagnetic radiation.

Rather than a treatment with a collection of formulas, the book emphasizes the physics of relativity as related to the radiative processes of light. Prior exposure to special relativity is not assumed. Many researchers and students undertake preliminary courses where the results of, for example, time dilation and length contraction for objects moving at speeds approaching the speed of light are introduced. However, when relativistic physics is needed for research or more advanced courses, the jump required in understanding can be a hurdle, partly due to the use in many texts of tensor field mathematics (as a way of paving the way for studies of general relativity). The treatment in this book sets out the ideas of special relativity assuming no prior relativistic study and uses US junior and European year 2 undergraduate mathematics. Initial chapters end with the idea of the Lorentz transformation of time and space between observers with different velocities. Electric and magnetic fields, particle acceleration, and the "beaming" effect of high velocity on the emission of light are then treated. Later chapters deal with the interaction of light with plasma and the use of magnetic fields for light production (with specific treatment of synchrotrons and free electron lasers). The final chapters present detailed treatments of plasma radiative processes, collisions, and relativistic optics. The aim is to present special relativity and the way special relativity impacts radiative processes in astrophysics and laboratory plasmas as a continuous and interconnected story using the mathematics commonly used by experimental physicists.[1]

---

[1] The mathematical knowledge assumed includes an understanding of vector and matrix operations, for example, the operations ∇^2, $\nabla \times$, and $\nabla \cdot$ on vectors. Standard calculus is also used extensively, for example,

An understanding of radiation physics is needed to interpret astrophysical observations and laboratory plasma diagnostics. Most of the observable universe is plasma and as the universe is expanding rapidly, special relativity changes light interactions with astrophysical plasmas which are often large distances apart with large relative velocities. Laboratory plasmas can also be heated to sufficiently high temperatures that relativistic effects are important. High-power laser systems focused onto solid, liquid, or gas targets can produce plasma material with the electrons moving at speeds approaching the speed of light. My earlier book [110] considered atomic and radiative processes in plasmas without dealing in any detail with the effects of special relativity. This book usually presents a nonrelativistic treatment as an introduction before dealing with relativistic effects. In some cases, relativistic "corrections" to nonrelativistic formulas are presented.

It is useful to potential readers to point out what is not covered in this book. Relativity theories are particularly important in three areas of modern physics with the labels of (i) special relativity, (ii) general relativity, and (iii) quantum electrodynamics. We treat the first of these. Special relativity deals with the physics of macroscopic bodies and particles where the quantum nature is not important. Special relativity departs from classical Newtonian physics when velocities relative to an observer are a significant fraction of the speed of light.

To understand gravity, it is necessary to invoke "general relativity." General relativity is not treated in this book. The theory of general relativity explains the gravitational attraction of masses as arising from the warping of space and time. Noninteracting massless photons propagate in straight lines in special relativity, but in general relativity, photon propagation paths follow space warped by the presence of mass. Time is also slowed near a large mass. The other relativity topic not covered is quantum electrodynamics which we can label as relativistic quantum mechanics. Relativistic quantum mechanics brings special relativity and quantum mechanics together and, for example, uses the Dirac equation rather than the Schrodinger equation to determine behavior on the atomic scale. Whereas a wavefunction for the Schrodinger equation is a complex variable of space and time, the Dirac equation requires wavefunctions with four elements and has components which are 4×4 matrices.

Numerous exercises are included at the end of each chapter. These are an essential part of the text as they allow self-tests of the material, illustrate the applications of the material discussed in the text, and show the wider relationships of the material. Some exercises are a bit of fun. For example, the twin paradox problem of

to sum up angular variations of light emission. The reader is not assumed to have prior knowledge of tensor analysis, which is introduced and briefly utilized (Sections 2.10, 3.1, and 3.2). Similarly, Lagrangian mechanics applied to relativistic kinetics is introduced (Section 5.1), but other approaches are used to evaluate, for example, the expression for the relativistic energy of a body (see Section 2.7).

Exercise 3.1 is an unlikely "thought experiment" with a detailed solution presented in Chapter 10. Originally posed more generally by Einstein, this exercise was set out as a problem question in 1962 in the textbook by Jackson [58]. A twin born in 1980 embarks on a rocket ship in the year 2000 and returns in 2020 (their time). According to everyone on Earth, the year of return is 2336 due to the relativistic time difference between the rocket and Earth. Other exercises have numerical solutions or, perhaps, a comment regarding the wider aspects of the exercise in brackets.

The book presents material using the International System of Units (SI). The convention used in particle and plasma physics is followed, where energies are expressed in units of electron volts (eV), where $1 \text{ eV} = 1.6 \times 10^{-19}$ joules. In plasma physics, temperatures T are often also expressed in energy units of $k_B T$ and quoted in electron volts. Here, k_B is Boltzmann's constant. A temperature of 1 eV corresponds to 11,605 degrees Kelvin.

This is a textbook rather than a research monograph. The book presents the development of the scientific discussion and the detailed derivation of formulas related to special relativity in more detail than is usually given in a research monograph. To emphasize the physics, simple expressions are often determined, but the presented mathematics shows each significant step. If a reader wants a formula or result without going through a derivation, it is possible to skip many of the mathematical steps presented, particularly in the later chapters.

As this is a textbook, the citations are representative and do not attempt to give the developmental history or the detailed current state of research on a topic. I apologize in advance to those whose contribution to knowledge is described but not properly cited. Some of the material included has been developed over several years in lecture courses presented at the University of York. Some material presented in published journal articles and in the classic textbooks by Clemmow and Dougherty [17], Feynman [40], Jackson [58], and Rybicki and Lightman [101] is described.

Following the historical development of ideas in physics is often not a good way to learn a subject. Subsequent work, well after a theory is accepted, usually follows a simpler and more logical development path to a necessary formula or other result. However, relativity is closely associated with Einstein and, unusually, his treatments are still close to the optimum way to present special relativity. There is a short biography of Albert Einstein at the start of Chapter 1 and some details of the lives of other relevant scientists are presented later, particularly where there is a physical effect named after a person.

I am grateful to the Department of Physics, University of York, for the opportunity to write this book. I would like to thank Christine for finding some interesting books on Einstein's life and for helping me to work on this book throughout the pandemic lockdowns that afflicted the world in the years 2020–2022.

1

A Brief Introduction to Special Relativity and Light

Special relativity is a theory that explains the space and time relationships between bodies moving at different velocities. Special relativity allows for situations where observers of the kinetics of bodies are moving at significantly different velocities to the bodies. A significantly different velocity needs to be usually within an order of magnitude or so of the very high velocity of light in vacuum.

The theory of special relativity supercedes the laws of motion developed by Isaac Newton (1643–1727). Bodies behave in a classical or Newtonian manner if the observer has a velocity close to that of the body and, for many applications, Newtonian physics is more than adequate. Newton's first law states that an object will remain at rest or in uniform motion in a straight line unless acted upon by an external force. His second law gives the manner of change of the velocity, or in other words, the acceleration of the body when acted upon by a force. When subject to a force $\mathbf{F}$, the acceleration $\mathbf{a}$ of a body of mass m is related to the force by $\mathbf{F} = m\mathbf{a}$. Newton's second law is often more generally expressed in terms of the momentum $\mathbf{p}$ of a body with the force $\mathbf{F}$ on the body related to the rate of change in time of the momentum $\mathbf{p}$ by $\mathbf{F} = d\mathbf{p}/dt$. Allowing $m\mathbf{a}$ or equivalently $d\mathbf{p}/dt$ to be a force, Newton's third law states that forces occur in equal but oppositely directed pairs.

Special relativity is based on the postulates that: (a) the laws of physics are the same for all observers,[1] and (b) the speed of light in a vacuum is the same for all observers regardless of their relative motion or the motion of the light source. Consequences of the constant speed of light are that to a stationary observer, the passing of time on a moving body appears to slow (known as time dilation) and distances in the direction of the moving body velocity appear shorter (known as length contraction). The logical progression to show that the constant speed of light

---

[1] There is a caveat for both Newtonian mechanics and special relativity regarding the constancy of the laws of physics for all observers. An accelerating observer experiences "fictional" forces such as the centrifugal force which depend on their acceleration. The laws of physics are taken to be the same for all observers if the observer is not subject to fictional forces.

leads to time dilation and length contraction is first outlined in Section 1.2 and further explored for different scenarios throughout this book.

There are theoretical deductions and a vast array of experimental measurements showing that the speed of light in vacuum is constant. For example, Maxwell's equations were developed to summarize relationships between electric charges and electric and magnetic fields. Electromagnetic waves are predicted by Maxwell's equations to have a velocity of propagation in vacuum solely dependent on some constants of proportionality between electric charges and oscillating electric and magnetic fields (the dielectric constant ϵ_0 and magnetic permeability μ_0). The electromagnetic wave velocity proportionality to electrical and magnetic quantities gives a *primae facie* case for the velocity to be a constant of nature (see Section 1.1).

This book deals particularly with the relativistic aspects of light interactions with plasma material, where "light" refers to any electromagnetic wave. The word "light" is popularly used to refer to visible electromagnetic waves (wavelengths 400−700 nm) and sometimes ultraviolet and infrared electromagnetic radiation. Other forms of electromagnetic radiation such as X-rays, microwaves, and radio waves exhibit many identical properties in vacuum, including the velocity of propagation. Differences between electromagnetic waves anywhere on the spectrum from X-ray to radio waves mainly arise when the waves interact with plasma or other states of matter. A plasma is created when a gas is heated and produces negatively charged free electrons and positively charged ions. Electromagnetic waves interact strongly with the free electrons in particular.

Most of the observable universe is made up of plasma material. Densities range from stellar interiors to the low density of intergalactic space (with approximately one ion and one free electron per cubic meter). The astrophysical and cosmological properties of the universe have been determined using the emission, absorption, scattering, and interaction of light with plasma. We consider these processes in this book with a special emphasis on the effects of special relativity. Special relativistic effects associated with the interaction of light with astrophysical plasmas are important when astrophysical objects are hot and when objects have large relative velocities. The expanding universe produces important relative velocity effects on cosmological length scales, for example, between galaxies.

Laboratory plasmas created when high-power lasers are focused onto solid, liquid, or gas targets produce expanding plasmas with densities ranging from many times solid state material densities down to densities obtained with laboratory vacuum equipment. Lasers can accelerate electrons to high velocity where relativistic effects are important. Some of the phenomena seen in astrophysical plasmas can be replicated in laboratory plasmas, enabling laboratory investigation of astrophysical phenomena [16], [78]. Relativistic laboratory plasmas are also useful sources of high-energy particles and high photon energy radiation [13], [73], [112]. Inertial confinement fusion driven by lasers seeks to compress the deuterium (D) and tritium (T) isotopes of hydrogen to high densities. Compression and heating of DT mixtures may produce "fusion-burning" plasmas exhibiting some of the relativistic effects seen in the early universe [99].

The theory of special relativity was developed by Albert Einstein in the early 20th century. Einstein is probably the most significant physicist of all time, though some may still credit Newton with this accolade. Einstein is certainly the most readily recognized physicist of all time. Einstein's name and image are instantaneously identified by almost everyone. Anyone with a passing interest in science has heard of Einstein's famous equation relating the mass m of a body with an energy: $E = mc^2$. Einstein's equivalence of mass and energy arises because the momentum of a body moving at a velocity $\mathbf{v}$ measured by a stationary observer increases due to time dilation and length contraction to a value above the momentum $m\mathbf{v}$ predicted by Newtonian mechanics. The new relationship between energy and momentum shows that even with zero momentum, there is a residual energy mc^2 (see Section 2.8).

Albert Einstein was born on March 14, 1879, in Ulm, Germany. He was 26 in 1905 and employed by the Swiss Patent Office in Bern when he published two papers setting out the theory of special relativity [32], [35]. Another paper in that year helped confirm the idea that light comes in "quanta," now referred to as photons [34]. In his "photons paper," Einstein explained the photo-electric effect where a threshold light frequency is required to ionize an atom and release an electron. A fourth paper dealt with the Brownian motion of atoms and led physicists to accept the existence of atoms [33]. As these fundamental papers in the development of modern physics appeared in the same year, the year 1905 has been termed Einstein's "annus mirabilis" ("the miracle year" in English). Einstein became director of the Kaiser Wilhelm Physical Institute and professor at the University of Berlin in 1914. By 1916, in order to explain gravitational fields, Einstein had extended the theory of special relativity to the theory of general relativity. In 1917, he applied the general theory of relativity to model the structure of the universe.

The theories of relativity are the towering monuments to Einstein, but he also developed other ideas important to radiation and statistical mechanics. Einstein proposed the idea of stimulated emission and the relationships between radiation emission and absorption processes [36], [37]. We treat these concepts in Chapter 7. Stimulated emission stayed as a process sometimes important in radiation theory until the invention of the laser in 1960 when its practical use started to become evident. Laser is an acronym for light amplification by the stimulated emission of radiation. Later Einstein supported the Indian physicist Satyendra Bose (1894–1974) in the development of the distribution function for photons, now known as the Bose-Einstein distribution function (see Section 6.1). In 1924 Einstein translated

Bose's article on particle distributions into German for the *Zeitschrift für Physik* after it was rejected by an English-language journal, added his own supporting paper, and ensured publication of the work.

The Nobel prize in physics for 1921 was awarded to Albert Einstein with the citation "for his services to Theoretical Physics, and especially for his discovery of the law of the photoelectric effect." The prize rules do not allow an award unless it has been "tested by time" which meant ironically that the prize was not awarded for special and general relativity. The Nobel judges did not fully accept the theories of relativity despite measurements in 1919 by Arthur Eddington (1882–1944) showing that the light from a star was influenced by the solar gravity during a solar eclipse as predicted by the general theory of relativity [28]. Antisemitism from other German physicists may have also influenced the Nobel committee [12].

Einstein traveled extensively in the 1920s, visiting and lecturing in the United States, the United Kingdom, Japan, Palestine, and South America. He became a "celebrity" in the full modern sense of the term. Although celebrated in most of the world, in Germany the rise of the Nazi Party led to campaigns against Einstein and bizarrely his relativity theories. The scientific campaign against Einstein and relativity was orchestrated largely by earlier Nobel prize winners Philipp Lenard and Johannes Stark.[2] Adolf Hitler achieved sufficient electoral success that he was appointed chancellor of Germany in January 1933. Einstein left Germany in December 1932 to lecture as a visiting professor at the California Institute of Technology. He returned to Europe in March 1933 by ship, berthing at Antwerp and stayed first in Belgium and then the United Kingdom. Both of these countries organized armed protection of Einstein as there were fears he might be assassinated. In 1933–1935, the German government officially seized all of Einstein's assets in Germany [98].

In October 1933 Einstein moved permanently to the United States and took up a position as professor of theoretical physics at Princeton University.[3] He became an American citizen in 1940. On the eve of World War II, Einstein endorsed a letter to President Franklin D. Roosevelt alerting him to the potential German nuclear weapons program and recommending that the United States begin similar

---

[2] Philipp Lenard was awarded the Nobel prize for work on "cathode rays" (electron beams) in 1905 and Johannes Stark was given the award in 1919 for showing that electric fields split the spectral lines of atoms, now known as the Stark effect. The Stark effect, Stark line broadening (see Section 10.3 of Tallents [110]), and the motional Stark effect (see Section 3.2) all now carry Stark's name. In 1933, Stark become president of the Physical and Technical Institute of the German Reich and removed all Jewish people from academic posts. There have been calls for the Stark effect to be renamed and for Lenard and Stark to be stripped of their Nobel prizes [72].

[3] In April 1933 while in Belgium, Einstein received offers of professorships from the Collège de France in Paris; the University of Madrid; universities in Holland, India, and Turkey; and the new Institute for Advanced Studies in Princeton University. He accepted the Princeton offer provided he could fulfill his lecture commitments in England [98].

research.[4] Roosevelt authorized the creation of an advisory committee that led to the setting up of the Manhattan project and the development of weapons based on nuclear fission. In nuclear fission some of the mass of, for example, the uranium-235 isotope is converted to energy.[5]

Einstein was denied a security clearance to work on the Manhattan project due to a perceived security risk because of his pacifist leanings and celebrity status. In 1952 Einstein was offered the largely ceremonial post of president of Israel, but he declined. In the 1950s up until his death, Einstein was vocal in his criticism of Senator Joseph McCarthy who zealously chaired the Senate Un-American Activities Committee investigating unevidenced communist subversion in universities, the motion-picture industry and the U.S. federal government.

After 1925, Einstein's scientific interest centered on what he termed "unified field theory." This was an attempt to unify general relativity and electromagnetism. Einstein accepted the evidence for quantum mechanics, but had difficulty with the idea of probability distributions and action at a distance (as wavefunctions have spatial extent). In 1926 Einstein famously wrote that "*He* (God) *is not playing at dice.*" In 1952 Einstein wrote, "*The conviction prevails that the experimentally assured duality of nature (corpuscular and wave structure) can be realized only by ... a weakening of the concept of reality*" [98].

One of Einstein's approaches to a unified field theory survives, in a much modified form, in today's string theory [98]. String theory is a theoretical framework in which point-like particles are replaced by one-dimensional objects called strings with different vibrational states determining the mass, charge, and other properties of the particle as observed on distance scales much larger than the string. String theory has led to mathematical developments, but it seems unclear (at least at the time of writing this book) if it is useful for physics predictions due to the freedom in string theory to choose a range of parameters to fit experimental observations.

---

[4] In 1966 U.S. president Lyndon Johnson recognized the discoverers of nuclear fission with the Enrico Fermi prize awarded to Otto Hahn (1879–1968), Friedrich Strassmann (1902–1980), and Lise Meitner (1878–1968) who had been based in Germany. Awarded in 1945 after the end of the war in Europe, Hahn alone had received the 1944 Nobel chemistry prize for the discovery of nuclear fission. He received news of the Nobel award while held in England where secret tape recording of German scientists revealed details of the German wartime effort to develop a nuclear bomb [8]. Meitner, who like Einstein was of Jewish descent, had fled to Sweden from Germany in 1938, while Strassmann remained in Germany, but was blacklisted by the Nazi regime. Lise Meitner was the first woman to receive the Enrico Fermi prize.

[5] Uranium 235 has 92 protons and 143 neutrons and an atomic mass of 235. The uranium 235 nucleus disintegrates to less massive elements if excited, releasing a neutron which, in turn, collides with other uranium nuclei, causing them to disintegrate. The lighter elements proceed to decay to still lighter elements with more neutron production in a "chain reaction." There is a mass decrease of the product particles which is converted to kinetic energy of the particles following Einstein's equation $E = mc^2$. Developed during and after World War II, nuclear reactors [103] typically use a low concentration of uranium 235 mixed with the more common uranium isotope of atomic mass 238 which in a reactor transmutes to the readily fissionable plutonium 239. The fission rate of a reactor is controlled using neutron-absorbing media (to reduce the fission rate) and "moderators" such as "heavy water" D_2O or graphite which slow the neutrons closer to the maximum cross-section for fission (to increase the fission rate). Techniques used to make sure that fission in a nuclear weapon occurs rapidly and with complete fission of uranium or plutonium are understandably classified.

Einstein died in 1955 (aged 76) in Princeton, New Jersey, due to a ruptured abdominal aortic aneurysm [84]. His life and his brain have been extensively analyzed and – literally for his brain – dissected. Einstein's brain was removed and dissected shortly after his death by a pathologist who preserved it in pieces which were distributed to a number of medical researchers. The fate of Einstein's brain was discovered by reporters in 1979 and efforts were made to repatriate some of the brain to Einstein's grand-daughter [12], [86].

1.1 The Propagation of Light

Electromagnetic waves comprise oscillating electric and magnetic fields which are directed transversely to the direction of propagation of the wave. The frequency of the oscillation determines the spectral name of the radiation, with a range from low-frequency radio waves, through the infrared, the visible, ultra-violet, and X-rays to gamma rays occupying the highest frequencies. Common usage often restricts "light" to mean visible light (wavelength 400–700 nm) and perhaps ultraviolet light (wavelength 50–400 nm). However, as the different spectral regions for electromagnetic waves only become important when electromagnetic radiation interacts with matter, we will use the word "light" to describe electromagnetic radiation of any frequency, particularly when there is no radiation interaction with matter.

The propagation of light at all frequencies can be modeled using a wave equation obtained by a manipulation of Maxwell's equations. In this section, we derive the wave equation and show that the speed of light in a vacuum in the International System of Units (SI) is determined by two quantities (the dielectric constant ϵ_0 and the magnetic permeability μ_0) used with SI units to get the correct relationships in Maxwell's equations. The simple relationship between the speed of light and some electric and magnetic field constants strongly supports the idea that the speed of light is a "constant of nature." A constant value of the speed of light in vacuum is a fundamental assumption of the theories of both special and general relativity.

Maxwell's equations describe the observed behavior of electrical and magnetic fields. The propagation behavior of light in vacuum is developed using Faraday's law, which describes the way a magnetic field oscillating in time produces an electric field, and Ampere's law, which describes the way an electric field oscillating in time produces a magnetic field. We have respectively:

$$\nabla \times \mathbf{E} = -\frac{\partial \mathbf{B}}{\partial t}, \tag{1.1}$$

$$\nabla \times \mathbf{B} = \mu_0 \mathbf{J} + \epsilon_0 \mu_0 \frac{\partial \mathbf{E}}{\partial t}. \tag{1.2}$$

Magnetic **B** and electric **E** fields can be produced by temporal changes of each field (and equivalently a magnetic field can be created by a flow of current density **J**). Here ϵ_0 is the vacuum dielectric constant and μ_0 is the vacuum magnetic permeability.

Taking the curl ($\nabla \times$) of Faraday's law and substituting Ampere's law for $\nabla \times \mathbf{B}$ means that

$$\nabla \times \nabla \times \mathbf{E} = -\frac{\partial}{\partial t}\left(\mu_0 \mathbf{J} + \epsilon_0 \mu_0 \frac{\partial \mathbf{E}}{\partial t}\right). \tag{1.3}$$

In vacuum, the density of charge ρ_c and current flow **J** is zero. The third of Maxwell's equations, known as Gauss's law, describes the production of an electric field from a charge density. In vacuum, we can write

$$\nabla.\mathbf{E} = \frac{\rho_c}{\epsilon_0} = 0 \tag{1.4}$$

as the charge density ρ_c is zero in vacuum. Using the vector identity $\nabla \times \nabla \times \mathbf{E} = \nabla(\nabla.\mathbf{E}) - \nabla^2\mathbf{E}$, setting $\mathbf{J} = 0$ and $\nabla.\mathbf{E} = 0$ produces the wave equation for light. We have

$$\nabla^2\mathbf{E} = \epsilon_0 \mu_0 \frac{\partial^2 \mathbf{E}}{\partial t^2}. \tag{1.5}$$

The wave equation has solutions for the electric and magnetic fields which oscillate in time and space and which are known generally as electromagnetic radiation. Substituting into the wave equation (Equation 1.5) verifies that variations of electric field of the following form satisfy the wave equation at position **r**:

$$\mathbf{E} = \mathbf{E}_0 \exp[i(\mathbf{k}.\mathbf{r} - \omega t)], \tag{1.6}$$

where $\mathbf{E}_0$ is a field amplitude, **k** is a wavevector representing the rapid spatial variation of the field with position vector **r**, and ω is an angular frequency representing the rapid temporal variation of the field with time t. The angular frequency ω is the phase change of the field in radians per second and is related to the frequency ν in cycles per second (i.e. hertz) by $\omega = 2\pi\nu$.

Points of constant phase in the oscillating electric field solution of the wave equation (Equation 1.6) have $\mathbf{k}.\mathbf{r} - \omega t$ constant. If the wave propagates in the z-direction, points of constant phase on the wave have $kz - \omega t = $ constant. Differentiating $kz - \omega t = $ constant with respect to time yields $kdz/dt - \omega = 0$ and hence the phase velocity of the wave is $dz/dt = \omega/k$. The phase velocity of light in vacuum in the direction of **k** is given the symbol c, where

$$c = \frac{\omega}{k}. \tag{1.7}$$

The wavelength λ of the light is the distance in the direction of propagation z where the spatial phase term in $\exp(i\mathbf{k} \cdot \mathbf{r}) = \exp(ikz)$ does a complete cycle and changes by 2π. The wavelength $\lambda = 2\pi/k = 2\pi c/\omega = c/v$.

Using Equation 1.6 as a solution of the wave equation suggests a plane wave oscillation that is uniform in directions perpendicular to the wave vector $\mathbf{k}$ with, consequently, energy flow due to the wave in the direction of $\mathbf{k}$. Real examples of electromagnetic waves such as a beam of light have a finite width with spatial variations of $\mathbf{E_0}$ perpendicular to $\mathbf{k}$. Practical electromagnetic waves can be regarded as a superposition of many plane waves, with the electric field of the plane waves canceling outside the volume of the radiation field. To a good approximation, spatial variations of $\mathbf{E_0}$ perpendicular to $\mathbf{k}$ result in energy flow in the direction of $\mathbf{k}$, provided the spatial variation scalelength is large compared to $1/k$.

As well as verifying that Equation 1.6 is a solution of the wave equation, substituting Equation 1.6 into the wave equation (Equation 1.5) produces a relationship between the amplitudes of the wavevector k and frequency ω. We have

$$\nabla^2 \mathbf{E} = -k^2 \mathbf{E},$$

$$\frac{\partial^2 \mathbf{E}}{\partial t^2} = -\omega^2 \mathbf{E}$$

so that

$$k^2 = \epsilon_0 \mu_0 \omega^2.$$

It follows that the vacuum speed of light c is related to the vacuum dielectric constant ϵ_0 and vacuum magnetic permeability μ_0 by

$$c = \frac{\omega}{k} = \frac{1}{\sqrt{\epsilon \mu_0}}. \tag{1.8}$$

The wave equation (Equation 1.5) can be rewritten as

$$\nabla^2 \mathbf{E} = \frac{1}{c^2} \frac{\partial^2 \mathbf{E}}{\partial t^2}. \tag{1.9}$$

The vacuum velocity of light c depends on two constants (ϵ_0 and μ_0) used to get the correct proportionality between a changing electric and magnetic field. There is a strong suggestion here that the speed of light in a vacuum is a constant of nature.

The fundamental and constant value of the speed of light in vacuum has been recognized since 1983 by the specification of an exact speed of light in vacuum in SI units. The vacuum speed of light has an exact numerical value $c = 2.99792458 \times 10^8$ ms^{-1} because the meter unit of length is defined in terms of the vacuum speed of light as the distance traveled by light in $1/2.99792458 \times 10^{-8}$ seconds in vacuum.[6]

There is a corresponding oscillating magnetic field arising from the oscillating electric field for an electromagnetic wave in vacuum. Ampere's law (Equation 1.2) is given by

$$\nabla \times \mathbf{B} = \mu_0 \mathbf{J} + \epsilon_0 \mu_0 \frac{\partial \mathbf{E}}{\partial t} = -i\omega \epsilon_0 \mu_0 \mathbf{E} \qquad (1.10)$$

upon differentiating $\mathbf{E}$ and assuming vacuum propagation with $\mathbf{J} = 0$. Equation 1.10 requires that the magnetic field varies as $\mathbf{B} = \mathbf{B}_0 \exp[i(\mathbf{k} \cdot \mathbf{r} - \omega t)]$ in phase with the electric field variation. The definition of $\nabla \times$ in Cartesian coordinates determines that $\mathbf{B}$ is perpendicular to $\mathbf{E}$ and gives $-ikB = (-i\omega \epsilon_0 \mu_0)E$. Using the equivalences $1/c^2 = \epsilon_0 \mu_0$ and $c = \omega/k$, we have that

$$\frac{E}{B} = c. \qquad (1.11)$$

Taking the divergence of $\mathbf{E}$ in Equation 1.6.

$$\nabla \cdot \mathbf{E} = i\mathbf{k} \cdot \mathbf{E},$$

which from Equation 1.4 is identically equal to zero. With $\mathbf{k} \cdot \mathbf{E} = 0$ we have that the electric field $\mathbf{E}$ is perpendicular to the wavevector $\mathbf{k}$. A similar argument using the fourth of Maxwell's equations ($\nabla.\mathbf{B} = 0$) shows that the magnetic field $\mathbf{B}$ is perpendicular to $\mathbf{k}$. We have already determined in the discussion for Equation 1.11 that the magnetic field $\mathbf{B}$ is perpendicular to $\mathbf{E}$. The direction of energy propagation of the wave follows the $\mathbf{k}$ vector if spatial variations of the electric and magnetic fields have scalelengths $>> 1/k$.

The orientation of the electric field $\mathbf{E}$ in the plane perpendicular to $\mathbf{k}$ is the polarization direction of the electromagnetic wave. Any electromagnetic wave can be decomposed into polarization components in two orthogonal directions both perpendicular to the wavevector $\mathbf{k}$. Unpolarized light can be regarded as the independent supposition of two polarized beams with polarization directions at angle $\pi/2$ to each other. For unpolarized light there is a random phase $\mathbf{k}.\mathbf{r} - \omega t$ relationship between the polarization directions. Elliptically polarized light can be regarded as consisting of polarization components in two orthogonal directions where the phase $\mathbf{k} \cdot \mathbf{r} - \omega t$ between components in the two orthogonal directions always differs by $\pi/2$.

The classical energy density of an electric field is $(1/2)\epsilon_0 E^2$ and for a magnetic field is $(1/2)B^2/\mu_0$. These quantities for the energy density of the fields can be

---

[6] The second has been defined since 1967 by an atomic physics measurement – the frequency of the ground state hyperfine transition for cesium 133 atoms (see https://physics.nist.gov/cuu/Units/).

obtained by, for example, considering the charging of a capacitor (for the electric field) and the increase of current into a solenoid (for the magnetic field). For an electromagnetic wave at any instant of time, the energies in the electric and magnetic fields are equal. Using $E/B = c$ and $1/c^2 = \epsilon_0\mu_0$, we have $(1/2)\epsilon_0 E^2 = (1/2)B^2/\mu_0$. The total instantaneous energy density of an electromagnetic wave is consequently $\epsilon_0 E^2$ or equivalently B^2/μ_0. The electric field expression for the energy density of an electromagnetic wave is usually used.

The average of the sinusoidal variation of the electric and magnetic fields in time is given by the temporal average of the real and imaginary components of $\exp[i(\mathbf{k.r} - \omega t)]$, that is,

$$\frac{\int_{\omega t=0}^{2\pi} \cos^2(\omega t)d(\omega t)}{\int_{\omega t=0}^{2\pi} d(\omega t)} = \frac{\int_{\omega t=0}^{2\pi} \sin^2(\omega t)d(\omega t)}{\int_{\omega t=0}^{2\pi} d(\omega t)} = 1/2.$$

The time-averaged electric and magnetic energies are equal and together have a total energy density $(1/2)\epsilon_0 E_0^2$. Imagine a unit cross-section volume of unit area and length c along the direction $\mathbf{k}$ of propagation of the beam. In unit time, the energy from the volume of c will pass through the unit area. This energy passing per unit time per unit area normal to the direction of the beam is known as the intensity or irradiance I and is given by

$$I = \frac{1}{2}\epsilon_0 c E_0^2. \tag{1.12}$$

The proportionality of the intensity of light to the square of the electric field of the light enables the intensity of light to be significantly enhanced when the electric fields from different sources of light add coherently. Coherent addition requires the sources of light to have the same phase relationship, so that addition of the electric fields, rather than subtraction, occurs. With coherent addition of light, adding n sources of coherent light together causes an instantaneous n^2 increase in light intensity when all the sources are in phase.

This section has illustrated the wave-like properties of light. Light also has a particle-like nature.[7] Light can be shown to comprise discrete quanta of radiation energy known as photons. The particle nature of light is experimentally apparent

---

[7] Both light and matter exhibit wave-like and particle-like properties depending on the measurement being undertaken. This is the essence of the "complementarity" principle proposed by Neils Bohr (1885–1962). With the complementarity principle, wave or particle models are not just useful approximations for a physical situation, but intrinsic models of light and matter with one description or the other applicable depending on the measurement. With a particle treatment, position can be regarded as being known accurately, with a larger uncertainty on the velocity or momentum. With a wave treatment, the velocity/momentum is known accurately, but with a large uncertainty on the position. The uncertainty principle first developed by Werner Heisenberg (1901–1976) showed that measurements of complementary parameters such as position/momentum, frequency/time could not both be made, even in principle, with infinite precision. Many assume that the complementarity principle is equivalent to the uncertainty principle, but the complementary principle has a deeper meaning (see [31]).

in the photo-electric effect as discussed originally by Einstein [34]. In the photo-electric effect, light irradiated material only emits an electron when the energy of a photon of the light exceeds the binding energy of the material. Light is absorbed by matter in units of energy and if the unit of light energy is less than the quantum differences for the energies in the material irradiated, the light is not absorbed.

The term "photon" was introduced in a paper published in 1926 [66]. The quantization of light in photons of energy $h\nu$, where $h = 6.626 \times 10^{-34}$ Js is Planck's constant and ν is the frequency of the light, is discussed later in Section 2.9, where the wave-like property of matter and radiation is introduced, and in Section 6.4, where the equilibrium radiation field (black-body radiation) is considered. Calculations of the trajectories of light rays in vacuum as seen by observers with different velocities can be made using the treatments developed for particles by setting the particle velocity to the vacuum speed of light (see Section 2.5) and for propagation in a medium by setting the particle velocity to the phase velocity of light in the medium (see Chapter 9).

1.2 Time Dilation and Length Contraction

In this section, a "thought experiment" involving light propagating in a moving frame of reference is used to introduce two consequences of special relativity: (i) that time in a moving frame of reference runs slower as measured in the stationary or rest frame, and (ii) that lengths in a moving frame of reference are shorter in the direction of motion when measured in the stationary or rest frame. We will quantify more precisely the transformations between time and distance for different inertial frames of reference in Chapter 2.

Maxwell's equations lead to the wave equation which has solutions indicating that the vacuum speed of light c is constant (see Section 1.1). The theories of relativity take as an assumption that the vacuum speed of light c is constant in all frames of reference. The theory of special relativity applies for frames of reference which are "inertial." An inertial frame of reference is one where free particles move in straight lines at constant speed. An inertial frame of reference with free particles moving at constant velocity implies that there are no forces, known as "fictitious forces," which deviate the path of a body due to the acceleration of the frame of reference. Examples of fictitious forces include the centrifugal and Coriolis forces associated with a rotating body.

An inertial frame of reference is often specified as one which is not undergoing acceleration. However, it is possible to treat acceleration using the equations of special relativity as long as the acceleration is considered for an infinitesimally short period of time (see, e.g., Section 3.3).

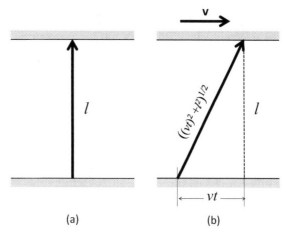

(a)   (b)

Figure 1.1   A thought experiment examining the time for light to propagate between two parallel mirrors in (a) a frame of reference moving with the mirrors, and
(b) a stationary frame of reference. The mirrors are moving parallel to the mirror
surfaces at a velocity **v** relative to the stationary observer. The distances for light
propagation as seen by (a) a moving observer in a frame of reference S' and (b) a
stationary observer in a frame of reference S are annotated.

We consider here a brief argument showing that to an observer in a stationary
frame of reference, time in a moving inertial frame of reference appears to run more
slowly. This is known as time dilation. Distances measured by a stationary observer
of lengths in the moving inertial frame of reference in the direction of the velocity
are also shown to be shortened.

Consider a frame of reference with a pulse of light reflecting between two parallel mirrors separated by a distance l which is moving at constant velocity v in
an orthogonal direction to the light beam (see Figure 1.1). We label this moving
frame of reference as S' and use primed symbols to designate time and distance in
the frame of reference. The light propagation between the two mirrors is also observed by a nominally stationary observer in a frame of reference which we label
as S with distances and time designated with unprimed symbols (see Figure 1.2).
In the S' frame of reference moving at velocity v, the light takes a time $t' = l/c$ to
pass from one mirror to the other. To the stationary observer, the light travels a
distance $\sqrt{l^2 + v^2 t^2}$ in a time given by $t = \sqrt{l^2 + v^2 t^2}/c$. The important issue now
is that to the observers in both the stationary S and moving S' frames of reference,
the speed of light c is the same. The only way to resolve the different distances that
the light travels according to the two observers is for the measurement of time to
be different in the two frames of reference. Substituting $l = ct'$ gives

$$t = \frac{\sqrt{c^2 t'^2 + v^2 t^2}}{c}.$$

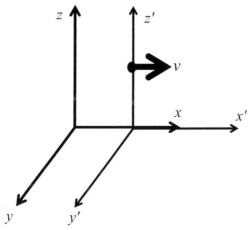

Figure 1.2   Two inertial frames of reference. The frame of reference S' with primed coordinates (x', y', z') has a velocity $\mathbf{v}$ in the x-direction relative to the frame of reference S with coordinates (x, y, z).

Squaring this expression and re-arranging gives

$$t^2 = t'^2 + \frac{v^2}{c^2}t^2,$$

which simplifies to

$$t = \frac{t'}{\sqrt{1 - v^2/c^2}}. \tag{1.13}$$

The observer in the stationary frame of reference S observes a longer time for the light to reflect from one mirror to the other. The numerator in Equation 1.13 is less than one, so time t in the stationary frame S is greater than time t' in the moving frame S'. We can imagine that the reflection of the light pulse acts as a clock. Time in the moving frame of reference as observed from the stationary frame of reference is running more slowly. A "moving" and a "stationary" observer are equally valid observers, so to a nominally moving observer, the stationary clocks also appear to run more slowly. In summary, the time measured in the two frames of reference is different because the speed of light is the same in the two frames of reference – but the light needs to travel different distances in the two frames of reference.

The slowing down of time to a stationary observer of the time in a frame of reference moving with velocity approaching the speed of light is a real effect. For example, muons created by cosmic rays[8] interacting with the Earth's upper atmosphere

---

[8] Cosmic rays striking the Earth's atmosphere are particles comprising 90% protons (hydrogen nuclei), 9% alpha particles (helium nuclei), and 1% electrons with a small number of higher atomic number nuclei and antiparticles. With particle numbers peaking at energies ≈ 0.3 GeV, cosmic rays interact with air molecules to produce a cascade of particles traveling close to the direction of the initial cosmic ray. Muons make up more than half the cosmic radiation at sea level with the remainder comprising mainly electrons, positrons, and

decay with a half-life of 1.56 μs and so would decay and not be observed on the Earth's surface if time in the muon frame of reference was the same as the passage of time on the Earth's surface [30]. Traveling close to the speed of light, high-energy muons would only travel close to 470 m in the time of their half-life from their creation, so almost all would have decayed before reaching the ground as they pass through 15 km or so of atmosphere. They are observed on the surface of the Earth because as observed on the ground, the muon time is passing much more slowly.

To quantify, let the distance x that a muon travels in the Earth's frame of reference be set so $x = vt$. The length of atmosphere moving past the muon in its rest frame is $x' = vt'$. We have

$$x' = vt' = vt\sqrt{1 - v^2/c^2} = x\sqrt{1 - v^2/c^2}.$$

Lengths of atmosphere x' as measured by the muon are reduced by a factor $\sqrt{1 - v^2/c^2}$ from the lengths x as measured in the Earth's frame of reference. For $v > 0.95c$, many muons created by cosmic rays reach the ground (see Exercise 1.8).

Distances in a frame of reference moving at velocity v as measured by stationary observers are reduced or contracted in the direction of the velocity. The muon in our example sees the Earth's atmosphere contracted by a factor $\sqrt{1 - v^2/c^2}$ as it moves past it at velocity v. This effect applies to any length measurement made in a stationary frame in the direction of velocity of a moving object, but only, of course, becomes significant when the velocity is a significant fraction of the speed of light.

As all observers in inertial frames of reference are equivalent, we can think of, for example, the length contraction of the Earth's atmosphere in the frame of reference of a muon even though the nominally stationary Earth may be regarded as a "better" frame of reference. To avoid confusion, the "proper time" of an event is used to mean the time measured by an observer in the frame of reference of the event. The "proper length" is the length of an object or the distance between two points measured by an observer in the same frame of reference as the object.

An observer in a stationary frame of reference measures times in a moving frame slower than the proper time and distances in the direction of the relative velocity of the two frames of reference shorter than the proper length. Distances measured orthogonal to the direction of the velocity are measured to have the same value in the two frames of reference. This change in time and measurement of distance is illustrated as a cartoon in Figure 1.3.

---

photons. Cosmic rays originate from solar flares and coronal mass ejections and from supernovae explosions in our galaxy and beyond. The sources and acceleration mechanisms for cosmic rays, particularly those with extreme energy, are the subjects of much research (see Section 5.2.3). Galactic cosmic ray energies up to 10^{15} eV and extragalactic up to 3×10^{20} eV have been observed [9], [22]. Cosmic rays with energies $> 10^{20}$ eV are scattered by background intergalactic photons (the cosmic microwave background) and so are rarely detected (see Exercise 6.15).

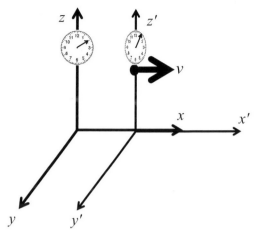

Figure 1.3   The two inertial frames of reference illustrated in Figure 1.2 with superimposed clocks as seen from the stationary frame *S* as the velocity **v** approaches
the speed of light *c*. Time appears slower, and distances in the direction of the velocity appear shorter in the moving frame as observed from the stationary frame.

We quantify more precisely the transformations between time and distance for
different inertial frames of reference in Chapter 2. We show that it is possible
to have a transformation between time and distance (the Lorentz transformation)
which ensures (i) that the speed of light is the same in different frames of reference,
and (ii) that with a simple change of sign of the relative velocity of the observers, a
"moving" observer sees dilated time and contracted lengths in the stationary frame,
and the stationary observer measures dilated time and contracted lengths in the
moving frame. The shortening of lengths in the moving frame as measured in the
stationary frame of reference is often referred to as Lorentz contraction.

1.3 Light Emission from Charge Acceleration

Emission and absorption of electromagnetic radiation occurs when charges are accelerated. A charged body of any mass radiates when accelerated, but the most
common emitters and absorbers are electrons as their low mass results in larger
acceleration. The acceleration is apparent for unbound particles such as free
electrons in a plasma, but is also present when dealing with the time-dependent
quantum mechanics of electrons bound in atoms and ions interacting with electromagnetic waves. In this section, we follow a relatively simple treatment by Purcell
[94] to determine the radiation produced by an accelerating charge.

   The production of electromagnetic radiation when a charge such as an electron
accelerates arises due to a dislocation of the radially symmetric electric field from
the charge. Immediately before an acceleration impulse, the electric field radiates

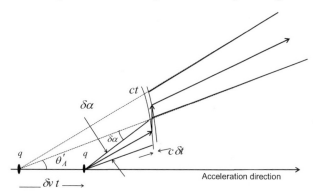

Figure 1.4   The electric field disruption at time t when a charge is accelerated to velocity δv over a time δt at $t = 0$ for an angle θ'_A to the acceleration direction in the frame of reference of the charge. All electric field lines from the new charge position within the angle $\delta \alpha$ need to move transversely at distance ct to link to the electric field lines established when the charge was in its initial position. The transverse electric field forms a pulse of electromagnetic radiation propagating away from the accelerated charge.

from the initial position in a reference frame moving with the initial charge velocity. Immediately after the acceleration, the electric field propagates in a radial direction from the new position in the reference frame. As the electric field lines from the charge are continuous, there is an electric field transverse jump propagating at the speed of light. For the purposes of evaluating an expression for the transverse electric field, we can assume that the velocity jump occurs on a very short timescale so that the frame of reference remains inertial.

We now consider more quantitatively the production of electromagnetic radiation by thinking about the electric field around a charged particle if it suffers an impulse of acceleration to a velocity δv in a time δt at some initial time $t = 0$. At a later time $t > 0$, the electric field lines associated with the initial position extend radially out from the initial position at distances greater than ct, which is the distance that light and the information about the disturbance in the electric field from the charge can travel in the time t. However, at the distance ct, there is a dislocation of the electric field associated with the charge having accelerated to a new position and the consequent need for a transverse movement of the electric field radially emanating from the new positions to join up with the electric field lines which started before time $t = 0$. The concept is schematically illustrated in Figure 1.4.

The acceleration of a charge q by a velocity δv in a time δt at time $t = 0$ causes a dislocation in the electric field distribution from the charge at a later time t at a distance ct from the charge and produces a transverse electric field. All electric field lines emanating from the later position of the charge move transversely within an annulus of thickness $c\delta t$ to match the field lines emitted before the charge acceleration. Provided that the velocity increase δv is much less than the speed of light c, the electric field lines from the charge in the reference frame of the charge are directed radially from the position of the charge, except at the distance ct. We can calculate the enhancement of the electric field transversely over the distance ct to $c(t + \delta t)$ by considering a small angular range $\delta\alpha$ for a particular angle θ'_A to the acceleration direction of the charge. Using Figure 1.4, by simple geometry

$$\frac{\sin \delta\alpha}{\delta v \, t} = \frac{\sin(\pi - \theta'_A - \delta\alpha)}{c\,t}.$$

If $\delta v << c$, the angle $\delta\alpha$ is small and we have that

$$\delta\alpha = (\delta v/c) \sin \theta'_A.$$

The enhancement of the electric field in moving transversely through a thickness $c\delta t$ rather than being spread emanating radially from the charge over an angle $r\,\delta\alpha$ is given by

$$\frac{r\,\delta\alpha}{c\delta t} = \frac{r\delta v}{c^2\delta t} \sin \theta'_A.$$

The radial electric field from the charge q at distance r is given by

$$E_r = \frac{q}{4\pi \epsilon_0 r^2},$$

so the transverse field at distance r is given by

$$E_t = \frac{q \sin \theta'_A}{4\pi \epsilon_0 c^2 r} \frac{\delta v}{\delta t}.$$

The quantity $\delta v/\delta t$ is the acceleration of the charge which we can write as dv'/dt' (with the primes indicating a measurement of acceleration in the frame of the charge). The transverse electric field can then be written as

$$E_t = \frac{q}{4\pi \epsilon_0 c^2 r} \left(\frac{dv'}{dt'}\right) \sin \theta'_A. \tag{1.14}$$

The acceleration direction of a charge oscillates or rotates in many situations. For example, charged particles orbit around magnetic fields in a plasma in Larmor or gyro-orbits. An electron approaching an ion undergoes a hyperbolic trajectory with a change of direction and hence change of acceleration direction due to the electron-ion Coulomb attraction. In these cases the electric field in a particular direction will oscillate sinusoidally (possibly at many frequencies) and consequently create sinusoidally oscillating magnetic fields (from Ampere's law). Sinusoidally oscillating electric and magnetic fields are simply electromagnetic radiation. The

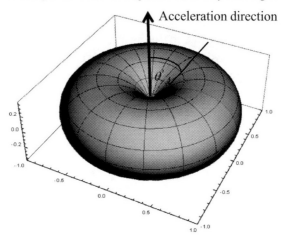

Figure 1.5 Three-dimensional profile of the power $dP'/d\Omega'$ per unit solid angle radiated by an accelerating charge in the frame of reference of the charge. The radiated power varies as $\sin^2\theta'_A$, where θ'_A is the angle to the acceleration direction.

power $dP'/d\Omega'$ radiated per unit solid angle is related to the intensity $I = \epsilon_0 c E_t^2$ at a distance r by $dP'/d\Omega' = Ir^2$, so we have that

$$\frac{dP'}{d\Omega'} = \frac{q^2}{4\pi\epsilon_0}\frac{1}{4\pi c}\left(\frac{dv'}{dt'}/c\right)^2\sin^2\theta'_A. \tag{1.15}$$

This expression for the radiation emitted by an accelerating charge is known as the Larmor formula after the Northern Irish physicist Joseph Larmor (1857–1942). Larmor was the first to determine the power radiated by an accelerated charge (Equation 1.15). He also developed the theoretical understanding of the splitting of spectral lines due to a magnetic field (now known as the Zeeman effect). Prior to Einstein's special relativity, Larmor worked on transformations similar to the relativistic Lorentz transformation (to be discussed in Section 2.2).

The radiation produced by an accelerating charge is polarized. The **k** vector of the radiation is directed radially away from the charge with the electric field directed transversely (in the plane formed by the **k** vector and the acceleration vector). The polarization of light is defined by the direction of the electric field so an accelerated charge produces light with a polarization in the plane formed by the light **k** vector and the charge acceleration vector. The power radiated by an accelerated charge q is proportional to the square of the acceleration and varies with angle θ'_A to the direction of acceleration (see Figure 1.5).

The power output (Equation 1.15) is valid in the frame of reference of the charged particle. If the charged particle has a velocity v approaching the speed of light c, we show later that there is a "beaming effect" in the stationary frame of reference with emission concentrated in the stationary frame in the direction of the charge velocity over a range of angles $1/\gamma$ to the velocity direction, where $\gamma = 1/\sqrt{1 - v^2/c^2}$ (see Section 3.4). Equation 1.15 is sometimes known as the "dipole" distribution of radiation as an oscillating electric dipole radiates with an identical angular variation ($\propto \sin^2 \theta_A'$).

Integrating Equation 1.15 for nonrelativistic emission over all solid angles gives the total radiated power

$$
\begin{aligned}
P &= \int \frac{dP'}{d\Omega'} d\Omega' \\
&= \int_0^\pi \frac{dP'}{d\Omega'} 2\pi \, \sin\theta_A' \, d\theta_A' \\
&= \frac{q^2}{4\pi\epsilon_0} \frac{1}{4\pi c} \left(\frac{dv'}{dt'}/c\right)^2 \int_0^\pi 2\pi \, \sin^3\theta_A' \, d\theta_A' \\
&= \frac{q^2}{4\pi\epsilon_0} \frac{2}{3c} \left(\frac{dv'}{dt'}/c\right)^2 .
\end{aligned}
\tag{1.16}
$$

The total power radiated in the frame of reference of the charged particle given by Equation 1.15 is the same as the total power detected in the stationary rest frame, so Equation 1.16 also applies to the total power emitted for charged particles moving at relativistic velocities provided the acceleration dv'/dt' is measured in the frame of the charged particle. The expression for the total power emitted by an accelerating charged particle after integration over 4π steradian when the acceleration is measured in the rest frame of reference is evaluated in Chapter 3 (see, e.g., Equation 3.44).

Exercises

1.1 The foot is a unit of length in the British Imperial and U.S. customary system of units defined as 0.3048 m exactly. How long in nanoseconds does it take for light to travel a distance of one foot? [1.016 ns]

1.2 How far does a pulse of light travel in one year? [9.46×10^{15} m]

1.3 Consider visible light of wavelength 500 nm (a cyan or light blue color). Determine the frequency and photon energy of light of wavelength 500 nm. [6×10^{14} Hz, 2.48 eV]

1.4 A spectral line at wavelength 500 nm has a spectral width of 1 nm. Determine the spectral width in frequency and photon energy of this spectral line. [1.2×10^{12} Hz, 5×10^{-3} eV]

1.5 Evaluate Equation 1.12 to show that the electric field E_0 and magnetic field B_0 associated with light of irradiance I measured in Wm^{-2} are given by

$$E_0 = 27.4 \; I^{1/2} \quad \text{Vm}^{-1}$$
$$B_0 = 9.13 \times 10^{-8} \; I^{1/2} \quad \text{tesla.}$$

1.6 The clocks illustrated in Figure 1.3 show that time in the moving frame S' as measured in the stationary frame S is running at half the rate of time measured in the frame S. Calculate the relative velocity of frame S' to frame S. [$0.86c = 2.6 \times 10^8$ ms^{-1}]

1.7 The average lifetime of a π meson in the frame of reference of the meson is 26.0 ns. If a π meson has a velocity of $0.95c$ in a laboratory experiment, determine the lifetime of the π meson in the laboratory frame of reference and the average distance the π meson travels in the laboratory before decaying. [83.3 ns, 24 m]

1.8 Muons decay with a half-life of 1.56 μs in their frame of reference. They are created in the Earth's atmosphere when cosmic rays interact with air. For muons created in the upper atmosphere with a velocity of $0.98c$, calculate the observed half-life of the muons and determine the percentage of an initial population of muons which do not decay after traveling 10 km through the atmosphere. [6.8 ns, 4.9%]

1.9 Laser cavities consist of two parallel mirrors as illustrated in Figure 1.1. In the frame of reference of the mirrors, the laser output consists of "modes" of frequencies separated by $\Delta v' = 2l/c$, where l is the distance between the mirrors. If the laser mirrors move at a velocity $\mathbf{v}$ parallel to the surface of the mirrors, show that the mode spacing Δv according to a stationary observer is given by

$$\Delta v = \frac{2}{c}(l^2 + (vt)^2)^{1/2}.$$

Hence, show that

$$\Delta v = \frac{\Delta v'}{(1 - v^2/c^2)^{1/2}}.$$

[The frequency spacing is Doppler shifted by an amount that is consistent with Equation 2.47 for light moving at angle $\pi/2$ to the velocity direction (as derived in Section 2.6).]

1.10 Evaluate Equation 1.16 to show that the power radiated by an accelerated electron is given by

$$P = 5.7 \times 10^{-30} \left(\frac{dv}{dt}\right)^2 \quad \text{watts,}$$

where the electron acceleration dv/dt is measured in units of m s^{-2}.

1.11 An isolated electron is accelerated by an electric field of 1 Vm^{-1} for a short time. Determine the power radiated by the electron. [1.76×10^{-7} watts]

1.12 Ignoring radiation losses for an isolated electron accelerated in a uniform electric field of 1 Vm^{-1}, determine the length of time needed to accelerate an isolated electron to a speed $c/100$, where c is the speed of light. [17 μs]

1.13 Considering the numerical answers for Exercises 1.11 and 1.12, discuss if it is valid to neglect radiation energy losses for an isolated electron accelerated to a speed of $c/100$ in an electric field. [The effect of radiated energy loss on the motion of charged particles is discussed in Section 7.12.]

1.14 A charge of one coulomb is accelerated at a rate c ms^{-2} for a short time. Determine the total instantaneous power of the electromagnetic radiation created by the acceleration. [20 watts]

2

An Introduction to Relativistic Kinematics, Kinetics and Energy

"Kinematics" describes and predicts the motion of bodies using the initial conditions of position, velocity and acceleration. The study of the effects of forces on bodies is known as "kinetics." The kinematic motion of bodies and particles can be accurately observed and predicted by Newtonian mechanics when the observer is moving with a velocity close to that of the objects or particles under study. The kinematics of special relativity need to be invoked when the observer of a body or particle is separated from the objects under study due to a large difference in the velocity of the objects and the observer. We show in this chapter that relativistic kinetics and relativistic energy concepts develop by extending the Newtonian treatment of forces (as the differential of momentum with respect to time) and by defining energy in terms of forces (as the scalar product of the force on a body with the displacement caused by the force).

We found in Chapter 1 that a nominally stationary observer measures that time in a rapidly moving body runs more slowly, while lengths in the direction of the velocity of the moving body are measured to be shorter by the stationary observer. These two effects are consequences of the constant value of the vacuum speed of light for all observers. Light propagates at the same velocity to an observer moving at speed and to a stationary observer. Light signals sent from one observer to another have the same velocity to the observer sending the signal and to the observer receiving the signal even when one observer is moving rapidly away (or towards) the other observer. Relativistic adaptions to Newtonian kinematics, kinetics and energy all arise as consequences of the constant speed of light.

Section 1.1 of this book examined how the propagation of light arises from Maxwell's equations relating electric fields, magnetic fields, charges, and currents. Light was shown to be composed of spatially and temporally oscillating electric and magnetic fields propagating with a velocity related to the constants needed to relate electric fields, magnetic field, charges, and currents. Such a result strongly suggests that the vacuum speed of light is constant for all observers as the vacuum electric and magnetic field constants must be the same everywhere. Over the 160 years since Maxwell proposed the electromagnetic nature of light,[1] a considerable body of experimental evidence has been produced to show that the speed of light in a vacuum is the same for all observers.[2]

In Section 1.2 we found a simple quantitative relationship for the relative rate of progress of time for different observers. The speed difference between the stationary and moving observers needs to be a significant fraction of the speed of light for a noticeable effect, but will always exist. We showed in Section 1.3 that accelerated charges produce light.

In this chapter, we develop the Lorentz transformation between observers. The Lorentz transformation formalizes and conceptually eases the calculation of changes in time and length associated with large velocity differences between observers. The Lorentz transformation is not confined to the consideration of time and length measurements and is extended to cover the transformation of velocities, particle momentum, and energy. The chapter presents a matrix transformation useful for calculating Lorentz transformations.

2.1 Galilean Transformations

We start the discussion of transformation between observers by considering the classical Galilean transformation of time and length, which is valid for observers with small differences in velocities. A Galilean transformation relates time and spatial coordinates for different observers within the constructs of Newtonian physics.

An inertial frame of reference is a coordinate system where free bodies move with constant speed and direction. An inertial frame of reference can be regarded as being at rest or moving at a constant velocity. A system in an inertial frame of reference does not experience forces external to the system.

In an inertial frame of reference, Newton's laws follow in a straightforward way. For example, a force $\mathbf{F}$ on a body causes the momentum $\mathbf{p}$ of the body to change in time such that

$$\mathbf{F} = \frac{d\mathbf{p}}{dt}.$$

The force $\mathbf{F}$ is the vector sum of forces on the body, such as forces arising from electromagnetic effects. A noninertial frame of reference can have additional forces

---

[1] James Clerk Maxwell (1831–1879) published an early version of the equations later known as Maxwell's equations in 1861 and 1862 and used them to propose that light is an electromagnetic phenomenon (see [76]).

[2] Though the speed of light is constant in a vacuum, the speed of light in a medium (and not in a vacuum) varies depending on the medium. In a medium, pulses of light propagate with a group velocity which can be different to the phase velocity and the vacuum velocity (see Section 4.4). The velocity of light in a medium also varies with the frame of reference of the observer (see Chapter 9).

associated with the acceleration of the frame of reference. For example, a rotating frame of reference has a centrifugal force on a body directed away from the center of rotation. A body moving with a velocity relative to a rotating frame of reference experiences a Coriolis force. The centrifugal and Coriolis forces are examples of forces referred to as "fictitious" forces. A fictitious force does not arise from any physical interaction between two objects, such as electromagnetism or contact forces, but rather from the acceleration of the noninertial reference frame.[3]

Newton conceived the intuitive notion of the transformation of coordinates between two inertial frames of reference. Consider a frame of reference with Cartesian coordinates (x', y', z') moving at velocity v along the x-axis relative to another frame of reference with Cartesian coordinates (x, y, z). If the two frames of reference measure the same time $t' = t$ and coincide in spatial position at some time $t' = t = 0$, the relation between the two coordinate systems is given by

$$t' = t,$$
$$x' = x - vt,$$
$$y' = y,$$
$$z' = z. \tag{2.1}$$

Though conceptualized by Newton, this transformation between the coordinate systems for the two frames of reference is known as the Galilean transformation after the Italian astronomer and physicist Galileo Galilei (1564–1642). It is a valid transformation between the frames of reference if the velocity v is much less than the speed of light. The Galilean transformation can be conveniently represented in matrix form:

$$\begin{bmatrix} t' \\ x' \\ y' \\ z' \end{bmatrix} = \begin{bmatrix} 1 & 0 & 0 & 0 \\ -v & 1 & 0 & 0 \\ 0 & 0 & 1 & 0 \\ 0 & 0 & 0 & 1 \end{bmatrix} \begin{bmatrix} t \\ x \\ y \\ z \end{bmatrix} \tag{2.2}$$

Evaluating this matrix equation gives the four relationships for time and distances as expressed in Equation 2.1.

---

[3] In a frame of reference rotating with angular momentum ω, assuming relativistic effects are small, the relationship between a force $\mathbf{F}'$ and acceleration $\mathbf{a}'$ becomes $\mathbf{F}' - m_0 \frac{d\omega}{dt} \times \mathbf{r}' - 2m_0\omega \times \mathbf{v}' - m_0\omega \times (\omega \times \mathbf{r}')$ $= m_0\mathbf{a}'$, where $\mathbf{r}'$ is the position vector relative to the rotating reference frame and $\mathbf{v}'$ is the velocity relative to the rotating reference frame. The three fictitious force terms due to rotation are known respectively as the Euler force, the Coriolis force, and the centrifugal force.

2.2 Lorentz Transformations

The special theory of relativity is based on two principles held to be true because they agree with observations: (i) The laws of physics are the same in any inertial frame of reference, and (ii) the vacuum speed of light is constant in all inertial frames of reference. As discussed in Section 2.1, an inertial frame of reference has no "fictitious" forces such as the centrifugal force due to the acceleration of the frame. The first principle assumed for special relativity requires, for example, that Newton's second law that a body subject to a force $\mathbf{F}$ undergoes a change of momentum $\mathbf{p}$ such that $\mathbf{F} = d\mathbf{p}/dt$ applies in all inertial frames. The second principle is a consistent extension of the first principle arising from experiments showing that the velocity of light in a vacuum does not change with the observer[4] and from treatments, such as presented in Section 1.1, showing that Maxwell's equations predicting the behavior of electric and magnetic fields also indicate a constant velocity for light propagation. The constant velocity of light has important consequences for time and spatial measurements in different inertial frames of reference if the relative velocity of the frames of reference are significant compared to the speed of light.

Consider two observers, one in a frame of reference S' with Cartesian coordinates (x', y', z') moving at velocity $\mathbf{v}$ along the x-axis relative to another observer in a frame of reference S with Cartesian coordinates (x, y, z) (see Figure 1.2). A pulse of light emitted from the common origin $(0, 0, 0)$ of both frames of reference at time $t' = t = 0$ will appear to both observers as an expanding sphere of light centered on the respective origins of the two frames of reference. The equations for the expanding spheres in each frame of reference are respectively

$$x^2 + y^2 + z^2 - c^2 t^2 = 0$$
$$x'^2 + y'^2 + z'^2 - c^2 t'^2 = 0, \tag{2.3}$$

where c is the speed of light. For these equations to both be valid, the time t' in the frame of reference S' is probably not equal to the time t in the frame of reference S, and the coordinates (x, y, z) and (x', y', z') may be different. We show below that the times t and t' and the distances x and x' change.

We want to determine the relationships between the S' coordinates (t', x', y', z') and S coordinates (t, x, y, z). At right angles to the x-direction (perpendicular to the

---

[4] The Michelson-Morley experiment in 1887 showed that light propagates at the same velocity in orthogonal directions at any orientation and time. Albert Michelson and Edward Morley, at what is now Case Western Reserve University in Cleveland, Ohio, undertook to measure any change in the velocity of light with direction using an interferometer which Michelson invented (see Exercise 2.2). The experiment helped to debunk the idea, current at the time, of the existence of a medium (known as the "luminiferous aether," "aether," or "ether") in which light propagates. It was previously thought that light propagated at a constant velocity relative to the aether, much like sound propagates at a constant velocity relative to air. The idea of the aether still survives in popular imagination, for example, when unintercepted radio messages are said to "disappear into the aether," but there is no evidence or need for its existence.

velocity **v**), the distances measured in the two frames of references are the same, as there is no velocity difference between the frames of reference in the y-and z-directions. We have

$$y' = y,$$
$$z' = z. \tag{2.4}$$

We need to find the relationship between (x, t) and (x', t'). We can always choose an initial time $t = 0$ and initial position $x = 0$, so that at this initial time and position $t' = 0$ and $x' = 0$. If space and time are homogeneous, derivatives of t and x should also remain constant with changes of the initial time and position. These conditions mean that the transform between (x, t) and (x', t') is linear and we can write

$$x' = Ax + Bt,$$
$$t' = Cx + Dt, \tag{2.5}$$

where the coefficients A, B, C, and D are to be determined.

Setting $x' = 0$ means that $(x, 0, 0)$ represents the origin of the S' frame of reference as measured in the S frame of reference. We have that

$$Ax + Bt = 0.$$

This can be written as

$$\frac{x}{t} = -\frac{B}{A}.$$

The value of x/t is the velocity of the origin O' as measured in S which is equal to v. We have that $B = -vA$, so can write

$$x' = A(x - vt). \tag{2.6}$$

It is possible to solve Equations 2.5 for x and t:

$$x = \frac{Dx' + vAt'}{AD - BC}, \tag{2.7}$$

$$t = \frac{At' - Cx'}{AD - BC} \tag{2.8}$$

These two equations can be manipulated as before. Consider the origin of the reference frame S where $x = 0$. Equation 2.5 with $x = 0$ can be rearranged to show that

$$\frac{x'}{t'} = -\frac{vA}{D}.$$

The frame S is moving at velocity $-v$ with respect to frame S' so $x'/t' = -v$ and hence $A = D$. Using $A = D$ and the earlier result that $B = -vA$, we can modify Equations 2.7 and 2.8:

$$x = \frac{A(x' + vt')}{A^2 + vAC} \tag{2.9}$$

$$t = \frac{A(t' - (C/A)x')}{A^2 + vAC} \tag{2.10}$$

We now compare this transformation from frame S' to S with the earlier transformation (Equation 2.5) from frame S to S'. With just a small modification and using Equation 2.6, Equations 2.5 become

$$x' = A(x - vt) \tag{2.11}$$

$$t' = A(t + (C/A)x). \tag{2.12}$$

Any difference between the transformation from frame S' to S and from frame S to S' arises because the velocity of S' with respect to S is v, while the velocity of S with respect to S' is $-v$. We can obtain Equations 2.11 and 2.12 by swapping the primed and unprimed variables and changing the sign of v in Equations 2.9 and 2.10. For this interchange to work, $A^2 + vAC = 1$ and C/A must be proportional to v. We have

$$A^2(1 + v(C/A)) = 1$$

which can be readily rearranged to give an expression for the coefficient A. We find

$$A = \frac{1}{\sqrt{1 + v(C/A)}}. \tag{2.13}$$

We gained some idea of the form of the dilation of time when measured in a different frame of reference in the discussion of Section 1.2. For events measured at the same position in space (e.g. $x = 0$), time dilates in the stationary frame as observed in the moving frame following $t' = t/\sqrt{1 - v^2/c^2}$ (adapting Equation 1.13). Similarly, the distance $x'_2 - x'_1$ measured for a moving frame of reference in a stationary frame appears as a contracted distance $x_2 - x_1 = (x'_2 - x'_1)\sqrt{1 - v^2/c^2}$. The swapping of frames of reference in comparing Equation 2.12 with Equation 2.10 shows that C/A is proportional to the velocity v. It is reasonable to propose a trial value of C/A, where $C/A = -v/c^2$. The transformation from frame S to frame S' is then given by

$$x' = \frac{x - vt}{\sqrt{1 - v^2/c^2}}, \tag{2.14}$$

$$t' = \frac{t - vx/c^2}{\sqrt{1 - v^2/c^2}} \tag{2.15}$$

in agreement with our ideas of time dilation and length contraction as discussed in Section 1.2. The reverse transformation from the moving frame S' (the proper frame) to the frame S is obtained by simply swapping the primed and unprimed symbols and changing the sign of the velocity v to $-v$:

$$x = \frac{x' + vt}{\sqrt{1 - v^2/c^2}}, \tag{2.16}$$

$$t = \frac{t' + vx'/c^2}{\sqrt{1 - v^2/c^2}} \tag{2.17}$$

The compete relationship between (t, x, y, z) and (t', x', y', z') is known as the Lorentz transformation after Hendrik Lorentz (1853–1928)[5] and can be written as:

$$t' = \gamma \left(t - \frac{vx}{c^2} \right)$$
$$x' = \gamma (x - vt)$$
$$y' = y,$$
$$z' = z, \tag{2.18}$$

where we define the Lorentz parameter

$$\gamma = \left(1 - \frac{v^2}{c^2} \right)^{-1/2}. \tag{2.19}$$

The velocity **v** of the moving frame S' is here taken to be positive v when directed away from the rest frame S in the positive x-direction.

Earlier, we considered how observers in frame S and S' would see an expanding pulse of light originating from their assumed common origin at time $t = 0$ (see Equations 2.3). At later times t, observers in both frames should see an expanding sphere of light centered on the S and S' origins such that the position coordinates (x', y', z') and (x, y, z) of the light pulses at times t' and t satisfy

$$x'^2 + y'^2 + z'^2 - c^2 t'^2 = 0$$

---

[5] Hendrik Lorentz was born in Arnhem, the Netherlands, in 1853. He was awarded the Nobel prize together with Pieter Zeeman in 1902 for the discovery of the magnetic field splitting of spectral lines, now known as the Zeeman effect. Between 1892 and 1895 Lorentz derived the relativistic transformation bearing his name as a way of transforming between different velocities relative to the luminiferous aether, but at the time, unlike Einstein, he did not make the logic leap to drop the idea of the aether.

and

$$x^2 + y^2 + z^2 - c^2 t^2 = 0.$$

Substitution of the Lorentz transformation (Equations 2.18) from frame of reference S to frame S' in the primed version of Equations 2.3 gives

$$x'^2 + y'^2 + z'^2 - c^2 t'^2 = \gamma^2 (x - vt)^2 + y^2 + z^2 - c^2 \gamma^2 \left(t - \frac{vx}{c^2} \right)^2$$

$$= x^2 + y^2 + z^2 - c^2 t^2 + (t^2 - x^2/c^2)(2 - v^2/c^2) = x^2 + y^2 + z^2 - c^2 t^2 = 0$$

upon specifically noting that in the S frame, the distance x light travels in time t is given by $x = ct$. This manipulation shows that both primed and unprimed expressions in Equations 2.3 are satisfied and that observers in frames S and S' do both see an expanding sphere of light. Our choice for the value $C/A = -v/c^2$ is justified as Equations 2.3 are both valid with this choice.

The Lorentz transformation can be represented in a matrix form:

$$\begin{bmatrix} t' \\ x' \\ y' \\ z' \end{bmatrix} = \begin{bmatrix} \gamma & -\gamma v/c^2 & 0 & 0 \\ -\gamma v & \gamma & 0 & 0 \\ 0 & 0 & 1 & 0 \\ 0 & 0 & 0 & 1 \end{bmatrix} \begin{bmatrix} t \\ x \\ y \\ z \end{bmatrix} \quad (2.20)$$

Operating in terms of ct and ct', the Lorentz transformation becomes:

$$\begin{bmatrix} ct' \\ x' \\ y' \\ z' \end{bmatrix} = \begin{bmatrix} \gamma & -\gamma\beta & 0 & 0 \\ -\gamma\beta & \gamma & 0 & 0 \\ 0 & 0 & 1 & 0 \\ 0 & 0 & 0 & 1 \end{bmatrix} \begin{bmatrix} ct \\ x \\ y \\ z \end{bmatrix}, \quad (2.21)$$

where $\beta = v/c$.

We expect an appropriate low velocity limit and an equivalence in the frames of reference to be apparent in the Lorentz transformation. When the velocity v of the moving frame of reference is much less than the speed of light c, we have that $\gamma = 1$ and $\beta = 0$. The Lorentz transformation (Equation 2.20) is then equivalent to the Galilean transformation (Equation 2.2).

Manipulating Equation 2.21 by inverting the 4×4 matrix we can determine an expression for the stationary coordinates (ct, x, y, z) as a function of the moving coordinates (ct', x', y', z'). We have

$$
\begin{bmatrix} ct \\ x \\ y \\ z \end{bmatrix} = \begin{bmatrix} \gamma & -\gamma\beta & 0 & 0 \\ -\gamma\beta & \gamma & 0 & 0 \\ 0 & 0 & 1 & 0 \\ 0 & 0 & 0 & 1 \end{bmatrix}^{-1} \begin{bmatrix} ct' \\ x' \\ y' \\ z' \end{bmatrix} = \begin{bmatrix} \gamma & \gamma\beta & 0 & 0 \\ \gamma\beta & \gamma & 0 & 0 \\ 0 & 0 & 1 & 0 \\ 0 & 0 & 0 & 1 \end{bmatrix} \begin{bmatrix} ct' \\ x' \\ y' \\ z' \end{bmatrix}.
$$

$$(2.22)$$

As the two inertial frames of reference are equivalent, in the dashed frame of reference, the undashed frame of reference is moving at a speed $-v$ so that $\beta = -v/c$ and Equation 2.22 is fully consistent with Equation 2.21. This manipulation exemplifies a general and useful way to change expressions involving Lorentz transformations. Formulas involving Lorentz transformations from frame S' (moving at velocity $\mathbf{v}$) to frame S (nominally stationary) can be changed to formulas involving Lorentz transformations from frame S (nominally stationary) to frame S' (moving at velocity $\mathbf{v}$) by simply swapping primed symbols for unprimed symbols and unprimed symbols for primed symbols and then reversing the sign of the velocity so that v becomes $-v$ in the Lorentz transformation matrix.

2.3 The Transformation of Velocities

The Lorentz transformation enables a straightforward method for determining the transformation of a velocity measured in one frame of reference into the velocity observed in another frame of reference. The transformation of a velocity $\mathbf{u}'$ from a frame S' to a frame S can be determined by considering small increments of space and time in the Lorentz transformation (Equation 2.20). As before we assume that frame S' is moving along the x-axis at a velocity $\mathbf{v}$ relative to the frame of reference S, with positive v implying that the velocity $\mathbf{v}$ of the moving frame S' is directed away from the rest frame S in the positive x-direction. Writing the Lorentz transformation in differential form:

$$
\begin{aligned}
dt &= \gamma \left(dt' + \frac{v}{c^2} dx' \right) \\
dx &= \gamma (dx' + v dt') \\
dy &= dy', \\
dz &= dz'
\end{aligned}
$$

$$(2.23)$$

The velocities along the three axes x, y, and z in the frame S are related to the velocities in frame S' by

$$
u_x = \frac{dx}{dt} = \frac{\gamma (dx' + v dt')}{\gamma (dt' + v dx'/c^2)} = \frac{u'_x + v}{1 + v u'_x/c^2},
$$

$$(2.24)$$

$$u_y = \frac{dy}{dt} = \frac{dy'}{\gamma(dt' + vdx'/c^2)} = \frac{u'_y}{\gamma(1 + vu'_x/c^2)}, \tag{2.25}$$

$$u_z = \frac{dz}{dt} = \frac{dz'}{\gamma(dt' + vdx'/c^2)} = \frac{u'_z}{\gamma(1 + vu'_x/c^2)}. \tag{2.26}$$

These equations for the velocities in Cartesian coordinates in the x-, y- and z-directions assume the velocity $\mathbf{v}$ is parallel to the x-axis. When the frame S' is moving with an arbitrary velocity $\mathbf{v}$ relative to the frame S, we can treat the velocity transformation by considering the components of $\mathbf{u}$ parallel and perpendicular to $\mathbf{v}$. The parallel component of velocity ($u'_{||}$) transforms as shown in Equation 2.24, while the perpendicular component of velocity ($u'_\perp$) transforms as for Equations 2.25 and 2.26. We write

$$u_{||} = \frac{u'_{||} + v}{1 + vu'_{||}/c^2}, \tag{2.27}$$

$$u_\perp = \frac{u'_\perp}{\gamma(1 + vu'_{||}/c^2)}. \tag{2.28}$$

The directions of the velocities in the two frames can be related by considering how the angle θ' between the velocity $\mathbf{u}'$ in the S' frame and the relative velocity $\mathbf{v}$ of the frames relates to the angle θ between the velocity $\mathbf{u}$ in the S frame and the relative velocity $\mathbf{v}$. We have that $u'_\perp = u' \sin\theta'$ and $u'_{||} = u' \cos\theta'$ and $\tan\theta = u_\perp/u_{||}$. This last relation gives an expression known as the "aberration formula," which relates the velocity direction for a particle as measured in two different reference frames. Angles θ' to the velocity $\mathbf{v}$ of a moving frame as measured in the moving frame transform to the angle θ to the velocity $\mathbf{v}$ in the stationary frame by

$$\tan\theta = \frac{u_\perp}{u_{||}} = \frac{u' \sin\theta'}{\gamma(u' \cos\theta' + v)}. \tag{2.29}$$

A feature of the Lorentz transformation mentioned in Section 2.2 is the general way of moving from a Lorentz transformation from frame S' (moving at velocity $\mathbf{v}$) to frame S (nominally stationary) to a Lorentz transformation from frame S (nominally stationary) to frame S' (moving at velocity $\mathbf{v}$). Formulas involving Lorentz transformations from frame S' (moving at velocity $\mathbf{v}$) to frame S (nominally stationary) can be changed to formulas involving Lorentz transformations from frame S (nominally stationary) to frame S' (moving at velocity $\mathbf{v}$) by (i) simply reversing the sign of the velocity so that v becomes $-v$ and (ii) interchanging primed and unprimed designations on parameters. The velocity transformations from frame S' to S presented for Equations 2.27 and 2.28 can be rewritten as transformations from frame S to S':

$$u'_{||} = \frac{u_{||} - v}{1 - vu_{||}/c^2},$$ (2.30)

$$u'_{\perp} = \frac{u_{\perp}}{\gamma(1 - vu_{||}/c^2)}$$ (2.31)

Similarly the aberration formula (Equation 2.29) for the angle of a particle velocity can be changed to give the transformation from frame S to frame S':

$$\tan\theta' = \frac{u\sin\theta}{\gamma(u\cos\theta - v)}$$ (2.32)

2.4 The Relativistic Rocket Equation

Vehicles that are accelerated using thrust supplied by expelling part of the mass of the vehicle are termed "rockets." Rockets are propelled by expelling mass in the opposite direction of desired travel so that the forward force on the rocket due to the backward momentum of expelled mass causes forward motion of the remaining rocket mass. The forward velocity of a rocket is related to the initial and final mass of the rocket by a "rocket equation."

Considering measurements in the frame of reference of a rocket, the classical rocket equation can be obtained. As before, we use primed symbols to indicate the "proper" frame of reference where the observer is moving with the rocket. Let the rate of change of the momentum of the rocket exhaust be $v'_{ex}dM_{ex}/dt'$, where v'_{ex} is the velocity of the exhaust in the frame of the rocket and dM_{ex}/dt' is the rate of rest mass flow in the exhaust. The force $v'_{ex}dM_{ex}/dt'$ is the "thrust" of the rocket. The rate of change of the momentum of the remaining rest mass M of the rocket can be written as Mdv'_R/dt', where v'_R is the rocket velocity. The rocket mass changes at a rate dM/dt' such that $dM_{ex}/dt' = -dM/dt'$. Equating the force from the exhaust $v'_{ex}dM_{ex}/dt'$ and the force on the rocket Mdv'_R/dt', we have

$$M\frac{dv'_R}{dt'} = -v'_{ex}\frac{dM}{dt'}.$$

Rearranging, we can write that

$$\frac{dM}{M} = -\frac{1}{v'_{ex}}dv'_R.$$ (2.33)

Integrating the left-hand side from an initial rest mass M_0 to a final rest mass M_F and the right-hand hand from a velocity of zero to a final velocity v'_F, we have

$$\int_{M_0}^{M_F} \frac{dM}{M} = -\frac{1}{v'_{ex}}\int_0^{v'_F} dv'_R.$$

Undertaking the integrations with a small rearranging gives the nonrelativistic rocket equation for the rocket velocity v'_F assuming a constant exhaust velocity v'_{ex}:

$$v'_F = v'_{ex} \ln\left(\frac{M_0}{M_F}\right), \tag{2.34}$$

where the final rocket mass is M_F and the initial mass is M_0.

The relativistic version of the rocket equation is obtained by applying the relativity theory examined in Section 2.3. We use the relationship (Equation 2.27) between the velocity v'_R of an object measured by an observer in the frame of the rocket moving at speed v and the velocity v_R of the same object measured by a stationary observer. We assume that the velocity v'_R of the object is parallel to the velocity $\mathbf{v}$ of the rocket. Rewriting Equation 2.27 in the notation we are using for the rocket motion, we have

$$v_R = \frac{v'_R + v}{1 + vv'_R/c^2}. \tag{2.35}$$

We now consider how the value of a small velocity increment dv'_R in the moving frame is measured in the stationary frame. Taking the differential of Equation 2.35 gives

$$dv_R = dv'_R \left(\frac{1}{1 + vv'_R/c^2} - \frac{v'_R + v}{(1 + vv'_R/c^2)^2}\frac{v}{c^2}\right)$$

$$= dv'_R \left(\frac{1 - v^2/c^2}{(1 + vv'_R/c^2)^2}\right).$$

In the limit where an object is moving with the rocket, the velocity v'_R of the object relative to an observer in the rocket is zero ($v'_R = 0$) and, furthermore, the velocity v_R seen by the stationary observer is such that $v_R = v$. The relationship between a small increment of velocity dv'_R in the moving frame and a small increment of velocity dv_R in the stationary frame is then given by

$$dv'_R = \frac{dv_R}{1 - v_R^2/c^2}. \tag{2.36}$$

We can substitute for dv'_R in the expression for dM/M obtained for the rocket equation in the frame of the rocket (Equation 2.33). Substituting and now integrating the right-hand side from a velocity $v_R = 0$ to a final velocity v_F in the stationary frame, we have

$$\int_{M_0}^{M_F} \frac{dM}{M} = -\frac{1}{v'_{ex}} \int_{v_R=0}^{v_F} \frac{dv_R}{1 - v_R^2/c^2}. \tag{2.37}$$

The integrations reduce to an expression for the relativistic rocket equation:

$$\ln\left(\frac{M_0}{M_F}\right) = \frac{c}{2v'_{ex}} \ln\left(\frac{1 + v_F/c}{1 - v_F/c}\right)$$

which simplifies to

$$\frac{M_0}{M_F} = \left(\frac{1 + v_F/c}{1 - v_F/c}\right)^{c/(2v'_{ex})}. \tag{2.38}$$

The asymptotic limit when $v_F \ll c$ is not apparent with the relativistic rocket equation in the form given in Equation 2.38. Rearranging Equation 2.38, we have that

$$\frac{v_F}{c} = \frac{(M_0/M_F)^{c/(2v'_{ex})} - 1}{(M_0/M_F)^{c/(2v'_{ex})} + 1}.$$

We now use $(M_0/M_F)^{c/(2v'_{ex})} = \exp\left(c/(2v'_{ex}) \ln(M_0/M_F)\right)$ and the identity for the hyperbolic tanh function:

$$\tanh x = \frac{e^{2x} - 1}{e^{2x} + 1}$$

Equation 2.38 can then be rewritten in a form that relates the rocket velocity v_F in the stationary frame to the exhaust velocity and initial M_0 and final rocket rest mass M_F:

$$\frac{v_F}{c} = \tanh\left(\frac{v'_{ex}}{c} \ln\left(\frac{M_0}{M_F}\right)\right) \tag{2.39}$$

The velocity v'_{ex} of the rocket exhaust is assumed constant in both the nonrelativistic (Equation 2.34) and relativistic (Equations 2.39 and 2.38) treatments and is measured in the frame of reference of the rocket. When the rocket exhaust velocity $v'_{ex} \ll c$, the final velocity v_F of the rocket in the stationary frame determined using Equation 2.39 is equal to values given by the nonrelativistic expression (Equation 2.34).

Our treatment to obtain the relativistic rocket equation is an example where we use special relativity to model an accelerating frame of reference. In the limit of an infinitesimally small duration of time when the velocity increment dv'_R is infinitesimally small, we can use special relativity as the frame of reference is inertial over an infinitesimally small duration of time. We use this argument to work out the transformation of acceleration from a moving to rest frame later in Section 3.3.

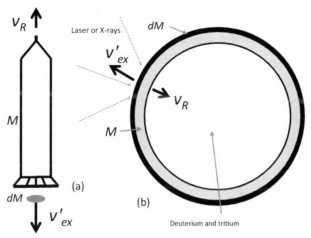

Figure 2.1   Schematic illustrations of (a) a rocket of mass M propelled by expelling mass dM, and (b) the rocket-like compression of deuterium and tritium
in inertial fusion where indirect x-rays (created in a black-body cavity) or direct
laser beams ablate a mass dM and propel shell material of mass M to compress the
deuterium and tritium.

The rocket equations are applicable to the laser-ablated capsules containing deuterium (D) and tritium (T) fuel utilized in inertial fusion research [82]. In inertial
fusion, optical lasers or X-rays ablate a spherical plastic shell containing frozen DT
on the inside of the shell and gas within the shell (see Figure 2.1). The ablated material expands outwards from the shell and drives the shell inwards, compressing
the D and T fuel so that fusion to helium (and energetic neutrons) occurs. The in-
ward velocity of the capsule shell is governed by the rocket equations. The ablation
rate of the shell material and the expanding plasma velocity are equivalent to dM
and the exhaust velocity v'_{ex}, respectively, as featured in the rocket equations. A
general form including relativistic effects of the rocket equation relevant to inertial
fusion can be found by modifying Equation 2.37. An integration of the relative rate
of ablation dM/M gives the final implosion velocity v_F:

$$\int_{M_0}^{M_F} \frac{dM}{M} = -\int_{v_R=0}^{v_F} \frac{dv_R}{v'_{ex}(1 - v_R^2/c^2)}, \qquad (2.40)$$

where v'_{ex} is the outward velocity of ablated material as a function of the implosion
velocity. If the outward velocity v'_{ex} of ablated material is constant, Equation 2.39
then gives the final implosion velocity v_F as a function of the initial mass M_0 of the
shell plus fuel and the final mass M_F of the DT fuel.

2.5 The Transformation of the Propagation Direction of Light

The complementarity principle discussed as a footnote in our initial wave treatment of light (Section 1.1) states that both light and matter exhibit wave-like and particle-like properties depending on the measurement or interaction being considered. A photon is the quantum of electromagnetic radiation (for further discussions on the quantization of light, see Sections 2.9 and 6.4).

In relativistic calculations, photons can be treated as zero-mass particles moving at the speed of light c in different frames of reference. This means that calculations of the trajectories of light rays seen by observers with different velocities can be made using relativistic treatments developed for particles by setting the particle velocity to c. The angular variation of light in a moving frame of reference as observed in a stationary frame is examined in the section using the formulas for velocity transformation between frames as determined in Section 2.3.

As an example of the effect on the observed direction of light propagation in different frames of reference, we set the velocity $u' = c$ in Equation 2.29 to get a relationship between the angle θ' of light propagation to the velocity $\mathbf{v}$ in a moving frame as observed (at angle θ to the velocity $\mathbf{v}$) in a nominally stationary frame:

$$\tan \theta = \frac{\sin \theta'/\gamma}{\cos \theta' + v/c} \tag{2.41}$$

A narrowing of the angular range of radiation observed in the stationary or laboratory frame for light emitted at a range of angles in a moving frame is illustrated in Figure 2.2, where results for Equation 2.41 are plotted. At high velocities where the Lorentz parameter $\gamma = 1/(1 - (v/c)^2)^{1/2}$ is large, the emission in the stationary frame is concentrated in the direction parallel to the velocity $\mathbf{v}$ of the moving frame.

To obtain values of $\sin \theta$ and $\cos \theta$ for light emission in the laboratory frame, we need to evaluate the sum of the squares of the numerator and denominator of Equation 2.41. We have

$$\left(\left(\cos \theta' + \frac{v}{c} \right)^2 + \frac{\sin^2 \theta'}{\gamma^2} \right)^{1/2} = 1 + \frac{v}{c} \cos \theta'$$

upon substituting $1/\gamma^2 = 1 - (v/c)^2$. Expressions for $\sin \theta$ and $\cos \theta$ then follow:

$$\sin \theta = \frac{\sin \theta'/\gamma}{1 + (v/c) \cos \theta'}, \tag{2.42}$$

$$\cos \theta = \frac{\cos \theta' + v/c}{1 + (v/c) \cos \theta'} \tag{2.43}$$

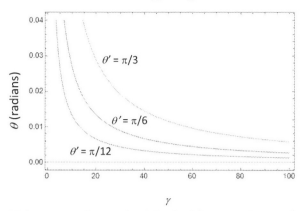

γ

Figure 2.2 Light propagation angles θ in the laboratory frame of reference for light emitted in the moving frame at angles θ' as labeled as a function of the Lorentz γ-parameter of the moving frame. Angles are measured relative to the velocity **v** of the moving frame.

For light emitted in the moving frame S' perpendicular to the direction of the velocity we have $\theta' = \pi/2$, so that the angle of the light in the stationary S frame is given by

$$\sin\theta = \frac{1}{\gamma}. \tag{2.44}$$

When the frame of reference is moving with velocity v close to the speed of light c, Equation 2.42 shows that θ becomes small and $\theta \approx 1/\gamma$ for all angles of emission θ' close to $\pi/2$. If photons are emitted isotropically in frame S' moving close to the speed of light, then half of them (with $\theta' < \pi/2$) are found in the stationary frame S within $\theta < 1/\gamma$. Very few photons are found in the stationary frame with $\theta \gg 1/\gamma$. This collection of light in the stationary frame around the direction of motion of the emitter is known as "beaming." The beaming effect dominates, for example, the photon emission angles observed in synchrotrons where electrons are accelerated with velocities close to the speed of light.

2.6 The Doppler Effect

The Doppler effect is the change in frequency of a wave detected by an observer when the source of the wave is moving relative to the observer. With sound waves, the Doppler effect arises solely due to the change in the time for the wave to travel to the observer as the source moves towards (or farther from) the observer. The Doppler effect is named after Christian Doppler (1803–1853), who lived and worked in the Austrian Empire at the Prague Polytechnic (now the Czech Technical University in Prague) and universities now in Hungary and Vienna.

Doppler explained the colors observed in binary stars in terms of the effect that now bears his name. Emission of light arises from the acceleration of a charged particle such as an electron or a quantum mechanical transition in an ion or atom (see Section 1.3).[6] As with the Doppler effect for sound waves, there is a change in the time of flight of photons from a moving source as the source moves closer to (or farther from) a stationary observer. With the Doppler effect for light with source velocities approaching the speed of light, an additional effect becomes evident. We have seen from consideration of the Lorentz transformation (Section 2.2) that a periodic phenomenon in a moving frame S' has a longer period by factor of the Lorentz parameter γ when viewed by observers S in a stationary frame. Time dilation means that the change in the period of emitted light between the moving S' and stationary S frames is given by

$$\Delta t = \gamma \Delta t'.$$

The reducing distance for the light to travel in a time Δt is $v \Delta t \cos \theta$, where $v \cos \theta$ is the component of the velocity v along the line of sight of the observer. Here θ is the angle to the velocity $\mathbf{v}$ from the light source to the observer in the observer's frame of reference. We assume that v is positive for velocities towards the observer. The period Δt_S of the light recorded by the observer is

$$\Delta t_S = \Delta t - \frac{v \Delta t \cos \theta}{c} = \Delta t \left(1 - \frac{v \cos \theta}{c} \right).$$

The observed frequency of light seen by the observer is

$$\omega = \frac{2\pi}{\Delta t_S} = \frac{2\pi}{\Delta t \, (1 - v \cos \theta / c)} = \frac{2\pi}{\gamma \, \Delta t' \, (1 - v \cos \theta / c)} = \frac{\omega'}{\gamma \, (1 - v \cos \theta / c)}.$$

$$(2.45)$$

where ω' is the frequency of light in the moving emitting frame of reference.

The Doppler shift of frequency from the stationary frame S to the moving frame S' (Equation 2.45) can be written as

$$\omega' = \omega \gamma \left(1 - \frac{v}{c} \cos \theta \right). \qquad (2.46)$$

The inverse of this equation produces a similar expression for the Doppler shift from the moving frame S' into the stationary frame of reference S for a velocity v moving towards the stationary observer. We interchange the frames of reference by interchanging primed and unprimed symbols and substituting $-v$ for v in Equation 2.46 to obtain

---

[6] A transition between discrete quantum states in an atom or ion also arises due to the acceleration of charge, but the evaluation of the probability of a transition needs to be undertaken using quantum mechanics.

$$\omega = \omega'\gamma \left(1 + \frac{v}{c}\cos\theta'\right),\tag{2.47}$$

where θ' is the angle to the velocity $\mathbf{v}$ from the light source to the observer in the frame of reference of the light source.

For velocities $v \ll c$, the Doppler shift in frequency $\Delta\omega = \omega - \omega'$ is directly proportional to the velocity component $v_{||} = v\cos\theta$ parallel to the line of sight. We have from Equation 2.47 that

$$\frac{\Delta\omega}{\omega} \approx \frac{v_{||}}{c}\tag{2.48}$$

as $\omega' \approx \omega$, $\gamma \approx 1$, and $\theta \approx \theta'$. The proportionality of the shift of frequency to the velocity of the emitter with Doppler broadening at lower velocities is used to measure the velocities of ions in plasmas as they emit light. The profile of ion velocity distribution along the observer's line of sight is replicated in the spectral shape of the observed emission (see Section 7.3).

The scatter of light from a moving object with velocity $\mathbf{v}$ is affected by the Doppler effect. Scatter of light usually means that in the frame of the scattering particle, the incoming and scattered light frequencies are the same (though Compton scatter of high-frequency photons does involve a frequency shift; see Section 4.8). If the light is incident along a direction with unit vector $\hat{\mathbf{i}}$, the component of the velocity along the direction of the incident light is $\hat{\mathbf{i}}.\mathbf{v}$. Scatter along a direction with unit vector $\hat{\mathbf{j}}$ has a component $\hat{\mathbf{j}}.\mathbf{v}$ along the direction of the velocity. There is a Doppler shift in both the absorption and re-emission processes such that the frequency ω_s observed after scattering is related to the incoming frequency ω by

$$\omega_s = \omega\gamma^2\left(1 - \hat{\mathbf{i}}\cdot\mathbf{v}/c\right)\left(1 + \hat{\mathbf{j}}\cdot\mathbf{v}/c\right) = \omega\gamma^2\left(1 + (\hat{\mathbf{j}} - \hat{\mathbf{i}})\cdot\mathbf{v}/c\right).\tag{2.49}$$

Equation 2.49 is obtained by first applying Equation 2.46 for the shift from frame S to S' and then Equation 2.47 for the shift from frame S' back to frame S. Velocities along the direction $\hat{\mathbf{j}} - \hat{\mathbf{i}}$ cause shifts in frequency of the scattered light. Again as for Equation 2.48 if $v \ll c$, there is a simple proportionality of the shift in frequency $\Delta\omega_s = \omega_s - \omega$ on scattering such that

$$\frac{\Delta\omega_s}{\omega} \approx \frac{(\hat{\mathbf{j}} - \hat{\mathbf{i}}).\mathbf{v}}{c}.\tag{2.50}$$

In astrophysics and cosmology, starlight and other light emission are Doppler shifted according to Equation 2.47 with the velocity v directed precisely away from the observer such that $\theta' = \pi$. The observed frequencies ω in the stationary frame on Earth are shifted from the moving frequencies ω' emitted by the star such that

$$\omega = \omega' \gamma \, (1 - v/c) = \omega' \frac{1 - v/c}{(1 - v^2/c^2)^{1/2}} = \omega' \sqrt{\frac{(1 - v/c)^2}{(1 + v/c)(1 - v/c)}}$$

$$= \omega' \sqrt{\frac{1 - v/c}{1 + v/c}}. \tag{2.51}$$

The shift of observed frequencies in astrophysics and cosmology to lower frequencies than those emitted is usually referred to as the "redshift" with a symbol z employed to give a quantitative redshift value. The redshift z is defined by

$$1 + z = \frac{\omega'}{\omega} = \sqrt{\frac{1 + v/c}{1 - v/c}}. \tag{2.52}$$

For small redshift $(v << c)$, $z \approx v/c$. As the velocity of a distant cosmic object relative to the Earth approaches the speed of light, the value of z approaches infinity. In astrophysics and cosmology, the largest redshift z of observable radiation in the expanding universe is that of the cosmic microwave background which is shifted by a value $z \approx 1100$ [108]. However, some of the shift of the cosmic microwave background occurs due to general relativistic effects associated with the expansion of the universe. For large cosmic redshifts, the redshift $z + 1$ is most accurately given by the ratio of the cosmic scalelength of the universe now and the cosmic scalelength when the light was emitted.

2.7 Relativistic Momentum

Measurements of time and distance change with the frame of reference of the observer, so measurements of momentum change with the frame of reference of the observer. Special relativity theory has as an axiom that the laws of physics are the same in all inertial frames of reference, so we require expressions for momentum that preserve for all observers the same relationship between force and momentum. In classical Newtonian physics, the experimentally observed relationship between force $\mathbf{F}$ and momentum $\mathbf{p}$ is

$$\mathbf{F} = \frac{d\mathbf{p}}{dt}. \tag{2.53}$$

In the frame of reference S' of a moving body, the force $\mathbf{F}$ is determined by a straightforward evaluation of momentum $\mathbf{p} = m_0 \mathbf{u}$, where m_0 is known as the rest mass and $\mathbf{u}$ is the velocity of the body. The familiar Newtonian equation relating force and acceleration is

$$\mathbf{F} = m_0 \frac{d\mathbf{u}}{dt}, \tag{2.54}$$

where $d\mathbf{u}/dt$ is the acceleration of the body.

A relationship between the velocities u_y and u'_y of a body perpendicular to the velocity **v** in the respective frames of reference S and S' is obtained from Equation 2.25. The conversion of velocity components perpendicular to the relative frame velocity **v** from frame S' to frame S is given by

$$u_y = \frac{dy}{dt} = \frac{u'_y}{\gamma(1 - vu'_x/c^2)}. \tag{2.55}$$

Assuming that the S' frame moves with the body in the x-direction, then $u'_x = v$ and Equation 2.55 yield

$$u_y = \gamma u'_y.$$

The apparent momentum of a body traveling in the y-direction measured in the stationary frame of reference S is given by

$$p_y = m_0 u_y = \gamma m_0 u'_y, \tag{2.56}$$

where $m_0 u'_y$ is the momentum of the body measured in the proper frame of reference S'.

The momentum directed in the x-direction (parallel to **v**) requires some consideration to check if the increase of momentum of a body in a speeding frame of reference also applies for momentum directed parallel to **v**. For any arbitrarily directed momentum, we can always take components parallel and perpendicular to the velocity **v** of the moving frame of reference, so showing the same momentum increase in going from the moving to stationary frame of reference in directions perpendicular and parallel to **v** means that the change in momentum applies generally.

Equation 2.27 gives an expression for the conversion of velocity components parallel to **v** from frame S' to frame S:

$$u_{\parallel} = \frac{u'_{\parallel} + v}{1 + vu'_{\parallel}/c^2} \tag{2.57}$$

Consider the effect of incremental velocity changes of a body in frame S' in the directions parallel to the velocity **v** in the stationary frame S. An observer in frame S allows for the velocity of the moving frame S' by subtracting the frame velocity in a measurement of the incremental change in velocity $du_{\parallel}$ of a body parallel to the velocity **v**. Considering the differential change in velocity parallel to the velocity **v**, Equation 2.57 gives that

$$du_{\parallel} = \frac{du'_{\parallel} + v}{1 + v\,du'_{\parallel}/c^2} - v = \frac{du'_{\parallel}(1 - v^2/c^2)}{1 + v\,du'_{\parallel}/c^2} \approx du'_{\parallel}(1 - v^2/c^2). \tag{2.58}$$

The right-hand expression in Equations 2.58 becomes exact when $du'_{\parallel} \to 0$ in the denominator.

Consider the force $F'_{\parallel}$ measured in the moving S' frame required to produce the velocity change $du'_{\parallel}$. The component of force parallel to the velocity $\mathbf{v}$ is related to the acceleration $du_{\parallel}/dt$ in the rest frame by

$$F'_{\parallel} = m_0 \frac{du'_{\parallel}}{dt'} = m_0 \frac{du_{\parallel}}{dt'} \frac{1}{1 - v^2/c^2} = m_0 \frac{du_{\parallel}}{dt} \frac{1}{(1 - v^2/c^2)^{3/2}} \qquad (2.59)$$

as time dilation means that $dt' = dt\sqrt{1 - v^2/c^2}$. Time dt' is the proper time as it is measured in the moving frame S'. In Section 3.3, we show that force parallel to the velocity $\mathbf{v}$ does not vary between frames of reference, that is, force parallel to $\mathbf{v}$ is Lorentz invariant.[7] This means that

$$F_{\parallel} = F'_{\parallel} = m_0 \frac{du_{\parallel}}{dt} \frac{1}{(1 - v^2/c^2)^{3/2}}. \qquad (2.60)$$

Changes of velocity $du_{\parallel}$ are parallel to the velocity $\mathbf{v}$ so that $du_{\parallel} = dv$. Equation 2.60 is consequently equivalent to

$$F_{\parallel} = \frac{d}{dt} \left(\frac{m_0 v}{\sqrt{1 - v^2/c^2}} \right) = \frac{dp_x}{dt}, \qquad (2.61)$$

where

$$p_x = \frac{m_0 v}{\sqrt{1 - v^2/c^2}} = \gamma m_0 v.$$

By considering the momentum parallel (Equation 2.61) and perpendicular (Equation 2.56) to the velocity $\mathbf{v}$ of the moving frame of reference, the increase in momentum of a body with velocity $\mathbf{v}$ when measured in a stationary frame of reference is seen to be general. We can write

$$\mathbf{p} = \frac{1}{\sqrt{1 - v^2/c^2}} m_0 \mathbf{v} = \gamma m_0 \mathbf{v}. \qquad (2.62)$$

The momentum increase shown in Equation 2.62 is sometimes said to be equivalent to a mass increase from m_0 to γm_0. While m_0 is referred to as the "rest mass," γm_0 is referred to as the "relativistic mass." However, while convenient and commonly used, the idea of the relativistic mass can be misleading in some situations as it changes with the frame of reference. The rest mass m_0, however, is an intrinsic property of a body or particle, much as the charge of particles (e.g. the charge of electrons and protons) is an intrinsic property.

---

[7] An alternative approach is to show that assuming $F_{\parallel} = F'_{\parallel}$ results in momentum in the direction of the moving frame velocity $\mathbf{v}$ given by $p_x = \gamma m_0 c$ as was found for transverse momentum $p_y = \gamma m_0 c^2$. If momentum changes between frames are independent of the direction of the momentum, then $F_{\parallel} = F'_{\parallel}$.

2.8 Relativistic Energy

The mechanics of relativistic bodies extends the Newtonian version of mechanics. At low velocities, the relativistic equations also need to be equivalent to the Newtonian equations. The relativistic force $\mathbf{F}$ on a body is defined in terms of the change of momentum $\mathbf{p}$ of the body by

$$\mathbf{F} = \frac{d\mathbf{p}}{dt}. \tag{2.63}$$

The relativistic work dW produced by a force $\mathbf{F}$ during a small displacement $d\mathbf{r}$ of a body is defined by

$$dW = \mathbf{F} \cdot d\mathbf{r}. \tag{2.64}$$

The rate at which a force $\mathbf{F}$ does work is the power P given by

$$P = \mathbf{F} \cdot \mathbf{u}, \tag{2.65}$$

where $\mathbf{u} = d\mathbf{r}/dt$ is the velocity of the object. The notion of the relativistic kinetic energy of a body can be developed by viewing the work done by a force $\mathbf{F}$ as a contribution towards the relativistic kinetic energy T_K of the body. We write

$$P = \frac{dT_K}{dt} = \mathbf{F} \cdot \mathbf{u}. \tag{2.66}$$

We can manipulate Equation 2.66 as follows:

$$\frac{dT_K}{dt} = \mathbf{F} \cdot \mathbf{u} = \mathbf{u} \cdot \frac{d\mathbf{p}}{dt} = \mathbf{u} \cdot \frac{d}{dt} \left(\frac{m_0 \mathbf{u}}{\sqrt{1 - u^2/c^2}} \right)$$

upon substituting for the momentum $\mathbf{p}$ using Equation 2.62. Differentiating and noting that $\mathbf{u} \cdot \mathbf{u} = u^2$ and $\mathbf{u} \cdot d\mathbf{u}/dt = u \, du/dt$, we obtain

$$\frac{dT_K}{dt} = u \frac{du}{dt} \frac{m_0}{\left(1 - u^2/c^2\right)^{3/2}}$$

which is equivalent to the expression

$$\frac{dT_K}{dt} = \frac{d}{dt} \left(\frac{m_0 c^2}{\sqrt{1 - u^2/c^2}} \right). \tag{2.67}$$

Integrating with respect to time t, we have for the relativistic kinetic energy T_K that

$$T_K = \frac{m_0 c^2}{\sqrt{1 - u^2/c^2}} - m_0 c^2, \tag{2.68}$$

where we have added a constant of integration of value $-m_0c^2$ as we require that the kinetic energy $T_K = 0$ when $u = 0$.

Further evidence that the value of the constant of integration $-m_0c^2$ is valid in Equation 6.62 is obtained by considering the limit of velocities much less than the speed of light. When the velocity is small so that $u << c$, the binomial expansion

$$\left(1 - u^2/c^2\right)^{-1/2} \approx 1 + \frac{u^2}{2c^2}$$

is accurate. The kinetic energy from Equation 2.68 becomes

$$T_K \approx m_0c^2(1 + u^2/2c^2) - m_0c^2 = \frac{1}{2}m_0u^2 \tag{2.69}$$

in agreement with the classical Newtonian expression for the kinetic energy of a body of mass m_0 moving with a velocity **u**.

We can define a total relativistic energy W given by

$$W = T_K + m_0c^2 = \frac{m_0c^2}{\sqrt{1 - u^2/c^2}} = \gamma m_0c^2. \tag{2.70}$$

As the total relativistic energy has a term m_0c^2 added to the kinetic energy, a body or particle at rest can be regarded as having a rest mass energy associated with the rest mass m_0. Indeed, the rest mass energy m_0c^2 can be partly converted to kinetic energy, for example, when large mass particles decay to lower mass particles in nuclear and particle physics reactions. The rest mass energy is the total relativistic energy E_R of a mass m_0 at rest.

The expression for the rest mass energy $E_R = m_0c^2$ is the most famous and well-known equation in physics.[8] Dropping the subscripts on E_R and m_0 for a moment, we have the equation usually used for the mass-energy equivalence:

$$E = mc^2 \tag{2.71}$$

We have defined the relativistic γ factor by $\gamma = 1/\sqrt{1 - u^2/c^2}$. It is clear from Equation 2.70 that an equivalent and useful relationship specifying γ is in terms of the energy W of a body or particle relative to the rest mass energy:

$$\gamma = \frac{W}{m_0c^2} = 1 + \frac{T_K}{m_0c^2}. \tag{2.72}$$

The kinetic energy T_K for highly relativistic particles with velocities close to the speed of light can be much greater than the rest mass energy, so that $\gamma \approx T_K/(m_0c^2)$.

---

[8] Einstein developed the famous relationship between energy and mass in his 1905 papers on special relativity (e.g. "Does the Inertia of a Body Depend upon its Energy-Content?" [32]). Others with different thought experiments had previously obtained a proportionality between energy and mc^2, but had not gotten the correct expression [100].

It is possible to use Equation 2.70 to obtain a relationship between the relativistic energy and the momentum of a body. Squaring Equation 2.70 gives

$$W^2 = \frac{m_0^2 c^4}{1 - u^2/c^2} = \frac{m_0^2 c^4 (1 - u^2/c^2 + u^2/c^2)}{1 - u^2/c^2} = m_0^2 c^4 + \frac{m_0^2 u^2}{1 - u^2/c^2} c^2.$$

Using Equation 2.62 for the momentum, we have

$$p^2 = \mathbf{p} \cdot \mathbf{p} = \frac{m_0^2 u^2}{1 - u^2/c^2}$$

which means that the relativistic energy W is given by

$$W = \sqrt{p^2 c^2 + m_0^2 c^4}. \tag{2.73}$$

The kinetic energy T_K is the total relativistic energy minus the rest mass energy:

$$T_K = \sqrt{p^2 c^2 + m_0^2 c^4} - m_0 c^2 \tag{2.74}$$

Equation 2.73 is particularly important as it shows how momentum p and energy W are related in different frames of reference. We can rewrite Equation 2.73 as

$$(m_0 c^2)^2 = W^2 - (pc)^2. \tag{2.75}$$

The rest mass energy $m_0 c^2$ is constant in all frames, so the energy/momentum quantity $W^2 - (pc)^2$ is constant in all frames of reference. The quantity $W^2 - (pc)^2$ is said to be Lorentz invariant; as in a Lorentz transformation between different frames of reference, $W^2 - (pc)^2$ remains unchanged. A range of Lorentz invariant quantities, including $W^2 - (pc)^2$, the rest mass m_0, and rest mass energy $m_0 c^2$, are listed together for convenience in Appendix A.

We have established relationships for energy and momentum for bodies moving at relativistic velocities as measured in frames of reference not moving with the bodies. We can now differentiate the relativistic energy W (Equation 2.73) with respect to momentum p:

$$\frac{dW}{dp} = \frac{p\, c^2}{(p^2 c^2 + m_0^2 c^4)^{1/2}} \tag{2.76}$$

Substituting the relativistic expression for momentum as a function of velocity v (see Equation 2.62) gives a useful simplification of Equation 2.76. We have that

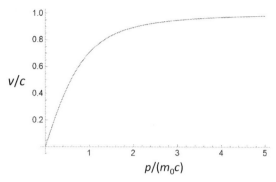

Figure 2.3   The velocity v/c of a particle as a function of the momentum p in units of m_0c, where both velocity and momentum are measured in the rest frame. The velocity increases approximately linearly with momentum up to $p \approx m_0c$ and then asymptotically approaches the speed of light c with increasing p.

$$\frac{dW}{dp} = \frac{m_0 v\, c^2}{(1 - v^2/c^2)^{1/2}(m_0^2 v^2 c^2/(1 - v^2/c^2) + m_0^2 c^4)^{1/2}}$$

$$= \frac{v\, c}{(v^2 + c^2(1 - v^2/c^2))^{1/2}} = v. \qquad (2.77)$$

For all velocities v up to and including $v = c$, the velocity of a body is given by the differential of the energy with respect to momentum. As the rest mass energy m_0c^2 is constant for a body, we can differentiate the total energy W or the kinetic energy T_K as $dW/dp = dT_k/dp = v$.

Rearranging Equation 2.74 we can write for the kinetic energy

$$T_k = m_0 c^2 \left[\left(1 + \left(\frac{p}{m_0 c}\right)^2\right)^{1/2} - 1 \right]. \qquad (2.78)$$

The velocity v of a particle of rest mass m_0 is then related to the momentum p by

$$v = \frac{dT_k}{dp} = \frac{p}{m_0}\left[1 + \left(\frac{p}{m_0 c}\right)^2\right]^{-1/2}. \qquad (2.79)$$

The velocity v/c of a particle (in units of the speed of light) is plotted as a function of the momentum $p/(m_0 c)$ using Equation 2.79 in Figure 2.3.

Radiation fields are quantized in energy with the unit of energy termed the "photon." The quantization of the radiation field is treated in some detail in Section 6.4 where the equilibrium (black body) radiation field is examined. Evidence for radiation quantization in photons also arises from a quantum mechanical treatment of free particles (see Section 2.9) and from an examination of the energy/momentum relationship of particles with zero mass. Photons have zero rest mass and an energy hv, where v is the frequency of the light and h is Planck's constant. Putting $m_0 = 0$

in Equation 2.75 indicates that a pulse of light with a large number n_p of photons has an energy $W = n_p h\nu$ and a momentum of magnitude

$$p = \frac{W}{c}. \tag{2.80}$$

If we set the photon velocity $v = c$ in Equation 2.77, we have that $dW/dp = c$, where c is the constant velocity of light. Integrating energy and momentum increases also gives the relationship $W = pc$ for the photon. Indeed, $W \approx pc$ is a useful approximation for all particles moving at velocities close to the speed of light.

As $p = W/c$ for light, each photon has a momentum of magnitude $h\nu/c$ which is directed in the direction of energy flow of the light field, that is, in the direction of the wavevector $\mathbf{k}$. Using $c = 2\pi\nu/k$, where k is the amplitude of the wavevector of light, we see that the momentum of each photon in vacuum is given by $\hbar\mathbf{k}$, where $\hbar = h/2\pi$.

In radiative transfer, an important parameter is the intensity of light. Intensity involves the light power per unit area and is discussed further in Chapters 3 and 6. Light with a power I per unit area in a beam incident on a surface at an angle of incidence α to the surface has an intensity per unit area on the surface of $I\cos\alpha$. The component of the momentum of a beam of light of energy W on a surface incident at an angle of incidence α normal to the surface is given by $p = W\cos\alpha/c$. The pressure of particles on a surface is the momentum change per unit area per unit time, so from Equation 2.80 a beam of light incident at an angle of incidence α to a surface exerts a pressure on the surface of

$$P_L = (1+R)I\cos^2\alpha/c, \tag{2.81}$$

where R is the reflectivity of the surface. For example, if incident at angle α and fully reflected ($R = 1$) at the surface, the direction of photon momentum reverses and the pressure on the surface is $P_L = 2I\cos^2\alpha/c$.

2.9 The Quantum Nature of Matter and Radiation

Depending on the measurement, matter and radiation both exhibit wave-like and particle-like properties. A quantum mechanical treatment of matter examining the wave-like properties can start with the time-dependent Schrodinger equation.[9] In

---

[9] Erwin Schrodinger (more correctly Schrödinger in German) was born in Vienna in 1887 and obtained a doctorate at the University of Vienna. He had subsequent employment at universities in Jena, Breslau (now Wroclaw, Poland), and Zurich. Schrodinger succeeded Max Planck in 1927 at the Friedrich Wilhelm University in Berlin, but opposed the Nazi party and in 1933 left for the University of Oxford and later Princeton University [79]. While at Oxford he shared the 1933 Nobel prize with Paul Dirac "for the discovery of new productive forms of atomic theory" (as cited by the Nobel committee). Schrodinger returned to Austria in 1936, but after the political union of Austria and Germany in 1938, his position became difficult and he left Austria for neutral Ireland. In 1940 Schrodinger became the first member of the School of Theoretical Physics

the time-dependent Schrodinger equation, the wavefunction of a material body determines the probability amplitude in space and time of finding the body. The square of the modulus of the wavefunction specifies the probability of finding the body as a function of space and time. The time dependent Schrodinger equation is written as

$$\hat{H}\Psi = i\hbar \frac{d\Psi}{dt},$$
(2.82)

where $\hat{H}$ is the quantum mechanical Hamiltonian (or energy operator) and Ψ is the time-dependent wavefunction. The time-dependent wavefunction Ψ can be expressed as a product of two functions, one of which is a function of time and the other a function of spatial coordinates:

$$\Psi = \psi(\mathbf{r})\phi(t)$$

Substituting into the time-dependent Schrodinger equation:

$$\frac{1}{\psi(\mathbf{r})}\hat{H}\psi(\mathbf{r}) = \frac{i\hbar}{\phi(t)}\frac{d\phi(t)}{dt}$$

upon dividing throughout by Ψ. The left-hand side of this equation is a function of spatial coordinates only, and the right-hand side is a function of time only, so the two sides must be separately equal to a constant, which we will denote by W. We obtain two equations:

$$i\hbar \frac{d\phi(t)}{dt} = W\phi(t)$$
(2.83)

and

$$\hat{H}\Psi(\mathbf{r}) = W\Psi(\mathbf{r}).$$
(2.84)

The equation for the temporal variation (Equation 2.83) of a wavefunction readily integrates to give

$$\phi(t) = \exp(-i\,\omega_{DB}t) = \exp\left(-i\frac{Wt}{\hbar}\right)$$
(2.85)

assuming the normalization that the modulus of $\phi(t)$ is unity. Equation 2.84 has the form of an energy eigenvalue equation and is generally known as the time-independent Schrodinger equation. From the form of Equation 2.84, W represents

of the Institute of Advanced Studies in Dublin, where he remained until retirement in 1955. He died in Vienna in 1961 at the age of 73.

energy. From Equation 2.85, it is then clear that the frequency ω_{DB} of oscillation of a time-dependent wavefunction is related to the energy W by

$$W = \hbar \omega_{DB}. \tag{2.86}$$

The energy W of a moving particle (including the rest mass energy) is related to the Lorentz factor $\gamma = 1/(1 - v^2/c^2)^{1/2}$ by $W = \gamma m_0 c^2$, so that $\omega_{DB} = \gamma m_0 c^2/\hbar = \gamma \omega_S$, where $\omega_S = m_0 c^2/\hbar$ can be considered to be the wavefunction frequency of a free particle in the frame of the particle. The period of oscillation $1/\omega_S = \hbar/(m_0 c^2)$ of, for example, the wavefunction of an electron has an extremely short duration numerical value of 1.3×10^{-21} second so cannot be directly observed.[10]

The Hamiltonian or energy operator is obtained by adding a kinetic energy operator and potential energy operator. For nonrelativistic bodies, the Hamiltonian can be written as

$$\hat{H}_0 = -\frac{\hbar^2}{2m_0} \nabla^2 + V(\mathbf{r}), \tag{2.87}$$

where m_0 is the mass of a body and $V(\mathbf{r})$ is the potential energy. Free bodies have wavefunctions with properties readily deduced by solving the time-independent Schrodinger equation (Equation 2.84) with a zero potential $V(\mathbf{r})$ term. For a free body traveling in the z-direction, the time-independent Schrodinger equation in the nonrelativistic limit becomes

$$\left(-\frac{\hbar^2}{2m_0} \frac{\partial^2}{\partial z^2} \right) \psi = T_K \psi, \tag{2.88}$$

where T_K is the particle kinetic energy. Solutions of Equation 2.88 for a body moving free of a potential are represented by traveling waves with wavefunction of form $\psi = \exp(i\, k_{DB} z)$, where k_{DB} is a wavevector. Substituting $\psi = \exp(i\, k_{DB} z)$ into Equation 2.88 shows that the kinetic energy T_K of a freely moving body is $T_K = \hbar^2 k_{DB}^2/(2m_0)$. As a nonrelativistic kinetic energy T_K is related to the momentum p by $T_K = p^2/(2m_0)$, the momentum p is given by $\hbar k_{DB}$.

The relationship $p = \hbar k_{DB}$ between the momentum p of a body and the wavenumber k_{DB} is valid for relativistic and nonrelativistic bodies. The universality of $p = \hbar k_{DB}$ can be evidenced by considering a highly relativistic body with momentum

---

[10] Indirect experimental evidence for rapid oscillation of the electron wavefunction does arise due to rapid rotation of the instantaneous position of an electron at a frequency $2\omega_S = 2m_0 c^2/\hbar$ with a radius $r_Z = c/(2\omega_S) = 1.9 \times 10^{-13}$ m: an effect known by the German word *Zitterbewegung*. In the steep potential gradient near an atomic nucleus, the zitterbewegung blurs the potential experienced by the electron, resulting in, for example, the Darwin correction for hydrogen-like s-orbital fine structure energies as the s-orbitals have nonzero values of wavefunction at the nucleus (see Section 7.4.1 of Tallents [110]). Perhaps more importantly, the angular momentum $m_0 c\, r_Z$ of an electron with phase velocity c (see Equation 2.93) produces a spin $\hbar/2$ of the electron in the electron frame [50].

$p = W/c$, where W is the total energy of a body. The energy operator $-\hbar^2/(2m_0)\nabla^2$ in Equation 2.84 becomes $\hat{p}\, c = (-i\,\hbar\, c)\nabla$ after substituting for the momentum operator $\hat{p} = -i\,\hbar\nabla$. For a free body traveling in the z-direction, the time-independent Schrodinger equation in the highly relativistic limit becomes

$$-i\hbar\, c\frac{\partial\psi}{\partial z} = W\psi. \tag{2.89}$$

Solutions of Equation 2.89 are represented by traveling waves with wavefunction of form $\psi = \exp(i\, k_{DB}z)$. Substituting $\psi = \exp(i\, k_{DB}z)$ into Equation 2.89 shows that the energy W of a freely moving body is $W = \hbar k_{DB}\, c$. As the highly relativistic energy W is related to the momentum p by $W = p\, c$, the momentum p is given by $\hbar k_{DB}$.

The wavelength λ_{DB} associated with a freely moving body is known as the De Broglie wavelength. The De Broglie wavelength is related to the wavevector k_{DB} and the momentum p by

$$\lambda_{DB} = \frac{2\pi}{k_{DB}} = \frac{h}{p}. \tag{2.90}$$

Allowing for a relativistic increase of the momentum p, the De Broglie wavelength of a body in a nominally stationary laboratory frame can be written in terms of the velocity v, rest mass m_0, and Lorentz factor γ for the body as

$$\lambda_{DB} = \frac{h}{\gamma\, m_0 v} = \frac{h(1 - v^2/c^2)^{1/2}}{m_0 v}. \tag{2.91}$$

A phase velocity v_p and the frequency ω_{DB} for the wavefunction of a free body are related in the usual way. We have that

$$\lambda_{DB}\,\omega_{DB} = 2\pi\, v_p.$$

Hence, the phase velocity v_p of the wavefunction is related to the frequency ω_{DB} and wavevector k_{DB} by

$$v_p = \frac{\omega_{DB}}{k_{DB}}. \tag{2.92}$$

Multiplying the numerator and denominator by $\hbar$ and setting an energy $W = \hbar\omega_{DB}$ (including the rest mass energy), we have for the phase velocity of the wavefunction of a particle moving outside any potential:

$$v_p = \frac{W}{p} = c\left(1 + \frac{m_0^2 c^2}{p^2}\right)^{1/2} \tag{2.93}$$

upon using Equation 2.73.

A group velocity for any traveling wave represents the velocity of the amplitude envelope formed when waves over a small wavevector range interfere. For waves of frequency ω and wavevector k, the group velocity is given by $d\omega/dk$, while the phase velocity is simply ω/k. For a more complete treatment of the group velocity, see Section 4.4 where the group velocity of light passing through a plasma medium is examined. The velocity v of a particle was shown in Section 2.8 to be equal to the differential of the total energy W with respect to the particle momentum p (i.e. $v = dW/dp$; see Equation 2.77). The particle velocity v is then analogous to a group velocity for the wavefunction given by $v = dW/dp = d\omega_{DP}/dk_{Db}$. Using Equation 2.76, the velocity of a particle is given by

$$v = \frac{dW}{dp} = c\left(\frac{1}{1 + m_0^2 c^2/p^2}\right)^{1/2}. \tag{2.94}$$

From Equations 2.93 and 2.94, we have a relationship between the phase velocity v_p and the particle velocity v:

$$v_p\, v = c^2 \tag{2.95}$$

For nonzero mass particles, the phase velocity v_p is greater than the speed of light c, while the particle velocity v, as expected, is always less than the speed of light c. The particle velocity v is equivalent to the group velocity and represents the speed of the movement of information. Information and energy can be conveyed at the group velocity by, for example, the detection of a particle. Energy is not moved at the wavefunction phase velocity, so the phase velocity $v_p > c$ does not violate any principle of relativity. For non-zero mass particles, but where $p \ll m_0 c$, Equation 2.94 predicts velocities $v = p/m_0$, in agreement with the expected classical Newtonian relationship.

Photons can be regarded as particles with zero mass ($m_0 = 0$) and energy $\hbar\omega$, where ω is the frequency of electromagnetic oscillation. For photons with $m_0 = 0$, Equations 2.93 and 2.94 predict, respectively, a phase and group velocity equal to the speed of light c in vacuum. The momentum $\mathbf{p}$ of a photon in vacuum is also given by $\mathbf{p} = \hbar\mathbf{k}$, where $\mathbf{k}$ is the wavevector for the corresponding electromagnetic wave.

2.10 Four-Vectors

Our treatment of special relativity has considered a frame of reference S' moving at a velocity $\mathbf{v}$ in the direction of the Cartesian x-coordinate relative to a nominally stationary frame of reference S. The Lorentz transformation introduced in Section 2.2 transforms coordinates (ct', x', y', z') in the S' frame of reference to the coordinates (ct, x, y, z) in the S frame of reference while preserving (i) the constant speed of light in both frames and (ii) the back transform from frame S to S' with a simple change of velocity to $-\mathbf{v}$. In Section 2.2 we considered a pulse of light originating from the origin $(0, 0, 0)$ of the coordinate systems S and S' at time $t = 0$. In any inertial frame of reference, it was verified that the Lorentz transform enables $-c^2t^2 + x^2 + y^2 + z^2$ to be invariant. In other words, the pulse of light originating at the origin at time $t = 0$ propagates as an expanding sphere of light intensity in all frames of reference.

It is a useful mathematical procedure to package the spacetime coordinates into a quantity known as a "four-vector" represented by a single symbol with a tensor notation for operations on the four-vectors. We introduce this notation and use it a little in this section and Sections 3.1 and 3.2 as a way of introducing the tensor notation used in relativity literature.

In the four-vector notation, an index μ with value $\mu = 0, 1, 2,$ or 3 is used to denote respective values of, for example, ct, x, y, and z such that the four-vector for spacetime is defined by

$$
\begin{aligned}
X^\mu &= ct && (\mu = 0), \\
&= x && (\mu = 1), \\
&= y && (\mu = 2), \\
&= z && (\mu = 3).
\end{aligned}
\tag{2.96}
$$

The Greek superscripts indicate that the vector is referred to as being *contravariant*. A contravariant vector is written out as an array of one column and four rows $\mu = 0, 1, 2,$ and 3. In relativity theory, quantities such as $-c^2t^2 + x^2 + y^2 + z^2$ are constant, so it is useful to also define another type of vector which is referred to as *covariant*. Covariant vectors are indicated with Greek subscripts, for example, X_μ, and defined so that the zeroth (e.g. the time) variable is calculated in operations as the negative of the listed value. A covariant vector is written out as an array of one row and four columns $\mu = 0, 1, 2,$ and 3. The four-vector for spacetime given by Equation 2.96 has a scalar product of the contravariant X^μ and covariant X_μ representations of form

$$
X^\mu X_\mu = \begin{bmatrix} -ct \\ x \\ y \\ z \end{bmatrix} \begin{bmatrix} -(-ct) & x & y & z \end{bmatrix} = \begin{bmatrix} -c^2t^2 \\ x^2 \\ y^2 \\ z^2 \end{bmatrix}.
\tag{2.97}
$$

The addition of the elements of the contravariant vector product $X^\mu X_\mu$ is often required, as, for example, the addition represents a Lorentz invariant quantity. Summing up the elements of $X^\mu X_\mu$, we have

$$\sum_{\mu=0}^{3} X^\mu X_\mu = -c^2 t^2 + x^2 + y^2 + z^2. \tag{2.98}$$

A Lorentz transformation is a linear map from four-vector X to four-vector X' using a 4×4 tensor Λ^μ_ν such that

$$X'^\mu = \Lambda^\mu_\nu X^\nu, \tag{2.99}$$

where elements of the tensor are in the νth column and μth row. A 4×4 tensor with 16 elements such as Λ^μ_ν is known as a first rank tensor, while column contravariant vectors and row covariant vectors can be referred to as zero rank tensors. The appropriate Lorentz transformation first rank tensor Λ^μ_ν is obtained from Equation 2.21:

$$\Lambda^\mu_\nu = \begin{bmatrix} \gamma & -\gamma v/c & 0 & 0 \\ -\gamma v/c & \gamma & 0 & 0 \\ 0 & 0 & 1 & 0 \\ 0 & 0 & 0 & 1 \end{bmatrix}, \tag{2.100}$$

where the Lorentz factor $\gamma = 1/\sqrt{1 - v^2/c^2}$.

We introduce a tensor known as the "Minkowski[11] metric" defined as

$$\eta_{\mu\nu} = \begin{bmatrix} -1 & 0 & 0 & 0 \\ 0 & 1 & 0 & 0 \\ 0 & 0 & 1 & 0 \\ 0 & 0 & 0 & 1 \end{bmatrix}. \tag{2.101}$$

The Minkowski metric is a diagonal array so $\eta_{\mu\nu} = \eta_{\nu\mu}$. The Minkowski metric makes it straightforward to write down quantities that are invariant under changes of the frame of reference. We have seen, for example, that the distance from the origin to a point in spacetime is invariant. This invariant distance is given by

$$\sum_\mu \sum_\nu \eta_{\mu\nu} X^\mu X^\nu = -c^2 t^2 + x^2 + y^2 + z^2. \tag{2.102}$$

The Minkowski metric represented by Equation 2.101 is applicable to so-called flat spacetime as envisaged for special relativity where light in vacuum propagates

---

[11] Hermann Minkowski (1864–1909) was born in what is now Lithuania and held teaching positions in Konigsberg (now the Russian enclave Kaliningrad); at universities in Germany; and in Zurich, where he taught Einstein. Immediately after Einstein's relativity publications in 1905, Minkowski became an influential ally who mathematically formalized Einstein's relativity despite his description of Einstein when a student as "a lazy dog" [12].

in a straight line. Other metrics are applicable for different distributions of matter in general relativity where the propagation of light depends on the distribution
of mass.

Einstein originated the "summation convention," dropping the summation symbols in any single term containing a repeated Greek index, so that Equation 2.102
can be written simply as

$$\eta_{\mu\nu}X^\mu X^\nu = -c^2 t^2 + x^2 + y^2 + z^2. \tag{2.103}$$

The summation convention indicates that a summation is implied over an index
with values 0, 1, 2, and 3 when a term contains a repeated Greek index. Referring
back to Equation 2.98, the scalar product of a contravariant and covariant vector
can also be written using the Einstein convention as

$$X^\mu X_\mu = -c^2 t^2 + x^2 + y^2 + z^2. \tag{2.104}$$

Quantities other than spacetime can undergo a Lorentz transformation. The
Lorentz transformation of the four-momentum p such that $(p^0, p^1, p^2, p^3) = (W/c, p_x, p_y, p_z)$ from the stationary frame of reference S to the four-momentum p'
such that $(p'^0, p'^1, p'^2, p'^3) = (W'/c, p'_x, p'_y, p'_z)$ in the moving frame of reference S'
is evaluated by the multiplication of the Lorentz transformation matrix (Equation
2.100) by the vector $(E/c, p_x, p_y, p_z)$. The energy and momentum transformations
simplify to

$$W' = \gamma(W - v p_x), \tag{2.105}$$

$$p'_x = \gamma(-vW/c^2 + p_x), \tag{2.106}$$

with $p'_y = p_y$ and $p'_z = p_z$.

The Lorentz transformation from frame S' to S is found using the inverse of
the Lorentz transformation matrix (see Equation 2.22). The energy and momentum
directed in the direction of $\mathbf{v}$ are related by

$$W = \gamma(W' + v p'_x), \tag{2.107}$$

$$p_x = \gamma(vW'/c^2 + p'_x). \tag{2.108}$$

Equations 2.107 and 2.108 can alternatively be found by swapping the primed and
unprimed symbols in Equations 2.105 and 2.106 and changing the sign of the velocity from v to $-v$. Such a change is consistent with a move of the nominally
stationary observer from a frame S to the frame S'. An observer in S' sees the frame
S moving away at a velocity $-\mathbf{v}$. We need Equations 2.105 and 2.107 in calculations where, for example, the energy of a particle moving at relativistic velocities
is transformed from the particle's frame of reference into a stationary or laboratory
frame.

Relativistic theories at the more fundamental level are built up by finding variables that are conserved or transformed in useful ways under Lorentz transformations. If a parameter or multiplication/division of parameters remains the same in different frames of reference, the parameter or grouping of parameters is said to be "Lorentz invariant." Parameters or groupings of parameters that are Lorentz invariant are examined in different chapters of this book, with a summary presented in Appendix A.

The four-momentum of a particle with relativistic energy W and momentum $\mathbf{p} = (p_x, p_y, p_z)$ represented by $p = (p^0, p^1, p^2, p^3) = (W/c, p_x, p_y, p_z)$ as discussed above yields a useful Lorentz invariant. Following Equation 2.102 using the Minkowski metric, an invariant of the momentum four-vector is given by

$$\eta_{\mu\nu} p^\mu p^\nu = -W^2/c^2 + p_x^2 + p_y^2 + p_z^2 = -W^2/c^2 + |\mathbf{p}|^2 \qquad (2.109)$$

upon using the Einstein convention. If the invariant value in Equation 2.109 is equal to $-m_0^2 c^2$, the energy E_R of a body of mass m_0 with zero momentum is $m_0 c^2$. We then obtain a previously determined result (Equation 2.73) for the relationship between the momentum p and relativistic energy W of a body:

$$W^2 = p^2 c^2 + m_0^2 c^4$$

The invariance of the energy/momentum four-vector under a Lorentz transformation requires that the rest mass energy is $E_R = m_0 c^2$.

Exercises

2.1 Consider a rod of length L' in a frame of reference moving at velocity v along the x-axis in a Cartesian coordinate system. If the rod is tilted at an angle θ' to the x-axis in the moving frame of reference, show that to a stationary observer the rod has a length given by

$$L = L' \left(1 - \frac{v^2}{c^2} \cos^2 \theta' \right)^{1/2},$$

with an angle θ to the x-axis given by

$$\tan \theta = \gamma \tan \theta'.$$

2.2 Consider the Michelson-Morley experiment of 1887 discussed as a footnote in Section 2.2. Light was sent on two paths of distance $2L'$ perpendicular to each other and the path length difference was measured using interferometry. If the whole apparatus moves at a velocity v at an angle θ' to one path in the frame of the apparatus, show that the length difference ΔL measured in the laboratory for the light to travel distances $2L'$ parallel and perpendicular to each other is given by

$$\Delta L = 2L' \left(\left(1 - \frac{v^2}{c^2} \cos^2 \theta'\right)^{1/2} - \left(1 - \frac{v^2}{c^2} \sin^2 \theta'\right)^{1/2} \right).$$

(In practice the Michelson-Morley experiment always shows a constant path-length difference independent of angle consistent with the central idea of relativity that light propagates at a constant velocity relative to the observer and is not constant relative to the "aether.")

2.3   Consider two spaceships directly approaching each other with velocities **v** of equal magnitude and opposite direction as measured on Earth. Use the velocity transformation between frames of reference given in Section 2.3 to show that the relative velocity u of approach of the other spacecraft measured by each spaceship is given by

$$u = \frac{2v}{1 + v^2/c^2}.$$

2.4   In the rocket equation, the exhaust velocity v'_{ex} is measured in the frame of the rocket. Using the velocity transformations developed in Section 2.3, show that the exhaust velocity v_{ex} measured in a stationary frame of reference is given by

$$v_{ex} = \frac{v'_{ex} - v_R}{1 - v_R v'_{ex}/c^2},$$

where v_R is the velocity of the rocket relative to the stationary frame.

2.5   When relativistic effects are small for a relativistic rocket, show that the rocket velocity v_F in the stationary frame including relativistic effects is related to the rocket velocity v'_F neglecting relativistic effects by the relationship

$$v_F \approx \frac{v'_F}{1 + (v'_{ex}/c) \ln(M_0/M_F)},$$

where M_0 is the initial rest mass of the rocket, M_F is the final rest mass, and v'_{ex} is the rocket exhaust velocity.

2.6   Consider a spectral line arising from a distant galaxy and detected by a telescope. If the spectral line is identified as arising from an atomic transition of wavelength λ' and is measured on Earth to have a wavelength λ, show that the relative velocity v of the galaxy to the Earth is given by

$$v = c\frac{\lambda^2 - \lambda'^2}{\lambda^2 + \lambda'^2}.$$

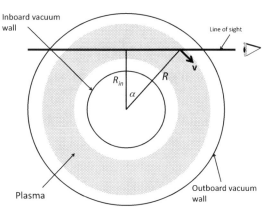

Figure 2.4 A schematic plan view of a tokamak showing the toroidal plasma with a spectrometer line of sight. The velocity **v** at a major radius R with a line of sight passing at a minimum major radius R_{in} is shown. The figure illustrates the geometry for Exercise 2.7.

2.7 A spectrometer used in a tokamak has a line of sight passing close to the "inboard" vacuum wall with a minimum major radius denoted by R_{in} (see Figure 2.4). Plasma rotation without shear often occurs such that the bulk velocity v of the plasma varies with angle α from the center of the torus such that $v = R(d\alpha/dt)$, where R is the major radius. Assuming $v \ll c$, show that the Doppler shift δv of line emission of frequency v_0 in the rest frame of reference due to plasma rotation is given by

$$\delta v = \frac{R_{in} v_0}{c} \frac{d\alpha}{dt}.$$

2.8 The cosmic microwave background (CMB) was created when hydrogen ions recombined with free electrons and black-body radiation was no longer being absorbed or scattered by free electrons and hydrogen ions. It is believed that the CMB was created at distances with redshift $z = 1,100$. Determine the apparent velocity relative to earth of the last CMB scattering event assuming the redshift is solely due to the Doppler effect. [$0.999998c$]

2.9 An isolated neutron decays by the weak interaction conversion of a down quark to an up quark with a half-life of 10.3 minutes producing a proton, an electron and an electron antineutrino. The proton rest mass is $1.6726219 \times 10^{-27}$ kg, the neutron rest mass is $1.674927471 \times 10^{-27}$ kg, the electron rest mass is 9.109×10^{-31} kg, and the neutrino has a rest mass $< 1 \text{eV}/c^2$. Calculate the total energy yield produced by neutron decay. [0.782 MeV]

2.10 Consider a neutron with a velocity of $0.9c$ which decays such that the electron takes all the energy yield from the decay with a velocity directed parallel to the original neutron velocity. Determine the velocity of the electron in the laboratory frame of reference. [$0.99c$]

2.11 An atom traveling at a velocity v in a laboratory experiment emits light perpendicular to the direction of the velocity in the atom frame of reference. Show that in the laboratory frame of reference, the light propagates at an angle θ to the velocity, where θ is given by the following equivalent expressions:

$$\tan\theta = \left(\frac{c^2}{v^2} - 1\right)^{1/2}$$

$$\tan\theta = \frac{1}{\left(\gamma^2 - 1\right)^{1/2}},$$

where γ is the Lorentz parameter.

2.12 According to cosmological models, the early universe had a temperature of 100 keV three minutes after the Big Bang. Look up the rest masses or rest mass energies and then evaluate the Lorentz γ parameter for electrons and protons with kinetic energy equal to this temperature $k_B T = 100$ keV. [Electrons $\gamma = 1.19$, protons $\gamma = 1.0001$]

3

Relativity and Electromagnetic Fields

We now further examine how quantities associated with light transform between frames of reference. Relativity theory is based on the observation that the speed of light is constant in all frames. The speed of light is said to be Lorentz invariant. In Section 2.6 we showed how the frequency of light transforms between frames of reference (a transformation known as the Doppler effect). In this chapter, we first show that the phase of the electric and magnetic field oscillations associated with the propagation of light is constant in different frames. We then examine the transform of electric and magnetic fields between frames of reference. An electric field can manifest as a magnetic field in another frame of reference and vice versa. For example, if a field is purely electric ($\mathbf{B} = 0$) in one frame of reference, there can be a mixture of electric and magnetic fields in another frame of reference.

The acceleration of particles is closely associated with light as acceleration of charged particles produces light (see Section 1.3). The transformation of particle acceleration between frames is examined in this chapter because of the association of particle acceleration with light emission. Acceleration to "relativistic velocities" (velocities approaching the speed of light) also creates interesting time differences between frames of reference with the possibility of colocation at the start and end of a period of time, with large time differences between frames of reference. This is the "twin paradox" problem considered as an exercise (Exercise 3.1) which is answered in full in Section 10.1.

Finally, we consider changes in the propagation direction of light in moving from one reference frame to another. There is a general "beaming" effect whereby light emitted at a range of angles in a moving frame is clustered around the direction of the velocity of the moving frame in the stationary (laboratory) frame. In this chapter, we allow for the transformation of the acceleration of particles between frames of reference in the deduction of the light emission in the stationary frame.

3.1 Revisiting the Propagation of Light

Maxwell's equations show that light is a directed electric and magnetic field oscillation in space and time (see Section 1.1). The oscillating electric and magnetic fields have magnitudes proportional to $\exp(i(\mathbf{k} \cdot \mathbf{r} - \omega t))$, where the wavevector $\mathbf{k}$ represents the rapid spatial oscillation of the wave with distance $\mathbf{r}$, and the frequency ω represents the rapid temporal variation with time t. The "phase" of the wave is given by $\mathbf{k} \cdot \mathbf{r} - \omega t$. The periodic vanishing of the electric and magnetic fields at "node" points when $\mathbf{k} \cdot \mathbf{r} - \omega t = 0$ occurs with some wavevector and frequency in all frames with a phase spacing of π.

The phase of an electromagnetic wave can, in principal, be measured by counting the node points of zero electric field passing an observer. Quantities that can be counted are invariably Lorentz invariant in different frames of reference as the quantity is either present or not. We expect that the phase $\mathbf{k} \cdot \mathbf{r} - \omega t$ of an electromagnetic wave is a quantity that is constant under a Lorentz transformation.

We can examine the Lorentz invariance of the phase of a light beam in more detail by considering the following "thought experiment." Imagine that an observer in a frame S is equipped to record the number of wave nodes passing their position P. If an initial wave node at time $t = 0$ is the first wave node recorded (when it reaches the observer at P), by time t the observer will record $(1/\pi)(\mathbf{k} \cdot \mathbf{r} - \omega t)$ wave nodes. Imagine another observer at position P' in the frame of reference S' moving with a velocity $\mathbf{v}$ parallel to the x-axis with an origin coincident with that of S at time $t = t' = 0$. The observer in S' begins to count nodes when the initial wave node passes them and continues counting until a later time t'. The observer at P' will count $(1/\pi)(\mathbf{k}' \cdot \mathbf{r}' - \omega' t')$ wave nodes. If the position P' at the end of the counting period coincides with position P, both observers count the same number of wave nodes. We have a constant phase for the light in both frames S and S' in this "thought experiment" with

$$\mathbf{k} \cdot \mathbf{r} - \omega t = \mathbf{k}' \cdot \mathbf{r}' - \omega' t.$$

If the phase is constant for this example scenario, a Lorentz transformation of the phase of light generally must give the same phase.

The appropriate four-vector for the phase of an electromagnetic wave is given by $(k^0, k^1, k^2, k^3) = (\omega/c, k_x, k_y, k_z)$ when the wavevector has Cartesian components k_x, k_y, and k_z. Using the Lorentz transformation (Equation 2.21) from the stationary frame S to the frame S' moving at velocity $\mathbf{v}$ in the x-direction, we obtain for the moving frame:

$$\omega'/c = \gamma(\omega/c - k_x v/c), \tag{3.1}$$

$$k'_x = \gamma(-v\omega/c^2 + k_x),$$
$$k'_y = k_y,$$
$$k'_z = k_z \tag{3.2}$$

For electromagnetic wave propagation directed at an angle θ to the velocity $\mathbf{v}$,

$$k_x = |\mathbf{k}|\cos\theta = k\cos\theta = (\omega/c)\cos\theta. \tag{3.3}$$

The wavevector $\mathbf{k}$ in different frames of reference is given by

$$|\mathbf{k}'|\cos\theta' = \gamma\left(k\cos\theta - \frac{v\,\omega}{c^2}\right). \tag{3.4}$$

The Lorentz transformation for the angle θ of propagation of light is given by

$$\tan\theta' = \frac{k'_y}{k'_x} = \frac{k_y}{\gamma(k_x - v\omega/c^2)}.$$

Using $k_x = (\omega/c)\cos\theta$ and $\tan\theta = k_y/k_x$, this expression simplifies to

$$\tan\theta' = \frac{\sin\theta/\gamma}{\cos\theta - v/c} \tag{3.5}$$

in agreement with the expression that we found in Section 2.3 (Equation 2.41). Equation 3.1 with the Equation 3.3 substitution reduces to the Doppler frequency shift expression found earlier (see Equation 2.46):

$$\omega' = \omega\gamma\left(1 - \frac{v}{c}\cos\theta\right) \tag{3.6}$$

Following Equation 2.102 using the Minkowski metric η, an invariant of the light phase four-vector is given by

$$\eta_{\mu\nu}k^\mu k^\nu = -\frac{\omega^2}{c^2} + k_x^2 + k_y^2 + k_z^2 = -\frac{\omega^2}{c^2} + |\mathbf{k}|^2 \tag{3.7}$$

upon using the Einstein summation convention. Setting the invariant value here equal to zero means that the speed of light $c = \omega/k$ (as we found in Section 1.1). The invariance of the four-vector for the phase of an electromagnetic wave under a Lorentz transformation requires that $C_0 = k^2 - \omega^2/k^2$, where C_0 is a constant. We can quickly verify that the constant C_0 must be zero. The speed of light $c = (\omega'^2/k'^2 + C_0 c^2/k'^2)^{1/2} = (\omega^2/k^2 + C_0 c^2/k^2)^{1/2}$ which for constant c and C_0 in all frames is only possible if $C_0 = 0$.

3.2 The Lorentz Transformation of the Electric and Magnetic Fields

Maxwell's equations are the same in all frames of reference, but the values of the electric and magnetic fields in different frames of reference are not the same. We show in this section that the idea of a pure electric or magnetic field is not valid for all frames of reference. In a different frame of reference, an electric field can manifest as a magnetic field and vice versa. We determine the relationships between electric and magnetic fields in different frames of reference in this section.

The relationships between the values of the electric and magnetic fields in different frames of reference are determined by the Lorentz transformation. Instead of directly using the vectors for the electric field $\mathbf{E}$ and magnetic field $\mathbf{B}$, it is simpler to treat electromagnetic field changes between frames of reference using the vector potential $\mathbf{A}$ and the scalar potential V. The vector potential $\mathbf{A}$ is related to the magnetic field by

$$\mathbf{B} = \nabla \times \mathbf{A}. \tag{3.8}$$

The scalar potential V represents a contribution to the electric field due to electrostatic effects associated with electric charges. The total electric field is the electrostatic contribution $(-\nabla V)$ plus the electric field $(-\partial \mathbf{A}/\partial t)$ created by a changing magnetic field, so that

$$\mathbf{E} = -\nabla V - \frac{\partial \mathbf{A}}{\partial t}. \tag{3.9}$$

Another condition known as the "Lorenz gauge"[1] is required to uniquely[2] define the vector potential $\mathbf{A}$ and the scalar potential V. The Lorenz gauge condition is that

$$\nabla \cdot \mathbf{A} + \frac{1}{c^2} \frac{\partial V}{\partial t} = 0. \tag{3.10}$$

Gauss's law $\nabla \cdot \mathbf{E} = \rho_c/\epsilon_0$, where ρ_c is the charge per unit volume, can be expressed in terms of the vector $\mathbf{A}$ and scalar V potentials using Equation 3.9:

$$\nabla \cdot \left(-\nabla V - \frac{\partial \mathbf{A}}{\partial t} \right) = \frac{\rho_c}{\epsilon_0}$$

which can be written as

$$\nabla^2 V + \frac{\partial}{\partial t} \nabla \cdot \mathbf{A} = -\frac{\rho_c}{\epsilon_0}. \tag{3.11}$$

Ampere's law expressed as Maxwell's equation ($\nabla \times \mathbf{B} = \mu_0 \mathbf{J} + \epsilon_0 \mu_0 \partial \mathbf{E}/\partial t$, see Equation 1.2) can be rearranged using $c = 1/(\epsilon_0 \mu_0)^{1/2}$ as

---

[1] The Lorenz gauge is name after the Danish physicist Ludvig Lorenz (1829–1891). The Lorentz transformation is named after the Dutch physicist and 1902 Nobel prize winner Hendrik Lorentz (1853–1928).

[2] Other "gauge" conditions are sometimes used to deal with different electromagnetic investigations. For example, the Coulomb gauge has $\nabla \cdot \mathbf{A} = 0$.

$$c^2 \nabla \times \mathbf{B} - \frac{\partial \mathbf{E}}{\partial t} = \frac{\mathbf{J}}{\epsilon_0}.$$

Substituting the equations defining the vector $\mathbf{A}$ and scalar V potentials in terms of electric and magnetic fields (Equations 3.8 and 3.9) gives

$$c^2 \nabla \times (\nabla \times \mathbf{A}) - \frac{\partial}{\partial t}\left(-\nabla V - \frac{\partial \mathbf{A}}{\partial t}\right) = \frac{\mathbf{J}}{\epsilon_0}.$$

Simplifying with the vector identity

$$\nabla \times (\nabla \times \mathbf{A}) = \nabla(\nabla \cdot \mathbf{A}) - \nabla^2 \mathbf{A}$$

leads to an expression

$$-c^2 \nabla^2 \mathbf{A} + c^2 \nabla(\nabla \cdot \mathbf{A}) + \frac{\partial(\nabla V)}{\partial t} + \frac{\partial^2 \mathbf{A}}{\partial t^2} = \frac{\mathbf{J}}{\epsilon_0}.$$

The Lorenz gauge condition (Equation 3.10) reduces this expression to

$$\nabla^2 \mathbf{A} - \frac{1}{c^2}\frac{\partial^2 \mathbf{A}}{\partial t^2} = -\frac{\mathbf{J}}{\epsilon_0 c^2}. \tag{3.12}$$

Using the Lorenz gauge condition (Equation 3.10), Equation 3.11 obtained from Gauss's law $\nabla \cdot \mathbf{E} = \rho_c/\epsilon_0$ also simplifies to

$$\nabla^2 V - \frac{1}{c^2}\frac{\partial^2 V}{\partial t^2} = -\frac{\rho_c}{\epsilon_0}. \tag{3.13}$$

There are four right-hand side quantities represented by Equations 3.12 and 3.13: the charge density ρ_c, and the components J_x, J_y, and J_z of the current density $\mathbf{J}$.

In all frames of reference, conservation of charge requires that the outflow of current $\mathbf{J}$ per unit volume corresponds to a decrease of charge density ρ_c so that

$$\nabla \cdot \mathbf{J} = -\frac{\partial \rho_c}{\partial t}. \tag{3.14}$$

Charge conservation (Equation 3.14) is valid for all frames of reference, so $\nabla \cdot \mathbf{J} + \partial \rho_c/\partial t = 0$ for all frames of reference. Using the notation introduced in Section 2.10, we can define the four-vector $j_\mu = (c\rho_c, J_x, J_y, J_z)$ and an operation

$$\nabla_\mu j_\mu = \frac{\partial \rho_c}{\partial t} + \nabla \cdot \mathbf{J} = \frac{1}{c}\frac{\partial(c \rho_c)}{\partial t} + \nabla \cdot \mathbf{J}. \tag{3.15}$$

Due to conservation of charge (Equation 3.14), the value here of $\nabla_\mu j_\mu$ is identically equal to zero in all frames.

The scalar product operation $\nabla_\mu \nabla_\mu$ produces the operators used on the left-hand side of Equations 3.12 and 3.13. We have[3]

$$\nabla_\mu \nabla_\mu = -\frac{1}{c^2}\frac{\partial}{\partial t}\frac{\partial}{\partial t} + \left(\frac{\partial}{\partial x}\right)\left(\frac{\partial}{\partial x}\right) + \left(\frac{\partial}{\partial y}\right)\left(\frac{\partial}{\partial y}\right) + \left(\frac{\partial}{\partial z}\right)\left(\frac{\partial}{\partial z}\right)$$
$$= -\frac{1}{c^2}\frac{\partial^2}{\partial t^2} + \nabla^2. \tag{3.16}$$

The operation $\nabla_\mu \nabla_\mu$ on a vector represents a wave equation with wave velocity equal to the speed of light c. For example, the wave equation for the propagation of light in vacuum (Equation 1.9) has form

$$\nabla_\mu \nabla_\mu E_\mu = 0, \tag{3.17}$$

where the three-vector for electric field is $E_\mu = (E_x, E_y, E_z)$. Equations such as the wave equation where the operation $\nabla_\mu \nabla_\mu$ produces a scalar value are Lorentz invariant.

Writing A_x, A_y, and A_z for the vector potential $\mathbf{A}$ components in Cartesian coordinates, Equations 3.13 and 3.12 can be written in terms of the four-vectors $A_\mu = (V/c, A_x, A_y, A_z)$ and $j_\mu = (c\rho_c, J_x, J_y, J_z)$ as

$$\nabla_\mu \nabla_\mu A_\mu = -\frac{j_\mu}{\epsilon_0 c^2}. \tag{3.18}$$

Charge conservation requires that $\nabla_\mu j_\mu = 0$ (see Equation 3.15). We then have that $\nabla_\mu \nabla_\mu j_\mu = 0$ (a scalar quantity) which means that the four quantities $c\,\rho_c$, J_x, J_y, and J_z transform as a four-vector between frames of reference as the $\nabla_\mu \nabla_\mu$ operator is Lorentz invariant. The four-vector $A_\mu = (V/c, A_x, A_y, A_z)$ on the left-hand side of Equation 3.18 then also Lorentz transforms as a four-vector between frames as all the other terms in Equation 3.18 are Lorentz invariant (namely the $\nabla_\mu \nabla_\mu$ operator, the dielectric constant ϵ_0, and the speed of light c).

The Lorentz transformation of the four-vector $(V/c, A_x, A_y, A_z)$ from the stationary frame S to the frame S' moving at velocity $\mathbf{v}$ along the x-axis is given by

$$\begin{bmatrix} V'/c \\ A'_x \\ A'_y \\ A'_z \end{bmatrix} = \begin{bmatrix} \gamma & -\gamma\beta & 0 & 0 \\ -\gamma\beta & \gamma & 0 & 0 \\ 0 & 0 & 1 & 0 \\ 0 & 0 & 0 & 1 \end{bmatrix} \begin{bmatrix} V/c \\ A_x \\ A_y \\ A_z \end{bmatrix}, \tag{3.19}$$

where $\beta = v/c$. Multiplying the Lorentz transformation matrix by the vector for the scalar and vector potentials shows that the transformation to the moving frame is given by

---

[3] In a similar procedure as introduced for covariant four-vector multiplication (see Section 2.10), a scalar product such as $\nabla_\mu \nabla_\mu$ for covariant operations conventially has the zeroth element as the negative of the product of the zeroth elements of each operation.

$$V'/c = \gamma(V/c - \beta A_x),$$
$$A'_x = \gamma(A_x - \beta V/c),$$
$$A'_y = A_y,$$
$$A'_z = A_z. \tag{3.20}$$

We want to evaluate the Lorentz transformation of the electric and magnetic fields from the Lorentz transformation of scalar and vector potentials. Returning to the Lorentz transformation of time and distance supplies some partial derivatives needed in the evaluation of the electromagnetic transformation. The Lorentz transformation of time and distance is given by Equation 2.21 such that

$$t = \gamma t' + \frac{\gamma \beta}{c} x',$$

$$x = \gamma \beta c t' + \gamma x',$$

with $y = y'$, $z = z'$, $\beta = v/c$, and $\gamma = 1/(1 - v^2/c^2)^{1/2}$. Considering partial derivatives of these expressions shows that $\partial t/\partial x' = \beta \gamma/c$, $\partial x/\partial x' = \gamma$ and, of course, $\partial y = \partial y'$ and $\partial z = \partial z'$. We now use these partial derivatives in the evaluation of the electric and magnetic fields in different frames of reference.

The relationship between magnetic field **B** and vector potential **A** means that the magnetic field **B**′ in the moving S' frame of reference can be related to the field quantities in the stationary S frame using $\mathbf{B}' = \nabla \times \mathbf{A}'$ by substituting the values of Equation 3.20. In the z-direction:

$$
\begin{aligned}
B'_z &= \frac{\partial A'_y}{\partial x'} - \frac{\partial A'_x}{\partial y'} = \frac{\partial A_y}{\partial x'} - \frac{\partial}{\partial y'}(\gamma(A_x - \beta V/c)) \\
&= \frac{\partial A_y}{\partial x}\frac{\partial x}{\partial x'} + \frac{\partial A_y}{\partial t}\frac{\partial t}{\partial x'} - \frac{\partial}{\partial y}(\gamma(A_x - \beta V/c)) \\
&= \gamma\left(\frac{\partial A_y}{\partial x} - \frac{\partial A_x}{\partial y} + \frac{\beta}{c}\left(\frac{\partial A_y}{\partial t} + \frac{\partial V}{\partial y}\right)\right)
\end{aligned}
$$

The first two terms here are together equal to B_z. Equation 3.9 means that $E_y = -\partial A_y/\partial t - \partial V/\partial y$ so that

$$B'_z = \gamma(B_z - \beta E_y/c). \tag{3.21}$$

A similar calculation to Equation 3.21 for the magnetic field in the y-direction simplifies to

$$B'_y = \gamma(B_y + \beta E_z/c). \tag{3.22}$$

Applying $\mathbf{B}' = \nabla \times \mathbf{A}'$ in the x-direction (the direction of the velocity **v** of frame S' relative to frame S) leads to

$$B'_x = \frac{\partial A'_z}{\partial y'} - \frac{\partial A'_y}{\partial z'} = \frac{\partial A_z}{\partial y} - \frac{\partial A_y}{\partial z} = B_x.$$

The magnetic field in the direction of the moving frame velocity $\mathbf{v}$ is the same in both frames S' and S.

The relationship between the electric field $\mathbf{E}'$ in the moving frame S' and the field quantities in the stationary frame S is found by substituting the values of Equation 3.20 into $\mathbf{E}' = -\nabla V' - \partial \mathbf{A}'/\partial t'$. In the z-direction:

$$E'_z = -\frac{\partial V'}{\partial z'} - \frac{\partial A'_z}{\partial t'}$$

$$= -\gamma \left(\frac{\partial V}{\partial z'} - \beta c \frac{\partial A_x}{\partial z'} \right) - \frac{\partial A_z}{\partial t'}$$

$$= \gamma \left(-\frac{\partial V}{\partial z} - \frac{\partial A_z}{\partial t} + \beta c \frac{\partial A_x}{\partial z} \right)$$

as $\partial t = \gamma \partial t'$ and $\partial z = \partial z'$. Magnetic fields are related to the vector potential using $\mathbf{B} = \nabla \times \mathbf{A}$ so for the magnetic field in the y-direction:

$$B_y = -\frac{\partial A_z}{\partial x} + \frac{\partial A_x}{\partial z} = \frac{\partial A_x}{\partial z}$$

if we choose our coordinate axes so that $\mathbf{A}$ is in the xy-plane. The expression for the electric field in the frame S' then simplifies to

$$E'_z = \gamma (E_z + \beta c B_y). \tag{3.23}$$

A similar calculation to Equation 3.23 for the electric field in the y-direction simplifies to

$$E'_y = \gamma (E_y - \beta c B_z). \tag{3.24}$$

The electric field in the direction of $\mathbf{v}$ (the x-direction) is evaluated using $\mathbf{E}' = -\nabla V' - \partial \mathbf{A}'/\partial t'$ so that

$$E'_x = -\frac{\partial V'}{\partial x'} - \frac{\partial A'_x}{\partial t'}$$

$$= -\gamma \left(\frac{\partial V}{\partial x'} - \beta c \frac{\partial A_x}{\partial x'} \right) - \frac{\partial A'_x}{\partial t'}$$

$$= -\left(\frac{\partial V}{\partial x} - \beta c \frac{\partial A_x}{\partial x} \right) - \gamma \left(\frac{\partial A_x}{\partial t'} - \beta/c \frac{\partial V}{\partial t'} \right)$$

$$= -\left(\frac{\partial V}{\partial x} - \beta c \frac{\partial A_x}{\partial x} \right) - \left(\frac{\partial A_x}{\partial t} - \beta/c \frac{\partial V}{\partial t} \right)$$

$$= \left(-\frac{\partial V}{\partial x} - \frac{\partial A_x}{\partial t} \right) + \beta c \left(\frac{\partial A_x}{\partial x} + 1/c^2 \frac{\partial V}{\partial t} \right).$$

The terms within the right bracket here are together equal to zero as the equation within the right bracket is the Lorenz gauge condition (Equation 3.10). The terms in the left bracket are equal to the electric field E_x. Hence the electric field in the direction of the velocity of the moving frame S' is equal to the electric field in the stationary frame S:

$$E'_x = E_x \tag{3.25}$$

We can summarize the relationships between the field quantities in the moving frame S' as observed in the stationary frame S using vector notation. Delineating components of the electric and magnetic fields parallel and perpendicular to the velocity $\mathbf{v}$ of frame S' relative to frame S:

$$\mathbf{E}'_{||} = \mathbf{E}_{||} \,, \qquad \mathbf{B}'_{||} = \mathbf{B}_{||} \tag{3.26}$$

For the electric and magnetic field components perpendicular to the velocity $\mathbf{v}$:

$$\mathbf{E}'_{\perp} = \gamma \left(\mathbf{E}_{\perp} + (\mathbf{v} \times \mathbf{B})_{\perp} \right) , \qquad \mathbf{B}'_{\perp} = \gamma \left(\mathbf{B}_{\perp} - (1/c^2)(\mathbf{v} \times \mathbf{E})_{\perp} \right) , \tag{3.27}$$

where $(\mathbf{v} \times \mathbf{B})_{\perp}$ and $(\mathbf{v} \times \mathbf{E})_{\perp}$ represent the projection of $\mathbf{v} \times \mathbf{B}$ and $\mathbf{v} \times \mathbf{E}$ onto the plane perpendicular to the velocity $\mathbf{v}$.

A consequence of Equations 3.27 and, more generally, the treatment in this section, is to demonstrate that a pure electric or magnetic field is not Lorentz invariant. For example, if a field is purely electric ($\mathbf{B} = 0$) in one frame of reference, in another frame of reference, there can be a mixture of electric and magnetic fields.

As a magnetic field in a stationary frame appears as an electric field in a moving frame of reference (Equation 3.27), electric field effects on atomic emission can be used as a diagnostic of magnetic field strength and direction in tokamak plasmas [65], [117]. Tokamaks use magnetic fields to confine a hot plasma with the aim of producing energy from the fusion of deuterium and tritium isotopes of hydrogen. Beams of neutral hydrogen or deuterium atoms of well-defined velocity are directed into the plasma in order to heat the plasma or act as a diagnostic of plasma parameters. As the neutral beams have a large velocity, the magnetic field containing the plasma appears as an electric field in the frame of reference of the atoms in the beam. The electric field causes an effect known as the "motional Stark effect" with the electric field perturbing atomic energy levels and shifting the emission of spectral lines. The broadening and shifting of spectral lines due to the electric field in the atomic frame of reference provides a measure of the magnetic field strength and direction in the stationary frame of reference.

3.3 Acceleration and Force in a Moving Frame of Reference

Particles achieve relativistic velocities due to forces producing acceleration. Forces on charged particles typically arise due to interactions with electric and magnetic fields. The interaction of electric and magnetic fields with charged particles is examined in more detail in Section 5.1, while we have just dealt with the Lorentz transformation of electric and magnetic fields between frames of reference (Section 3.2). A relativistic treatment of acceleration is particularly important when the acceleration of charged particles produces light emission and the particles achieve velocities close to the speed of light. As acceleration of charged bodies produces electromagnetic radiation, the Lorentz transformation of charged particle acceleration enables the radiated power produced in the frame of reference of a moving and accelerating charged body to be related to the radiated power and acceleration observed by a stationary observer.

In order to determine how an acceleration measured in, say, the rest frame affects the radiation produced by a body with a velocity moving at relativistic velocities, we need to obtain expressions for the transformation of the acceleration between a moving and rest frame. It is sometimes written, particularly in popular accounts of relativity, that as special relativity applies to inertial frames of reference, special relativity cannot treat the acceleration of bodies moving at relativistic velocities. We show in this section that special relativity deals with acceleration using the idea of an "instantaneous" inertial frame of reference, where for an infinitesimally small interval of time, the frame of reference remains inertial. We use the expressions for the velocity of a body in a moving frame of reference as observed in a stationary frame (as found in Section 2.3) to determine the appropriate transform of acceleration from the moving to the rest frame.

The transformation of the velocity parallel ($u_{\parallel}$) and perpendicular ($u_{\perp}$) to the velocity $\mathbf{v}$ from the moving frame of reference S' to the stationary frame S is given by Equation 2.27 and 2.28, namely:

$$u_{\parallel} = \frac{u'_{\parallel} + v}{1 + vu'_{\parallel}/c^2}, \qquad (3.28)$$

$$u_{\perp} = \frac{u'_{\perp}}{\gamma(1 + vu'_{\parallel}/c^2)}, \qquad (3.29)$$

where the primed velocities are measured in the moving frame S' and the unprimed velocities in the stationary frame S. The accelerations can be found by differentiating the two velocities parallel and perpendicular to $\mathbf{v}$ with respect to time. The acceleration parallel to $\mathbf{v}$ in the rest frame S is related to velocities in the moving frame S' by

$$a_{||} = \frac{du_{||}}{dt} = \frac{d}{dt}\left(\frac{u'_{||} + v}{1 + v\,u'_{||}/c^2}\right) = \left(\frac{(1 + v\,u'_{||}/c^2) - (u'_{||} + v)v/c^2}{(1 + v\,u'_{||}/c^2)^2}\right)\frac{du'_{||}}{dt}$$

$$= \left(\frac{1}{\gamma^2(1 + v\,u'_{||}/c^2)^2}\right)\frac{du'_{||}}{dt}$$

as the Lorentz factor $\gamma = 1/\sqrt{1 - v^2/c^2}$. We need to relate differential intervals of time in the two reference frames. Using Equation 2.15 to transform time from the moving reference frame S' to the stationary frame S, we have

$$t = \frac{t' + v\,x'/c^2}{\sqrt{1 - v^2/c^2}},$$

where x' is distance in the direction parallel to $\mathbf{v}$. Taking differentials

$$dt = \frac{dt' + (v/c^2)dx'}{\sqrt{1 - v^2/c^2}} = \gamma\left(1 + \frac{v\,u'_{||}}{c^2}\right)dt' \qquad (3.30)$$

using $u'_{||} = dx'/dt'$. Substituting Equation 3.30 in the above equation for $a_{||}$ yields the transformation of acceleration from the moving frame S' to the stationary frame S:

$$a_{||} = \left(\frac{1}{\gamma^3(1 + v\,u'_{||}/c^2)^3}\right)\frac{du'_{||}}{dt'} = \left(\frac{1}{\gamma^3(1 + v\,u'_{||}/c^2)^3}\right)a'_{||} \qquad (3.31)$$

The acceleration perpendicular to $\mathbf{v}$ in the rest frame S is related to velocities in the moving frame S' by

$$a_{\perp} = \frac{du_{\perp}}{dt} = \frac{d}{dt}\left(\frac{u'_{\perp}}{\gamma(1 + vu'_{||}/c^2)}\right)$$

$$= \left(\frac{\gamma(1 + v\,u'_{\perp}/c^2)(du'_{\perp}/dt) - \gamma(v\,u'_{\perp}/c^2)(du'_{||}/dt)}{\gamma^2(1 + vu'_{||}/c^2)^2}\right)$$

$$= \left(\frac{1}{\gamma(1 + v\,u'_{||}/c^2)}\right)\frac{du'_{\perp}}{dt} - \left(\frac{u'_{\perp}v/c^2}{\gamma(1 + v\,u'_{||}/c^2)^2}\right)\frac{du'_{||}}{dt}.$$

Substituting Equation 3.30 in the above equation for $a_{\perp}$ yields the transformation of acceleration from the moving frame S' to the stationary frame S:

$$a_{\perp} = \left(\frac{1}{\gamma^2(1 + v\,u'_{||}/c^2)^2}\right)\frac{du'_{\perp}}{dt'} - \left(\frac{u'_{\perp}v/c^2}{\gamma^2(1 + v\,u'_{||}/c^2)^3}\right)\frac{du'_{||}}{dt'}$$

$$= \left(\frac{1}{\gamma^2(1 + v\,u'_{||}/c^2)^2}\right)a'_{\perp} - \left(\frac{u'_{\perp}v/c^2}{\gamma^2(1 + v\,u'_{||}/c^2)^3}\right)a'_{||} \qquad (3.32)$$

In an "instantaneous" frame of reference for a body undergoing acceleration, all velocities according to the body under acceleration are zero. "Instantaneous" here means that we consider a sufficiently small increment of time for the acceleration process so the frame of reference remains inertial. The frame of reference of the accelerating body still has a velocity $\mathbf{v}$ relative to a stationary observer. Setting the velocities $u'_{||}$ and $u'_{\perp}$ equal to zero in Equations 3.31 and 3.32 means that the moving frame of reference S' of the body has accelerations which are related to the stationary frame S by

$$a'_{||} = \gamma^3 a_{||}, \tag{3.33}$$

$$a'_{\perp} = \gamma^2 a_{\perp}. \tag{3.34}$$

As in all the above treatment, the acceleration in the moving frame of reference is the primed variable.

The force $\mathbf{F}$ producing an acceleration $d\mathbf{v}/dt$ when both force and acceleration are measured in the rest frame can be evaluated using our expression for relativistic momentum $\mathbf{p}$ (Equation 2.62). We have that

$$\mathbf{F} = \frac{d\mathbf{p}}{dt}$$

with the momentum in the rest frame of a body with velocity $\mathbf{v}$ and rest mass m_0 given by

$$\mathbf{p} = \frac{1}{\sqrt{1 - v^2/c^2}} m_0 \mathbf{v} = \gamma m_0 \mathbf{v}. \tag{3.35}$$

Differentiating with respect to time t, we find

$$\mathbf{F} = \frac{d(\gamma m_0 \mathbf{v})}{dt} = m_0 \mathbf{v} \frac{d\gamma}{dt} + m_0 \, \gamma \, \frac{d\mathbf{v}}{dt}. \tag{3.36}$$

The first term here is in the direction of the velocity, while the second term has components of acceleration $d\mathbf{v}/dt$ in directions parallel to the velocity and perpendicular to the velocity. Considering the component perpendicular to the velocity relates force $F_{\perp}$ perpendicular to the velocity to the perpendicular acceleration $a_{\perp}$:

$$F_{\perp} = m_0 \gamma \frac{dv_{\perp}}{dt} = m_0 \gamma \, a_{\perp} \tag{3.37}$$

Using Equation 3.34, we have that the force $F_{\perp}$ perpendicular to $\mathbf{v}$ is equal to $(1/\gamma)m_0 a'_{\perp}$. The force $F'_{\perp}$ in the moving frame is given by $m_0 a'_{\perp}$ so that

$$F_{\perp} = (1/\gamma)m_0 a'_{\perp} = (1/\gamma)F'_{\perp}. \tag{3.38}$$

The first term and a component of the second term of Equation 3.36 contribute to the force $F_{||}$ parallel to the velocity $\mathbf{v}$. We have

$$F_{||} = m_0 v (-1/2)(1 - v^2/c^2)^{-3/2}(-2v\, a_{||}/c^2) + m_0 \gamma\, a_{||}.$$

Collecting terms, the force parallel to the velocity **v** becomes

$$F_{||} = m_0 \gamma^3\, a_{||}. \tag{3.39}$$

The variation with the Lorentz parameter γ for the force/acceleration behavior perpendicular (Equation 3.38) and parallel (Equation 3.39) to the velocity **v** between frames means that a force measured in the rest frame produces different acceleration as measured in the rest frame depending on its direction. In addition, for forces applied to bodies moving at relativistic speeds, the acceleration vector in the rest frame is not parallel to the force vector in the rest frame unless the force vector is exactly parallel or perpendicular to the velocity **v**.

In the moving frame S' (the proper frame), the force $F'_{||}$ parallel to the velocity **v** is related to the acceleration $a'_{||}$ measured in frame S' by the Newtonian relationship

$$F'_{||} = m_0 a'_{||}. \tag{3.40}$$

Equation 3.33 relates the acceleration $a_{||}$ in the rest frame to the acceleration $a'_{||}$ in a moving frame with the relationship $a'_{||} = \gamma^3 a_{||}$. Substituting into Equation 3.40 means that

$$F'_{||} = m_0 \gamma^3 a_{||} = F_{||} \tag{3.41}$$

upon using Equation 3.39. The force directed parallel to the velocity **v** in the moving and rest frames is the same, or in other words, the force directed parallel to the velocity **v** is Lorentz invariant (see Appendix B).

The Lorentz invariance of a force applied parallel to the velocity **v** is consistent with our finding in Section 3.2 that an electric field parallel to **v** is Lorentz invariant (Equation 3.26). A charge q experiences a force $qE_{||}$ parallel to **v** in an electric field $E_{||}$ parallel to **v**. The charge of a body is an intrinsic Lorentz invariant quantity, so both the force and electric field parallel to **v** are Lorentz invariant.

3.4 Emission from Relativistic Particles

Relativistic transformations are needed to find the angular variation of power detected in the rest frame of reference by an emitting particle moving at relativistic speeds. Conservation of energy requires that the total power radiated into all angles in the frame of reference of the charged particle is the same as the total power detected over all angles in the stationary rest frame. Evaluations of the total power emitted in the proper frame of reference of emission integrated over all angles (see, e.g. Equation 1.16) are identical to the total power integrated over all angles detected in the rest frame. However, the angular variation of the radiated power detected in the rest frame can differ significantly from the angular variation of the power radiated in the frame of reference of the emitting body. The most significant relativistic effect when the emitting body frame of reference has a relative velocity close to the speed of light is to concentrate the detected power in the rest frame in the direction of the velocity of the emitting particle. This behavior is known as the "beaming" of the light emission and is considered in detail in Section 3.5.

Acceleration of a charged particle causes the emission of electromagnetic radiation (see Section 1.3). An accelerating frame of reference is by definition not an inertial frame of reference. However, it is possible to use special relativity to deal with the detection of radiation from relativistic particles. We use the idea of an "instantaneous inertial" frame of reference for the particle emitting radiation. In an instantaneous inertial frame of reference, the frame remains inertial by only considering a sufficiently small increment of time so that "fictitious" forces (as discussed in Section 2.1) due to the acceleration are infinitesimally small.

The electromagnetic power $dP'_\Omega/d\Omega'$ radiated per unit solid angle by an accelerated charged particle was evaluated in the instantaneous frame of reference of the particle in Section 1.3. Modifying Equation 1.15 for the emission due to the acceleration $\mathbf{a}'$ in the instantaneous frame of reference S', the power radiated is proportional to the square of the acceleration measured in the moving frame:

$$\frac{dP'_\Omega}{d\Omega'} = \frac{q^2}{4\pi\epsilon_0}\frac{1}{4\pi c}\left(\frac{\mathbf{a}'\cdot\mathbf{a}'}{c^2}\right)\sin^2\theta'_A,\tag{3.42}$$

where q is the charge of particle and θ'_A is the angle of the emission to the acceleration direction of the particle in the frame S'. The acceleration can be represented as components parallel $a'_\parallel$ and perpendicular $a'_\perp$ to the direction of the velocity $\mathbf{v}$ of S'. The acceleration term in Equation 3.42 then becomes

$$\mathbf{a}'\cdot\mathbf{a}' = a'^2_\parallel + a'^2_\perp.$$

Using the relationships (Equations 3.33 and 3.34) for converting accelerations in an instantaneous frame of reference to the stationary frame:

$$\frac{dP'_\Omega}{d\Omega'} = \frac{q^2}{4\pi\epsilon_0}\frac{1}{4\pi c}\gamma^4\left(\frac{\gamma^2 a^2_\parallel + a^2_\perp}{c^2}\right)\sin^2\theta'_A\tag{3.43}$$

As used throughout this book, parameters with primes are evaluated in the moving frame of reference. Equation 3.43 includes the angle θ'_A of the emitted radiation to the acceleration direction in the frame of reference of the emitting body and is evaluated per unit $d\Omega'$ of solid angle in the frame of the emitting body.

The total power P radiated into 4π steradian in the frame of reference of the charged particle is the same as the total power detected in 4π steradian in the stationary rest frame. Integrating Equation 3.43 over all angles in the moving frame (see Equation 1.16), we have for the total radiated power in the rest frame

$$P = \frac{q^2}{4\pi\epsilon_0} \frac{2}{3c} \gamma^4 \left(\frac{\gamma^2 a_{\parallel}^2 + a_{\perp}^2}{c^2} \right), \tag{3.44}$$

where the accelerations parallel $a_{\parallel}$ and perpendicular $a_{\perp}$ to the velocity $\mathbf{v}$ are evaluated in the rest frame of reference. There is an enhancement of the total radiated power due to the velocity of the emitter approaching the speed of light as $P \propto \gamma^4$ (acceleration perpendicular to the particle velocity) or $P \propto \gamma^6$ (acceleration parallel to the particle velocity). This Lorentz parameter γ dependence of the total emission (over all angles in the rest frame) arises due to the different measurements of acceleration in the moving and rest frames of reference.

3.5 "Beaming" of Emission from Relativistic Bodies

In this section, we start to consider the precise direction of radiation emission in the rest frame of reference when the emitting body is moving with a velocity approaching the speed of light. The angular variation of the radiated power detected in the rest frame can differ significantly from the angular variation of the power radiated in the frame of reference of the emitting body. The most significant relativistic effect when the emitting body frame of reference has a relative velocity close to the speed of light is to concentrate the detected power in the rest frame in the direction of the velocity of the emitting particle. This behavior is known as the "beaming" of the light emission. For a quantitative treatment, we need to relate solid angles as measured in the moving and rest frames and we need to take account of the transform of the radiated power as measured in the moving and rest frames.

The solid angle $d\Omega'$ in Equations 3.42 and 3.43 is related to the angle θ' of the emission to the velocity $\mathbf{v}$ by

$$d\Omega' = \sin\theta' d\theta' d\phi',$$

where $d\phi'$ is the incremental azimuthal angle (see Figure 3.1). Setting $\mu = \cos\theta$ and $\mu' = \cos\theta'$, we can find the relationship between $d\Omega'$ in the moving S' frame and $d\Omega$ in the stationary S frame. As $d\mu' = -\sin\theta' d\theta'$, we have that

$$d\Omega' = -d\mu' d\phi'.$$

The relationships between the angles of emission in the moving frame S' and the stationary frame S were evaluated in Section 2.3. We have from Equation 2.43 that

$$\cos\theta = \frac{\cos\theta' + v/c}{1 + (v/c)\cos\theta'}.$$

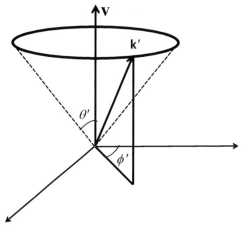

Figure 3.1   Spherical polar coordinates for a frame of reference moving at velocity **v**. Relativistic effects are uniform for all angles θ' to the velocity **v** and so do
not vary with azimuthal angle ϕ'.

Differentiating and substituting $\mu=\cos\theta$, $\mu'=\cos\theta'$, and $\gamma=1/\sqrt{1-v^2/c^2}$ gives

$$d\mu = \frac{d\mu'}{\gamma^2(1+\mu'v/c)^2}.$$

Relativistic transformations from frame S' to S only depend on the angle θ' to the
velocity **v** and so azimuthal angles ϕ are Lorentz invariant with $\phi'=\phi$. Consequently, azimuthal increments are also Lorentz invariant so that $d\phi'=d\phi$. The relationship between solid angles $d\Omega$ in the rest frame S and $d\Omega'$ in the moving frame
S' are consequently given by

$$d\Omega = \frac{d\Omega'}{\gamma^2(1+\mu'v/c)^2}. \tag{3.45}$$

Dealing with radiated power from a particle in a frame moving at relativistic
speeds requires the Lorentz transformation of increments of radiation energy. The
energy W measured in the stationary frame S is related to the energy W' of a body
in the moving frame S' by Equation 2.107. We have

$$W = \gamma(W' + vp_x').$$

Taking the differential of the energy relationship leads to

$$dW = \gamma(dW' + vdp_x') = \gamma\left(1 + \frac{v\cos\theta'}{c}\right)dW' = \gamma(1+\mu'v/c)dW', \tag{3.46}$$

as the momentum p_x' of light in the direction of the velocity **v** emitted at an angle of
θ' to the velocity **v** is $W'\cos\theta'/c$ (see Equation 2.80). Combining Equations 3.45

and 3.46 gives an equation which relates the energy per solid angle in the stationary S frame to that in the moving S' frame:

$$\frac{dW}{d\Omega} = \gamma^3(1 + \mu'v/c)^3\frac{dW'}{d\Omega'} \tag{3.47}$$

To evaluate the power radiated we can divide Equation 3.47 by appropriate differentials of time. In Section 2.3 we showed that Equation 2.23 relates intervals of time in the moving S' and stationary S frames such that we can write

$$dt = \gamma\left(dt' - \frac{v}{c^2}dx'\right) = \gamma(1 - \mu'v/c)dt', \tag{3.48}$$

where the velocity v of an emitting body is positive when directed towards the stationary observer. The second equality in Equation 3.48 replaces dx'/dt' by the component of the speed of photons along the x-direction. The speed of light emitted at angle θ' to the x-direction parallel to $\mathbf{v}$ is given by $dx'/dt' = c\cos\theta' = c\mu'$. We can change frames of reference for the time intervals by interchanging primed and unprimed symbols and changing the sign of the velocity $\mathbf{v}$:

$$dt' = \gamma(1 + \mu v/c)dt$$

so that

$$dt = (\gamma(1 + \mu v/c))^{-1}dt'. \tag{3.49}$$

These relationships between frames of reference of small increments of time (Equations 3.48 and 3.49) were presented earlier when discussing changes in the period of an electromagnetic wave due to the Doppler effect (see Section 2.6).

An expression for the relationship between the power $dP/d\Omega = dW/(dt d\Omega)$ detected per solid angle in the stationary S frame to the power $dP'/d\Omega' = dW'/(dt'd\Omega')$ radiated per solid angle by a single emitting body in the moving S' frame is obtained using Equation 3.47 and the relationship (Equation 3.49) between the time differentials dt and dt':

$$\frac{dP}{d\Omega} = \gamma^4(1 + \mu'v/c)^4\frac{dP'}{d\Omega'}, \tag{3.50}$$

where μ' is measured in the moving frame. It is useful to have the angular variation of power $dP/d\Omega$ in terms of the cosine μ of the angle θ to the velocity $\mathbf{v}$ measured in the stationary frame rather than μ' measured in the moving frame. Changing the frames of reference so that the primed symbols become unprimed and unprimed symbols are primed and the velocity $\mathbf{v}$ changes to $-\mathbf{v}$, we have

$$\frac{dP'}{d\Omega'} = \gamma^4(1 - \mu v/c)^4\frac{dP}{d\Omega}$$

and hence

$$\frac{dP}{d\Omega} = \frac{1}{\gamma^4(1-\mu\, v/c)^4}\frac{dP'}{d\Omega'}.\tag{3.51}$$

For isotropic or near-isotropic emission in the moving frame S', Equation 3.51 predicts emission in the stationary frame S with an angular variation which is highly peaked in the forward direction parallel to the velocity $\mathbf{v}$. Assuming that $dP'/d\Omega'$ is constant with angle θ' in the moving frame means that in the rest frame

$$\frac{dP}{d\Omega} \propto \frac{1}{\gamma^4(1-\mu\, v/c)^4}.\tag{3.52}$$

For a body moving close to the speed of light, the angle θ to the velocity in the rest frame is small and γ is large, so that we can expand $\mu = \cos\theta \approx 1 - \theta^2/2$ and use

$$\frac{v}{c} = \left(1 - \frac{1}{\gamma^2}\right)^{1/2} \approx 1 - \frac{1}{2\gamma^2}.$$

It follows that

$$1 - \mu\, v/c = 1 - \frac{v}{c}\cos\theta \approx 1 - \left(1 - \frac{1}{2\gamma^2}\right)\left(1 - \frac{\theta^2}{2}\right)$$

$$= \frac{1+\gamma^2\theta^2}{2\gamma^2} - \frac{\theta^2}{4\gamma^2} \approx \frac{1+\gamma^2\theta^2}{2\gamma^2}.\tag{3.53}$$

Substituting the approximation of Equation 3.53 into Equation 3.52, the power per steradian in the rest frame if emission in the moving frame is isotropic varies with the angle θ to the velocity of the emitter as

$$\frac{dP}{d\Omega} \propto \frac{1}{\gamma^4(1-\mu\, v/c)^4} \approx \left(\frac{2\gamma}{1+\gamma^2\theta^2}\right)^4.\tag{3.54}$$

Isotropic emission in the moving frame appears concentrated in the rest frame at angles θ close to the direction of the velocity of the moving frame. Equation 3.54 is a highly peaked function with an angular spread close to $\theta \approx 0$ of $1/\gamma$. For example, the power $dP/d\Omega$ radiated drops by a factor $2^4 = 8$ in moving from $\theta = 0$ to $\theta = 1/\gamma$ away from the direction of the velocity $\mathbf{v}$ (see Figure 3.2). This concentration of the radiated power as detected in the rest frame around the direction of the velocity of the emitter is known as the "beaming" effect.

Equation 3.51 uses Equation 3.49 to allow for the Doppler effect on the detected power due to the distance moved by the emitting body in the time dt'. In some evaluations, we need to ignore the effect of the Doppler shift of detected frequencies on the power. For example, if we measure radiation power in terms of the number of photons, we can ignore the Doppler energy shift of the photons. In

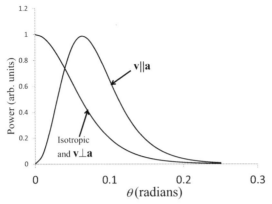

Figure 3.2   The angular variation of the power $dP/d\Omega$ in the stationary frame
S observed when the emission is from a frame S' with velocity $0.99c$ ($\gamma = 7.09$).
Power is shown as a function of the angle θ to the velocity. The emission profiles
are shown assuming (i) isotropic emission and emission where the acceleration is
perpendicular to the velocity ($\mathbf{v}\perp\mathbf{a}$) in the S' frame and (ii) the emission due to
acceleration of a charged particle where the acceleration is parallel to the velocity
($\mathbf{v}\|\mathbf{a}$). The plot shows that almost all emission is within $\theta < 1/\gamma = 0.14$ radians.
Emission with $\mathbf{v}\perp\mathbf{a}$ is for an azimuthal angle $\phi = \pi/2$.

this case, we only need to take account of time dilation in the moving frame, so
set $dt = \gamma dt'$, rather than $dt = (\gamma(1 + \mu'v/c))^{-1}dt'$ as used for Equation 3.51. The
number $dP_p/d\Omega$ of photons per unit time per steradian measured in the stationary
frame is then related to the number $dP'_p/d\Omega'$ of photons per unit time per steradian
emitted in the moving frame by

$$\frac{dP_p}{d\Omega} = \gamma^2(1 + \mu'v/c)^3\frac{dP'_p}{d\Omega'} = \frac{1}{\gamma^4(1 - \mu v/c)^3}\frac{dP'_p}{d\Omega'}, \tag{3.55}$$

where the subscript "p" has been added to indicate power P_p or P'_p in units of
photons per unit time. The two forms of Equation 3.55 are obtained by substitut-
ing $dt = \gamma dt'$ rather than $dt = (\gamma(1 + \mu'v/c))^{-1}dt'$ in Equation 3.47, with a similar
modification to Equation 3.51 producing the second equality.

3.6 Emission with Acceleration Parallel or Perpendicular to v

We now consider the particular examples of light emission as detected in the rest
frame when the acceleration of the charged body producing the emission is parallel or perpendicular to the relative velocity of the body. The general case where the
acceleration is at an arbitrary angle to the relative velocity of the body is mathematically difficult, so we concentrate on the two special cases where the acceleration
of the charged body is parallel or perpendicular to the relative velocity of the body.

The special cases illustrate the general behavior and are common geometries in evaluations of the emission of relativistic particles.

Returning to Equation 3.43, using Equation 3.51 the power radiated per unit solid angle in the stationary frame S is found to be

$$\frac{dP}{d\Omega} = \frac{q^2}{4\pi\epsilon_0} \frac{1}{4\pi c} \left(\frac{\gamma^2 a_\parallel^2 + a_\perp^2}{(1 - \mu v/c)^4 c^2} \right) \sin^2 \theta'_A. \tag{3.56}$$

If the acceleration of the charged particle is parallel to the velocity, angles relative to the velocity (θ') and acceleration (θ'_A) are equivalent ($\theta' = \theta'_A$). Adapting Equation 2.42, the angle θ'_A of the emission to the acceleration in the moving frame S' is related to the angle θ to the velocity in the stationary frame S by

$$\sin \theta'_A = \frac{\sin \theta / \gamma}{1 - (v/c) \cos \theta}. \tag{3.57}$$

Setting $a_\perp = 0$ in Equation 3.56, the radiated power as a function of angle θ in the stationary frame S for acceleration parallel to the velocity becomes

$$\frac{dP}{d\Omega} = \frac{q^2}{4\pi\epsilon_0} \frac{1}{4\pi c^3} a_\parallel^2 \left(\frac{1}{1 - v\cos\theta/c} \right)^6 \sin^2 \theta. \tag{3.58}$$

A similar argument presented to determine Equation 3.54 shows that the angular variation in θ of the bracketed quantities raised to the sixth power in Equation 3.58 peaks sharply around angles $\theta = 0$ in the direction of the velocity $\mathbf{v}$ (see Figure 3.2).

If the acceleration of the charged particle is perpendicular to the velocity, some three-dimensional geometry is needed to determine the relationship in the moving frame between the angle θ'_A of emission to the acceleration direction and the angle θ' of the emission relative to the velocity $\mathbf{v}$ (see Figure 3.3). From Figure 3.3, the cosine rule for a triangle formed between a vector $\mathbf{k}$ representing the light emission at angle θ' to $\mathbf{v}$ with azimuthal angle ϕ' and the acceleration direction perpendicular to $\dot{\mathbf{v}}$ has

$$\cos \theta'_A = \frac{b_2^2 - b_1^2 - k^2}{2kb_1},$$

with $b_1 = k \sin \theta' \cos \phi'$ and $b_2^2 = k^2 (\cos^2 \theta' + \sin^2 \theta' \sin^2 \phi')$. Substituting these values for b_1 and b_2, we find that

$$\cos \theta'_A = \frac{\cos^2 \theta' + \sin^2 \theta' \sin^2 \phi' - \cos^2 \phi' \sin^2 \theta' - 1}{2 \cos \phi' \sin \theta'} = \sin \theta' \cos \phi'.$$

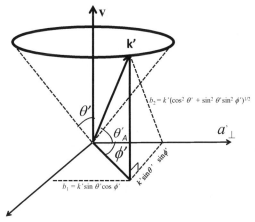

Figure 3.3   An illustration of the relationship of the angle θ'_A between the **k'**-vector of light emission and the acceleration direction for a charged particle and angles θ' between the velocity **v** and the **k'**-vector. The acceleration is assumed to be directed perpendicular to the velocity **v** in the moving frame S'.

We need to evaluate $\sin^2 \theta_A$ and substitute into Equation 3.56 with $a_{\parallel} = 0$. Using Equation 3.57,

$$\sin^2 \theta'_A = 1 - \cos^2 \theta'_A = 1 - \frac{\sin^2 \theta / \gamma^2}{(1 - v \cos \theta / c)^2}.$$

The power radiated per unit solid angle in the stationary frame S for acceleration perpendicular to the velocity in the moving frame S' is then given by

$$\frac{dP}{d\Omega} = \frac{q^2}{4\pi \epsilon_0} \frac{1}{4\pi c^3} a_{\perp}^2 \left(\frac{1}{1 - v \cos \theta / c} \right)^4 \left(1 - \left(\frac{\sin \theta \cos \phi}{\gamma (1 - v \cos \theta / c)} \right)^2 \right). \quad (3.59)$$

There is an azimuthal (ϕ) dependence on the power radiated when the acceleration is perpendicular to the velocity. The azimuthal $\cos^2 \phi'$ term in the moving S' frame transforms to $\cos^2 \phi$ in the stationary frame S as relativistic effects only depend on the angle to the velocity and are consequently independent of azimuthal angle.

As shown for isotropic emission (Equation 3.54) and emission where the acceleration is parallel to the velocity (Equation 3.58), when the acceleration is perpendicular to the velocity, Equation 3.59 indicates a sharply peaked emission in the stationary frame S for velocities v approaching the speed of light c. The emission is centered in the velocity **v** direction ($\theta \approx 0$) due to a term involving $1/(1 - v \cos \theta / c)$ which is raised to the power of four (for isotropic emission and when the acceleration is perpendicular to the velocity), or to the power of six (when the acceleration is parallel to the velocity). Figure 3.2 shows a sample angular variation of emission for the three scenarios considered: isotropic emission, emission where the particle acceleration is parallel to the velocity, and emission where the acceleration is perpendicular to the velocity.

In the extreme relativistic limit when $\gamma >> 1$, the quantity $(1 - v\cos\theta/c)$ is small in the forward direction where $\theta \approx 0$ as the radiation angular profile becomes extremely peaked in the direction of the velocity of emitting particles. In the extreme relativistic limit, it is sufficiently accurate and more suitable for numerical evaluation to use the approximation for $(1 - v\cos\theta/c)$ given in Equation 3.53. In the extreme relativistic limit, upon modifying Equation 3.58 the power radiated in the stationary frame for acceleration parallel to the velocity becomes

$$\frac{dP}{d\Omega} = \frac{q^2}{4\pi\epsilon_0}\frac{1}{4\pi c^3}a_\parallel^2\, \gamma^{10}\left(\frac{\gamma^2\theta^2}{(1+\gamma^2\theta^2)^6}\right). \tag{3.60}$$

In the extreme relativistic limit, upon modifying Equation 3.59 the power radiated in the stationary frame for acceleration perpendicular to the velocity becomes

$$\frac{dP}{d\Omega} = \frac{q^2}{4\pi\epsilon_0}\frac{1}{4\pi c^3}a_\perp^2\, \gamma^{8}\left(\frac{1 - 2\gamma^2\theta^2\cos(2\phi) + \gamma^4\theta^4}{(1+\gamma^2\theta^2)^6}\right) \tag{3.61}$$

upon using the identity $\cos(2\phi) = 1 - 2\cos^2\phi$. Equations 3.60 and 3.61 both depend on the angle θ of emission solely through the combination of $\gamma\theta$ which confirms that the emission for extremely relativistic particles ($\gamma >> 1$) is concentrated within an angle $\theta \approx 1/\gamma$.

Exercises

3.1   With an eye to the future, the 1967 edition of the textbook *Classical Electrodynamics* by J. D. Jackson [58] posed the following question: "Assume that a rocket ship leaves the Earth in the year 2000. One of a set of twins born in 1980 remains on Earth; the other rides in the rocket. The rocket ship is so constructed that it has an acceleration g in its own rest frame (this makes the occupants feel at home). It accelerates in a straight-line path for 5 years (by its own clocks), decelerates at the same rate for 5 more years, accelerates for 5 years, decelerates for 5 years, and lands on Earth. The twin in the rocket is 40 years old. (a) What year is it on Earth? (b) How far away from the Earth did the rocket ship travel?" [See an answer in Section 10.1.]

3.2   Assume that the rocket ship employed in Exercise 3.1 has a constant rest mass including occupants of 10^3 kg. We can imagine that the rocket is accelerated by some mass-conserving process, say, electron-positron annihilation so that the mass of the spaceship remains approximately constant. (a) Calculate a minimum energy needed for the described spaceflight. (b) Determine the power required to be supplied by the rocket in order to maintain the acceleration g at the end of the two accelerating phases of the flight. [See an answer in Section 10.1.]

3.3 The American drama television series *Manifest* premiered in 2018 featuring the passengers and crew of a commercial airliner who suddenly reappear after their aircraft is missing for 5 years. To the passengers and crew the flight time appears only slightly delayed. Assuming that some acceleration phenomenon to relativistic velocities is responsible for the different passage of time on the aircraft and the rest of the world, estimate the acceleration needed in the frame of reference of the aircraft if the aircraft uniformly accelerates for 1 hour, deaccelerates for 1 hour, and then undergoes the same acceleration in a return journey. [Use Equation 10.3 to get $\approx 10^6$ m s^{-2}. The aircraft would suffer more than the mild turbulence portrayed onscreen.]

3.4 A simpler variation of the question set by Jackson (Exercise 3.1) is often presented in special relativity courses. A twin in a spaceship is assumed to instantly accelerate to a constant high speed v and travel a distance x away from Earth. At distance x, the twin reverses direction and instantly accelerates up to speed v returning to Earth. Show that the time t on Earth is related to time t' in the spaceship by

$$t = \frac{t'}{\sqrt{1 - v^2/c^2}}.$$

3.5 Electrons are accelerated and oscillate, producing radiation when a high-power laser pulse propagates through a gas. The electron acceleration is known as "wakefield" acceleration (see e.g. [3], [73]). In their frame of reference, the electrons produce light close to the laser wavelength due to an oscillation perpendicular to the laser direction. If electrons of energy 300 MeV traveling parallel to the laser beam are produced, (a) determine the range of angles where you would expect to observe in the laboratory light emitted by the electrons. (b) Estimate the enhancement of the total radiated power of electron emission from the electron to laboratory frame due to the relativistic electron velocity. (c) Determine the fractional change of emitted wavelength from the electron to the laboratory frame. [(a) Light is emitted within ≈ 1.7 milliradians of the laser beam. (b) 1.2×10^{11} (c) $\approx 1.7 \times 10^{-3}$]

3.6 Neutral hydrogen atoms of kinetic energy 100 keV are injected into a tokamak plasma perpendicular to a 10 Tesla magnetic field. Calculate the electric field seen by the hydrogen atoms in their frame of reference. [4.4×10^7 V m^{-1}]

3.7   Consider a charge of velocity $\mathbf{v}$ such that $v \ll c$ which travels perpendicular to an electric field $\mathbf{E}$ and a magnetic field $\mathbf{B}$ where the electric and magnetic fields are perpendicular to each other. When the forces on the charge due to the electric and magnetic fields are balanced with $\mathbf{E} = -\mathbf{v} \times \mathbf{B}$ and hence the charge travels in a straight line, show that in the frame of reference of the charge, the magnetic field $\mathbf{B}'$ seen by the charge is given by

$$\mathbf{B}' \approx \mathbf{B} - \frac{1}{2} \frac{\mathbf{v} \times \mathbf{E}}{c^2},$$

while the electric field $\mathbf{E}'$ in the frame of the charge is zero.

4

Electromagnetic Wave Interactions with Plasma and Other Matter

A plasma can be defined as a collection of free electrons and ions where for distances much greater than the atomic scale, the average charge is zero. A gas becomes a plasma if heated sufficiently so that electrons are removed from atoms creating negatively charged free electrons and positively charged ions.

Electrons undergo greater acceleration in the presence of an electric field or electromagnetic wave as the magnitude of the electron charge is the same as that of a proton, while the electron mass is a factor 1,836 times smaller than the mass of a proton. Consequently, the electrons in a plasma can be accelerated much more readily to velocities approaching the speed of light and electrons exhibit relativistic effects at much lower particle energies than ions. Similarly, electrons can modify the passage of light much more effectively than ions because they accelerate at a much greater rate in an electromagnetic wave.

We examine single electron motion in an electromagnetic wave treating linear and elliptical polarization and a relativistic treatment in this chapter. A relativistic treatment of electron motion is needed when electron acceleration at high laser irradiances leads to the electron velocities approaching the speed of light. We extend the equations for the vacuum propagation of light (Section 1.1) to a medium with a refractive index and determine the refractive index for plasmas with a significant electron density. We continue with an examination of Cerenkov radiation produced when the velocity of a particle in a medium exceeds the phase velocity of light in that medium, and the processes of Thomson and Compton scatter. Thomson and Compton scatter can arise due to an isolated electron radiating light as the electron accelerates in an electromagnetic wave. Thomson scatter also occurs due to distributions of electrons if the scattered light electric field from different electrons adds coherently. Compton scatter is similar to single electron Thomson scatter, but is the name of the scattering process at high photon energies where the photon momentum is important.

4.1 Nonrelativistic Single Electron Motion: Linear Polarization

Due to the difference in mass of electrons and ions, an electromagnetic wave interacts with the electrons in a plasma and causes only an insignificant direct acceleration of the ions. In this section, we examine the motion of a single free electron
in an electromagnetic wave when the velocities reached by the electron as it accelerates in the field are much less than the speed of light c. All calculations here are
in the frame of reference of the electron, so that the initial electron velocity relative to an impinging electromagnetic wave is zero. At the high irradiances possible
with focused laser light, electrons can be accelerated to velocities approaching the
speed of light c. Modifications to the treatment of this section to take account of
relativistic effects are discussed later in Section 4.3.

   If an electromagnetic wave is linearly polarized, we can assume that the wave
propagates in the z-direction with the electric field directed in the x-direction. The
magnetic field of the electromagnetic wave is then directed in the y-direction. For
a plane wave uniform in the transverse direction to the light propagation direction,
movement or drift of an electron in the z-direction of light propagation due to the
incoming wave can be taken into account by undertaking calculations in a retarded
time τ, such that

$$\tau = t - \frac{z(t)}{c}, \tag{4.1}$$

where $z(t)$ represents the z-position of the electron at time t. The electron is assumed
to be at position $x = y = z = 0$ at time $t = 0$. For a constant light intensity, we show
in this section that the drift velocity $v_d = dz(t)/dt$ of an electron in the direction of
propagation of the wave is constant so that $d\tau = dt(1 - v_d/c) = dt$. We assume for
this section that any electron velocity $<< c$.

   Assuming an electric field of amplitude E_0 and angular frequency ω, the acceleration of an electron of charge $-e$ and mass m_0 is given by

$$\frac{d^2x}{dt^2} = -\frac{e}{m_0}E_0 \cos(\omega\tau). \tag{4.2}$$

Integrating from retarded time 0 to time τ, we obtain an expression for the velocity
of the electron in the electromagnetic field

$$\frac{dx}{dt} = -\frac{eE_0}{m_0\omega} \sin(\omega\tau). \tag{4.3}$$

Integrating again, we find the position of the electron relative to its starting position,
namely

$$x = +\frac{eE_0}{m_0\omega^2}(\cos(\omega\tau) - 1). \tag{4.4}$$

The electron oscillates or "quivers" in the electric field of the wave. We can calculate an average quiver energy which is often referred to as the "ponderomotive" energy. At any time the kinetic energy U of the electron is given by

$$U = \frac{1}{2}m_0\left(\frac{dx}{dt}\right)^2 = \frac{e^2E_0^2}{2m_0\omega^2}\sin^2(\omega\tau). \tag{4.5}$$

If we average the square of the sine function in time over an oscillation period of the electromagnetic wave, we can obtain a time-averaged quiver energy. Using

$$\frac{\int_0^{2\pi}\sin^2\theta\,d\theta}{\int_0^{2\pi}d\theta} = \frac{1}{2} \tag{4.6}$$

gives the time-averaged electron quiver energy $< U >$ in an electromagnetic field as

$$< U >= \frac{e^2E_0^2}{4m_0\omega^2}. \tag{4.7}$$

This average quiver energy of the electrons in an electromagnetic field is known as the ponderomotive potential. Evaluating constants, converting the square of the magnitude of the electric field (E_0^2) to irradiance (Equation 1.12) and the frequency ω to wavelength λ (using $\omega = 2\pi c/\lambda$) gives a numerical expression (in energy units of electron volts) for the ponderomotive potential

$$< U >\approx \frac{I\lambda^2}{10^{13}} \quad \text{eV}, \tag{4.8}$$

where the wavelength λ is measured in units of microns and the irradiance I is measured in units of Wcm^{-2}. For laser-plasma interactions at irradiances greater than 10^{14} $\text{Wcm}^{-2}\mu\text{m}^2$, the ponderomotive potential exceeds 10 eV and starts to become comparable to typical atomic ionization potentials (e.g. 13.6 eV for hydrogen atoms) and to the temperatures and hence kinetic energy per electron in the plasma.

The motion of the electron in an electromagnetic field is complicated by the presence of the magnetic field. The electron has a velocity $\mathbf{v}$ due to its acceleration in the electric field, which produces a Lorentz force $e\mathbf{v} \times \mathbf{B}$ pointing in the direction of propagation of the electromagnetic wave. We can write for the acceleration in the direction of wave propagation (our z-direction) that

$$\frac{d^2z}{dt^2} = -\frac{e}{m_0}\frac{dx}{dt}B,$$

where B is the magnetic field of the electromagnetic wave. The magnetic field B oscillates in phase with the electric field such that

$$B = \frac{E_0}{c} \cos(\omega\tau).$$

We can write for the electron acceleration in the z-direction due to the $e\mathbf{v} \times \mathbf{B}$ force that

$$\frac{d^2z}{dt^2} = \frac{e^2E_0^2}{m_0^2\omega c} \sin(\omega\tau)\cos(\omega\tau). \tag{4.9}$$

Integrating from retarded time 0 to time τ gives the velocity in the direction of propagation as

$$\frac{dz}{dt} = \frac{e^2E_0^2}{2m_0^2\omega^2 c} \sin^2(\omega\tau). \tag{4.10}$$

The position of the electron relative to its starting position is found by integrating again to give

$$z = -\frac{e^2E_0^2}{8m_0^2\omega^3 c} \sin(2\omega\tau) + \frac{e^2E_0^2}{4m_0^2\omega^2 c}\tau. \tag{4.11}$$

We have used here the integration result that

$$\int_0^{\omega\tau} \sin^2(\theta)d\theta = \frac{1}{2}\left(\omega\tau - \frac{1}{2}\sin(2\omega\tau)\right).$$

Equations 4.4 and 4.11 show that an electron in an electromagnetic field executes a figure-8 motion in the $xz-$plane with a superimposed electron drift in the z-direction (see Figure 4.1). The x-position of the electron is oscillating in position proportionally to $\cos(\omega\tau)$, while the z-position oscillates proportionally to $\sin(2\omega\tau)$. These equations for the electron position describe a figure-8 motion relative to a position moving with drift velocity v_d in the direction of electromagnetic propagation (the z-direction). Looking at the second term on the right-hand side of Equation 4.11, we see that the drift velocity $v_d = z(t)/t$ is given by

$$v_d = \frac{e^2E_0^2}{4m_0^2\omega^2 c}\left(\frac{\tau}{t}\right) = \frac{e^2E_0^2}{4m_0^2\omega^2 c}\left(\frac{t - z(t)/c}{t}\right) = \frac{e^2E_0^2}{4m_0^2\omega^2 c}\left(1 - \frac{v_d}{c}\right).$$

Rearranging to have all expressions involving the drift velocity v_d on the left-hand side, we have

$$\frac{v_d}{c} = \frac{e^2E_0^2}{m_0^2\omega^2 c^2}\left(\frac{1}{4 + e^2E_0^2/(m_0^2\omega^2 c^2)}\right). \tag{4.12}$$

Theoretical work on electron acceleration at high irradiance often uses the vector potential $\mathbf{A}$ description, where for an oscillating electric field $\mathbf{E}$, we have $\mathbf{E} = -d\mathbf{A}/dt$. The amplitude of the vector potential A_0 is employed in a reduced form

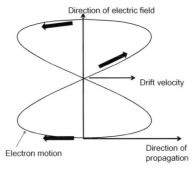

Figure 4.1   The trajectory of an electron in an electromagnetic wave. The electron exhibits a figure-8 motion as well as a superimposed drift in the direction of propagation of the wave.

$$a_o = \frac{eA_0}{m_0 c} = \frac{eE_0}{m_0 c \omega}. \tag{4.13}$$

Substituting for a_o in our expression for the drift velocity (Equation 4.12), we find that the electron drift velocity relative to the speed of light becomes a more readily memorized equation:

$$\frac{v_d}{c} = \frac{a_o^2}{4 + a_o^2} \tag{4.14}$$

The drift velocity v_d has no dependence on the phase of the electromagnetic field (there are no $\cos(\omega t)$ or $\sin(\omega t)$ terms in Equation 4.12) and, for a uniform electromagnetic field, no variation with z, so the drift velocity is not dependent on the z-position of the electron as it drifts.

We show in Section 4.3 that relativistic effects change the dependence of the drift velocity v_d with reduced vector potential a_o found in Equation 4.14. Allowing for relativistic effects, the maximum drift velocity v_d at high a_o becomes $c/3$ rather than the speed of light c (see Equation 4.37). Both the nonrelativistic (Equation 4.14) and relativistic (Equation 4.37) expressions have $v_d = a_o^2/4$ for small a_o.

The reduced vector potential a_o is related to the laser irradiance $I = (1/2)\epsilon_0 c E_0^2$ measured in Wcm^{-2} by

$$a_o = \sqrt{\frac{I\lambda^2}{1.37 \times 10^{18}}}, \tag{4.15}$$

where $\lambda = 2\pi c/\omega$ is the laser wavelength in microns. The drift velocity of the electrons in a radiation field becomes an important electron accelerating mechanism at high laser irradiances ($I > 10^{18}$ Wcm^{-2}). At $a_o > 1$, electron acceleration by the drift velocity starts to dominate other electron acceleration mechanisms. The electron acceleration and consequent heating of laser target material when the electrons collide, for example, with a solid target is commonly referred to as $\mathbf{J} \times \mathbf{B}$ "heating."

The velocity v_x of electrons accelerated in the direction of the electric field of the electromagnetic wave can be expressed in terms of the reduced vector potential a_o. An average of the square of the velocity $<v_x^2>$ is found by averaging the time-varying $\sin^2(\omega\tau)$ component of Equation 4.3 using $<\sin^2(\omega\tau)>= 1/2$. Assuming the retarded time τ is independent of a_o, we have

$$\frac{<v_x^2>}{c^2} = \frac{a_o^2}{2}. \tag{4.16}$$

This expression is valid when $a_o << 1$ as then the retarded time τ is independent of a_o. Similarly the average velocity $<v_z>$ in the direction of propagation of the electromagnetic wave in terms of a_o is found from Equation 4.10, again averaging over the $\sin^2(\omega\tau)$ component. We have

$$\frac{<v_z>}{c} = \frac{a_o^2}{4}. \tag{4.17}$$

This expression is again only valid when $a_o << 1$ as then the retarded time τ is independent of a_o. For $a_o << 1$, the time-averaged velocity $<v_z>$ in the direction z of the wave propagation is consistent with the drift velocity v_d (Equation 4.14).

In a nonuniform electromagnetic field such as a laser focus, the ponderomotive potential or time-averaged electron quiver energy $< U >$ given by Equation 4.7 varies spatially. Electrons will tend to move from high $< U >$ to lower values of this potential energy. For any potential field Φ, there exists a force $\mathbf{F}$ given by $\mathbf{F} = -\nabla\Phi$ which serves to homogenize potential energy variations. With a spatially varying ponderomotive potential, there is consequently a time-averaged force on each electron given by

$$\mathbf{F}_p = -\frac{e^2}{4m_0\omega^2}\nabla E_0^2 = -\frac{m_0 c^2}{4}\nabla a_o^2 \tag{4.18}$$

after using Equation 4.7 for the ponderomotive potential $<U>$ and the expression given (Equation 4.13) for the reduced vector potential a_o. The values of E_0^2 or a_o^2 are the spatially varying quantities in a nonuniform electromagnetic field if we assume relativistic effects on the electron momentum are small.

In a laser focus, there is a gradient of intensity and hence a gradient of E_0^2 perpendicular to the direction of laser propagation. On the microscopic level, electrons near the center of the focus are accelerated outwards away from the focal center by the electric field of the electromagnetic wave, and then accelerated back to the center when the oscillating electric field changes direction. However, the value of the electric field is smaller at a distance away from the focal center, so the electrons do not move back to the center completely. With many cycles, the net effect is a motion away from the high-intensity (high E_0^2) region.

The drift velocity v_d in the direction of light propagation is directly linked to the ponderomotive force. Our treatment has implicitly assumed a uniform electromagnetic wave in time, but in practice, light in laser-plasma or radio wave interactions arrives as a pulse. On the leading edge of an electromagnetic pulse, the drift velocity increases proportionally to the intensity I (and hence the square of the laser electric field E_0^2). The instantaneous spatial gradient of the ponderomotive potential $<U>$ of an electromagnetic pulse in the direction of propagation is $(1/c)\,d<U>/dt$. The instantaneous spatial gradient of the ponderomotive potential is equal to the ponderomotive force on an electron, so that the electron momentum change in the direction of the light pulse is given by

$$\frac{dp}{dt} = \frac{1}{c}\frac{d<U>}{dt}.$$

Assuming the electron motion in the direction of the light pulse is small, we can undertake a simple integration in time so that the electron momentum $p \approx <U>/c$. The electron velocity $v_d = p/m_0$ in the direction of the light propagation is then given by

$$v_d \approx \frac{<U>}{m_0 c} = \frac{e^2 E_0^2}{4 m_0 \omega^2 c} = \frac{a_o^2}{4}$$

upon using Equation 4.7 for the value of the ponderomotive potential $<U>$ and our definition of the reduced vector potential a_o (Equation 4.13). This value of the electron drift velocity is consistent with Equation 4.14 in the limit that the drift velocity is small. A large drift velocity changes the ponderomotive potential variation in time seen by an electron, so that the retarded time τ needs to be used and Equation 4.14 applies.

4.2 Nonrelativistic Single Electron Motion: Elliptical Polarization

At a fundamental level, a photon state is best represented by some combination of elliptically polarized components. Photons of light can be regarded as being composed of elliptically polarized components with the electric field rotating clockwise or anticlockwise and returning to the original direction after a wavelength. Such rotation is interpreted as the spin $\pm\hbar$ of the photon directed parallel or antiparallel to the **k**-vector of the light depending on the electric field rotation (clockwise or anticlockwise). Photons with linear polarization are then the linear superposition of two oppositely rotating (opposite spin) circular polarization states.

It is important to consider the interaction of elliptical polarized light with plasma as many laser-plasma experiments are undertaken with elliptical or circular polarization of the focused laser light. The treatment of electron motion in Section 4.1 assumes linear polarization of the electromagnetic wave. Laser light is often linearly polarized, but can be readily converted to elliptical or circular polarization. With circular polarization, the electric field direction of the electromagnetic wave rotates as the wave propagates, returning to the original direction after one wavelength. With elliptical polarization, the electric field magnitude varies within the period of the wave as the wave propagates. The polarization of a light field can be generally quantified using Stokes parameters (see Section 9.4).

We need not consider the fundamental aspects of a single photon polarization further as laser-plasma experiments are undertaken with typically more than 10^{16} photons (e.g. see Exercise 6.9). The electric field for an elliptically polarized electromagnetic wave can be represented by electric field components in the x- and y-directions which are $\pi/2$ out of phase with each other. For a wave propagation direction in the z-direction, we can write that

$$E_x(t) = E_x \cos(\omega\tau),$$
$$E_y(t) = E_y \sin(\omega\tau), \tag{4.19}$$

where E_x and E_y are the peak magnitudes of the electric fields in the x- and y-directions. For circular polarization $E_x = E_y$. We are using retarded time τ as defined by Equation 4.1. The acceleration of an electron in the x- and y-directions is given by

$$\frac{d^2x}{dt^2} = -\frac{e}{m_0}E_x \cos(\omega\tau),$$
$$\frac{d^2y}{dt^2} = -\frac{e}{m_0}E_y \sin(\omega\tau). \tag{4.20}$$

Assuming that the electromagnetic wave is incident for a long time prior to retarded time $\tau = 0$, we integrate the expressions for acceleration to give velocity components in the x- and y-directions:

$$\frac{dx}{dt} = -\frac{eE_x}{m_0\omega}\sin(\omega\tau),$$
$$\frac{dy}{dt} = -\frac{eE_y}{m_0\omega}(1 - \cos(\omega\tau)) \tag{4.21}$$

Assuming the electron at retarded time $\tau = 0$ is at a position $x = 0$ and $y = 0$, the x- and y-positions at subsequent times are given by:

$$x = +\frac{eE_x}{m_0\omega^2}(\cos(\omega\tau) - 1)$$

$$y = -\frac{eE_y}{m_0\omega^2}\sin(\omega\tau) \qquad (4.22)$$

The ponderomotive energy of an electron in the elliptically polarized wave is similar to that found for linear polarization (Equation 4.5). We have

$$U = \frac{1}{2}m_0\left(\left(\frac{dx}{dt}\right)^2 + \left(\frac{dy}{dt}\right)^2\right) \approx \frac{e^2}{2m_0\omega^2}(E_x^2 \sin^2(\omega\tau) + E_y^2 \cos^2(\omega\tau)). \qquad (4.23)$$

The average of the square of sine and cosine functions in time over an oscillation period of the electromagnetic wave are both half of the square of the peak of the sine or cosine values (see e.g. Equation 4.6). A time-averaged quiver or ponderomotive energy is then given by

$$<U> = \frac{e^2(E_x^2 + E_y^2)}{4m_0\omega^2}. \qquad (4.24)$$

Adapting Equation 1.12, the intensity I for elliptically polarized light is related to the electric field strengths by

$$I = \frac{1}{2}\epsilon_0 c(E_x^2 + E_y^2).$$

The ponderomotive energy $<U>$ is, consequently, related to the electromagnetic intensity I as it is for linear polarization (see Equation 4.8). The ponderomotive force due to the gradient of the ponderomotive energy is similarly identical to the ponderomotive force found for linear polarization upon setting $E_0^2 = E_x^2 + E_y^2$ (see Equation 4.18).

As we found for linear polarization, an electron has velocity $\mathbf{v}$ due to its acceleration in the electric field, which produces a Lorentz force $e\mathbf{v} \times \mathbf{B}$ pointing in the direction of propagation of the electromagnetic wave. For elliptical polarization, we can write for the acceleration in the direction of wave propagation (the z-direction) that

$$\frac{d^2z}{dt^2} = -\frac{e}{m_0}\left(\frac{dx}{dt}B_y + \frac{dy}{dt}B_x\right),$$

where B_y and B_x are the magnetic field components of the electromagnetic wave perpendicular to the x- and y-directions of the electron quiver velocities dx/dt and dy/dt. The magnetic fields oscillate in phase with the electric field such that

$$B_y = \frac{E_x}{c}\cos(\omega\tau),$$

$$B_x = \frac{E_y}{c} \sin(\omega\tau).$$

We can write for the electron acceleration in the z-direction due to the $e\mathbf{v} \times \mathbf{B}$ force that

$$\frac{d^2z}{dt^2} \approx \frac{e^2(E_x^2 - E_y^2)}{m_0^2\omega c} \sin(\omega\tau) \cos(\omega\tau).$$

Integrating from retarded time 0 to time τ gives the velocity in the direction of propagation as

$$\frac{dz}{dt} \approx \frac{e^2(E_x^2 - E_y^2)}{2m_0^2\omega^2 c} \sin^2(\omega\tau).$$

Including dropped terms from Equation 4.21, the position of the electron relative to its starting position is found by integrating again to give

$$z \approx -\frac{e^2(E_x^2 - E_y^2)}{8m_0^2\omega^3 c} \sin(2\omega\tau) + \frac{e^2(E_x^2 + E_y^2)}{4m_0^2\omega^2 c} \tau. \tag{4.25}$$

Substituting $E_0^2 = |E_x^2 - E_y^2|$, we have an oscillating motion similar to that found for linear polarization (Equations 4.12 and 4.14), but which vanishes for circular polarization.

4.3 Relativistic Single Electron Motion

When high laser intensities interact with free electrons in a plasma, the electrons can be accelerated by the electric field of the laser to velocities approaching the speed of light so that the Lorentz factor γ for the electron motion becomes much greater than one. The electromagnetic wave reduced vector potential for nonrelativistic particles $a_o = eE_0/(m_0 c\omega)$ takes a value a_0/γ for relativistic electron motion in the electromagnetic wave due to the relativistic increase of the electron mass m_0 to γm_0.

Assuming linear polarization and a significant relativistic mass increase, a modification to Equation 4.16 gives the average of the square of the electron velocity in the electric field direction:

$$\frac{<v_x^2>}{c^2} = \frac{a_o^2}{2 <\gamma^2>} \tag{4.26}$$

We employ here the square of the Lorentz factor $<\gamma^2>$ averaged over a cycle of the electromagnetic wave and neglect electron velocities in the direction of beam propagation (the z-direction) as such velocities are significantly less than velocities

v_x in the direction of the electric field of the wave (see the discussion below for Equation 4.37). An expression for the Lorentz factor in terms of the reduced vector potential a_o for linearly polarized light is found by setting

$$<\gamma^2> = \frac{1}{1 - <v_x^2>/c^2} = \frac{1}{1 - a_o^2/(2<\gamma^2>)}. \qquad (4.27)$$

For elliptically polarized light, the Lorentz factor $<\gamma^2>$ value averaged over a cycle of the electromagnetic wave is given by

$$<\gamma^2> = \frac{1}{1 - <v_x^2 + v_y^2>/c^2} = \frac{1}{1 - a_o^2/(2<\gamma^2>)} \qquad (4.28)$$

as we can define the reduced vector potential a_o for elliptical polarized light by

$$a_o^2 = \frac{e}{m_0 c\omega}(E_x^2 + E_y^2). \qquad (4.29)$$

For elliptically polarized light, $<v_x^2 + v_y^2>$ represents the cycle-averaged square of the electron velocities produced by electric fields of amplitude E_x and E_y in the directions x and y perpendicular to the light propagation direction (see Equation 4.21). Defining the reduced vector potential a_o as given by Equation 4.29 ensures that the relationship between a_o and laser intensity I and laser wavelength λ is given by Equation 4.15 for all polarizations. Rearranging Equation 4.27 or 4.28 gives an expression for the cycle average relativistic Lorentz factor for an electron in an electromagnetic wave of reduced vector potential a_o for both linear and elliptical polarization:

$$<\gamma^2> = 1 + \frac{a_o^2}{2} \qquad (4.30)$$

This relationship between the cycle-averaged square of the relativistic Lorentz factor (i.e. $<\gamma^2>$) and the reduced electric potential a_o is a useful expression as it enables a quick calculation of the effect of relativistic electron mass increase using the intensity of focused light (see Equation 4.15 for the conversion of intensity to a_o). We also utilize Equation 4.30 in our treatment of relativistic Thomson scatter (Section 4.6).

It is interesting to examine the peak Lorentz factor γ experienced by an electron in the field of an electromagnetic wave. The peak Lorentz factor γ_{max} can be found by dropping the sinusoidal field averaging process used to obtain Equation 4.30. We have

$$\gamma_{max} = (1 + a_o^2)^{1/2}. \qquad (4.31)$$

Electrons can be accelerated to kinetic energies $T_{kin} = (\gamma_{max} - 1)m_0 c^2 = (\gamma_{max} - 1)$ 511 keV. With high laser intensity irradiation in plasmas, this scaling is often equated to an electron temperature T_e:

$$k_B T_e = ((1 + a_o^2)^{1/2} - 1)\, m_0 c^2 \qquad (4.32)$$

Simulations show that superthermal temperatures with an intensity scaling as given by Equation 4.32 are produced in the plasmas created when focused high-intensity $(a_o > 1)$ lasers are incident onto targets [24], [85]. The temperature scaling for electrons created in high-intensity laser interactions given by Equation 4.32 is known as the Wilks scaling after Wilks et al. [114], who first found this relationship in a simulation study of laser-produced plasmas.

The ponderomotive potential for an electron accelerated to relativistic velocities by an electromagnetic wave is a cycle-averaged electron quiver energy $<U>$ determined by the average relativistic kinetic energy of the electron (see Equation 6.62). We have

$$<U> = m_0 c^2 \left((<\gamma^2>)^{1/2} - 1 \right). \qquad (4.33)$$

The ponderomotive force on an electron in an intense electromagnetic wave arises due to gradients of the ponderomotive potential and so is given by

$$\mathbf{F}_p = -m_0 c^2\, \nabla\left((<\gamma^2>)^{1/2} - 1\right) = -m_0 c^2\, \nabla\left((1 + a_o^2/2)^{1/2} - 1\right). \qquad (4.34)$$

For low-intensity laser or other interactions with a_o small, expanding the gradient term

$$\left((1 + a_o^2/2)^{1/2} - 1\right) \approx a_o^2/4$$

shows that the relativistic ponderomotive force at small a_o reduces to the nonrelativistic ponderomotive force given by Equation 4.18:

$$\mathbf{F}_p \approx -\frac{m_0 c^2}{4}\, \nabla a_o^2$$

For high-intensity laser interactions with a_o large, the ponderomotive force (Equation 4.34) becomes

$$\mathbf{F}_p \approx -\frac{m_0 c^2}{\sqrt{2}}\, \nabla a_o.$$

Substituting for $<\gamma^2>$ into Equation 4.26 shows that the cycle-averaged electron velocity with linear polarization in the electric field direction is given by

$$\frac{<v_x^2>}{c^2} = \frac{a_o^2}{2 + a_o^2}. \qquad (4.35)$$

At high a_o, the electron velocity v_x in the electric field direction approaches the speed of light c, while at low a_o, the cycle-averaged velocity is $a_o c/\sqrt{2}$. At low a_o

values, the scaling of the electron velocity in the direction of the electric field given by Equation 4.35 is identical to that found neglecting the relativistic effect on a_o (see Equation 4.16).

The electron drift velocity v_d arising when an intense electromagnetic wave is incident on an electron is modified when the electron velocity in the direction of the electric field approaches the speed of light. The effect of a relativistic momentum increase on the electron can be also taken into account by replacing the nonrelativistic reduced vector potential a_o by a_o/γ. From Equation 4.14, the relativistic drift velocity in the direction of the wave propagation becomes

$$\frac{v_d}{c} = \frac{a_o^2}{<\gamma^2> (4 + a_o^2/<\gamma^2>)}. \tag{4.36}$$

Using the value of $<\gamma^2>$ expressed in terms of a_o (Equation 4.30), we obtain for the electron drift velocity in the direction of propagation of an electromagnetic wave allowing for relativistic effects:

$$\frac{v_d}{c} = \frac{a_o^2}{4 + 3a_o^2} \tag{4.37}$$

The maximum electron drift velocity v_d for extreme laser intensities with high a_o is $c/3$, while at low a_o, the drift velocity varies as $a_o^2 c/4$.

The expression for the electron drift velocity given in Equation 4.37 differs from the scaling found assuming no relativistic effect on the reduced vector potential a_o (Equation 4.14). At low a_o values, relativistic and nonrelativistic evaluations both predict a drift velocity $v_d = a_o^2 c/4$. However, at high a_o relativistic calculations predict a maximum drift velocity of $c/3$, while the nonrelativistic treatment predicts a maximum drift velocity of c. The relativistic momentum increase of the electron due to the velocity parallel to the electric field of laser light reduces the electron drift velocity achieved in the direction of a high-power laser beam.

Equation 4.37 shows that the drift velocity in the direction of propagation of the electromagnetic wave is always significantly less than the cycle-averaged velocity in the electric field direction (Equation 4.35). Calculating $<\gamma^2>$ using only the velocity in the electric field direction (Equation 4.27) is, consequently, a good approximation.

Values of the reduced vector potential a_o can be regarded as small when $a_o << 1$ and large when $a_o >> 1$. Numerical values of a_o in terms of laser intensity I in units of Wcm^{-2} and laser wavelength λ are given in Equation 4.15. A value of $a_o = 1$ occurs at laser intensity 1.37×10^{18} Wcm^{-2} for laser wavelengths of 1 micron. The effects on electron acceleration in an electromagnetic wave as outlined in this section and in our nonrelativistic treatments (Sections 4.1 and 4.2) assume single electron motion, monochromatic light, a plane wave, and light pulse durations significantly longer than the period of the light. Departures from these constraints do not create significantly different results (see [95]). The effects when light interacts with a high density of electrons are examined in the next section (Section 4.4).

4.4 The Propagation of Light in a Plasma

A plasma is a globally neutral gas with significant ionization so that free electrons are present, but on longer length scales, the positive ionic and negative electron charges balance. Lengths longer than a quantity known as the Debye length have charges which average to zero. If the electrons have a temperature T and number density n_e, the Debye length λ_D is given by

$$\lambda_D = \left(\frac{\epsilon_0 k_B T}{n_e e^2} \right)^{1/2}. \tag{4.38}$$

For a derivation of this expression see, for example, Section 1.1 of Tallents [110].

We have examined the propagation of light in a vacuum in Chapter 1 (Section 1.1). The electric field $\mathbf{E}$ of the light is determined by Equation 1.3. The global neutrality of a plasma means that just as in vacuum, for wavelengths longer than the Debye length of the plasma, Gauss's law requires that

$$\nabla \cdot \mathbf{E} = \frac{\rho_c}{\epsilon_0} = 0, \tag{4.39}$$

where ρ_c is charge density. Using the vector identity $\nabla \times \nabla \times \mathbf{E} = \nabla(\nabla \cdot \mathbf{E}) - \nabla^2 \mathbf{E}$ and our result for $\nabla \cdot \mathbf{E} = 0$ leads to the wave equation

$$\nabla^2 \mathbf{E} = \mu_0 \frac{\partial \mathbf{J}}{\partial t} + \epsilon_0 \mu_0 \frac{\partial^2 \mathbf{E}}{\partial t^2}. \tag{4.40}$$

The free electrons in a plasma result in an electromagnetic wave inducing a current density $\mathbf{J}$ in the plasma which affects the propagation of the wave.

The wave equation has solutions for the electric and magnetic fields which can oscillate in time and space as we found for electromagnetic wave propagation in vacuum (Section 1.1). Provided the current density $\mathbf{J}$ variation in space and time is such that $\mathbf{J} \propto \mathbf{E}$, substituting into the wave equation (Equation 4.40) verifies that oscillations of electric field of the following form satisfy the wave equation at position $\mathbf{r}$:

$$\mathbf{E} = \mathbf{E}_0 \exp[i(\mathbf{k} \cdot \mathbf{r} - \omega t)], \tag{4.41}$$

where $\mathbf{E}_0$ is a field amplitude, $\mathbf{k}$ is a wavevector representing the rapid spatial variation of the field, and ω is an angular frequency representing the rapid temporal variation of the field. The angular frequency is the phase change of the field in radians per second and is related to the frequency in cycles per second (i.e. hertz) by $\omega = 2\pi\nu$.

Taking the divergence of $\mathbf{E}$ in Equation (4.41) gives

$$\nabla \cdot \mathbf{E} = i\mathbf{k} \cdot \mathbf{E}$$

which from Equation 4.39 is identically equal to zero. With $\mathbf{k} \cdot \mathbf{E} = 0$, we have that the wavevector $\mathbf{k}$ is perpendicular to the electric field $\mathbf{E}$ (provided Equation 4.39 is applicable). As found for vacuum light propagation, in a plasma the electric field of electromagnetic radiation is perpendicular to the propagation direction.

If the wavelength of the radiation is less than the plasma Debye length, it is possible for $\nabla \cdot \mathbf{E}$ to be nonzero so that the electric field $\mathbf{E}$ of the wave is not perpendicular to the wavevector $\mathbf{k}$. Waves where the electric field $\mathbf{E}$ has a component parallel to $\mathbf{k}$ are "longitudinal" plasma waves where the movement of plasma charges in the direction of propagation of the wave is important. Longitudinal plasma waves are not "electromagnetic" waves and involve treatments beyond the scope of this book (see e.g. Section 2.4.1 of Pert [89]).

The magnetic field $\mathbf{B}$ for an electromagnetic wave in a plasma is found using Equation 1.10 as considered for vacuum propagation of light in Chapter 1. As the current density $\mathbf{J}$ induced by the electric field is in the direction of the electric field, Equation 1.10 requires that the magnetic field vary in a similar fashion to Equation 4.41 with $\mathbf{B}$ and $\mathbf{B}_0$ replacing $\mathbf{E}$ and $\mathbf{E}_0$. The definition of $\nabla \times \mathbf{B}$ in Equation 1.10 determines that $\mathbf{B}$ is perpendicular to $\mathbf{E}$ and oscillates in phase with the electric field (i.e. $\mathbf{k} \cdot \mathbf{r} - \omega t$ is the same for both the magnetic and electric fields). The amplitude of the magnetic field $B = (k/\omega)E$ (see Equation 1.11). The orientation of the electric field $\mathbf{E}$ in the plane perpendicular to $\mathbf{k}$ is the polarization direction of the electromagnetic wave.

If we assume that the $\mathbf{k}$ vector is directed in the z-direction with the electric field $\mathbf{E}$ directed in the x-direction, then we can write that the solutions of the wave equation have the form $E = E_0 \exp[i(kz - \omega t)]$. The velocity v of electrons in a plasma due to the electric field of the light is found by integrating the acceleration $(-eE/(\gamma m_0))$ of the electrons due to the electric field to give

$$v = \frac{eE}{i\gamma m_0 \omega}.$$

We have included the relativistic Lorentz parameter $\gamma = 1/\sqrt{1 - v^2/c^2}$. Including the Lorentz factor allows for the possibility that the electric field is sufficiently large that a relativistic "mass increase" of the electrons needs to be taken into account because the electrons are accelerated to velocities approaching the speed of light. The current density in the rest frame is consequently directed in the x-direction with amplitude given by

$$J = -n_e e v = i \frac{n_e e^2}{\gamma m_0 \omega} E. \tag{4.42}$$

The wave equation then becomes

$$\frac{\partial^2 E}{\partial z^2} = \mu_0 i \frac{n_e e^2}{\gamma m_0 \omega} \frac{\partial E}{\partial t} + \epsilon_0 \mu_0 \frac{\partial^2 E}{\partial t^2}. \tag{4.43}$$

In vacuum (with $n_e = 0$), the first term on the right-hand side here vanishes and substituting Equation 4.41 gives that $\omega/k_0 = 1/\sqrt{\epsilon_0 \mu_0}$, where k_0 is the vacuum wavenumber. For constant phase $k_0 z - \omega t$, we see that the speed of light in vacuum $c = dz/dt = \omega/k_0$. Consequently, we have that

$$\epsilon_0 \mu_0 = \frac{1}{c^2}. \tag{4.44}$$

Substituting solutions $E = E_0 \exp[i(kz - \omega t)]$ into Equation 4.43 shows that

$$k^2 = \frac{\omega^2}{c^2} - \frac{n_e e^2}{\epsilon_0 <\gamma> m_0 c^2} = \frac{\omega^2 - \omega_p^2}{c^2}, \tag{4.45}$$

where we assume a value $<\gamma>$ for the Lorentz parameter averaged over a cycle of the electromagnetic oscillation. We have also introduced an expression known as the plasma frequency ω_p for the resonance oscillation of electrons relative to the ions in a plasma such that

$$\omega_p = \sqrt{\frac{n_e e^2}{\epsilon_0 <\gamma> m_0}}. \tag{4.46}$$

For electromagnetic radiation to propagate in the original forward direction requires $k \geq 0$. A value of $k < 0$ can be interpreted as the radiation being reflected backwards. For example, radiation incident into an increasing density of plasma only propagates up to where the laser frequency $\omega = \omega_p$. The corresponding maximum electron density for light penetration can be found using Equation 4.46 and is known as the critical density. Rearranging Equation 4.46 and setting $\omega = \omega_p$ shows that the critical electron density is given by

$$n_{crit} = \frac{\epsilon_0 <\gamma> m_0 \omega^2}{e^2}. \tag{4.47}$$

Evaluating Equation 4.47 numerically, we find that for light of wavelength $\lambda = 1.06$ μm and moderate- to low-intensity where $<\gamma> \approx 1$, the critical electron density $n_{crit} = 10^{21}$ cm^{-3} and scales proportionally to $1/\lambda^2$.

At extreme intensities ($>10^{18}$ Wcm^{-2}) available with focused high power lasers, the relativistic Lorentz parameter $<\gamma>$ increases above unity which increases n_{crit} proportionally and enables high-intensity laser pulses to penetrate high electron density plasma material. It is possible to penetrate the electron densities ($>10^{22}$ cm^{-3}) found in solid target material. We can estimate the intensities required for such intensity-induced transparency by modifying Equation 4.30. The relativistic Lorentz parameter for electron oscillation in an electromagnetic wave can be written as

$$<\gamma> \approx (<\gamma^2>)^{1/2} = \left(1 + \frac{a_o^2}{2}\right)^{1/2}. \tag{4.48}$$

Using Equation 4.15 for the relationship of the reduced electric potential a_o to electromagnetic intensity I, a useful expression for the critical density n_{crit} including the effect of intensity-induced transparency is given by

$$n_{crit} \approx \frac{10^{21}}{\lambda^2}\left(1 + \left(\frac{1}{2}\right)\frac{I\lambda^2}{1.37 \times 10^{18}}\right)^{1/2}, \tag{4.49}$$

where the light intensity I is in units of Wcm^{-2}, the wavelength λ is in units of microns, and the critical electron density n_{crit} is in units of cm^{-3}.

Instead of explicitly dealing with the plasma current density $\mathbf{J}$ in the wave equation (Equation 4.40), we can introduce the refractive index η to allow for plasma effects on the light propagation

$$\nabla^2 \mathbf{E} = \eta^2 \epsilon_0 \mu_0 \frac{\partial^2 \mathbf{E}}{\partial t^2} = \frac{\eta^2}{c^2}\frac{\partial^2 \mathbf{E}}{\partial t^2} \tag{4.50}$$

after using $\epsilon_0 \mu_0 = 1/c^2$ (see Equation 4.44). Substituting Equation 4.41, the phase velocity ω/k is found to be equal to c/η. The refractive index η is equivalent to the ratio of the speed of light in vacuum to the speed of light in the medium. As $\eta = kc/\omega$, using Equation 4.45 we have

$$\eta = \sqrt{1 - \frac{\omega_p^2}{\omega^2}} = \sqrt{1 - \frac{n_e}{n_{crit}}}. \tag{4.51}$$

The variation of the refractive index η with frequency ω of a medium is known as the "dispersion" of the medium. Light of different frequencies follows different dispersed paths in propagating through the medium.

From Equation 4.51 we have that refractive indices in plasmas are less than unity, while at the critical density, the refractive index decreases to zero. Radiation can no longer propagate at frequencies below the plasma frequency or at densities greater than the critical density. As the critical density n_{crit} can vary with intensity I at high intensities (Equation 4.49), plasma refractive indices η can be greater at high intensity than at low intensity. A focused laser pulse has a peak intensity at the beam center, so the refractive index can be higher at the beam center. The phase velocity of the laser light at the center of the beam is consequently slower than in the beam wings, leading to a focusing of a high-intensity beam in a plasma. Such an effect is known as relativistic self-focusing [93], [112].

The refractive indices of plasmas as given by Equation 4.51 are less than unity which means that the phase velocity of light is greater than the vacuum speed of light. The principle of relativity is not violated, however, as it is not possible to convey information or energy at the phase velocity. The light has to be modulated or pulsed to send a signal or transfer energy.

Rather than using just a single frequency ω and wavevector $\mathbf{k}$, sending a signal requires a pulse or packet of slightly different frequencies and wavevectors. We can suppose that a pulse of radiation has a total electric field $E(z, t)$ at position z and time t given by

$$E(z, t) = \int_{-\infty}^{\infty} E_0(k) \exp(i(kz - \omega t)) dk, \qquad (4.52)$$

where the integration is over different wavevectors which contribute an electric field $E_0(k)dk$ in the range k to $k + dk$. Assuming a range of frequencies ω around a central frequency ω_m associated with a range of wavevectors k centered on k_m, we can expand the frequency variation in a Taylor series:

$$\omega \approx \omega_m + (k - k_m)\frac{d\omega}{dk} + \frac{1}{2}(k - k_m)^2\frac{d^2\omega}{dk^2} + \dots \qquad (4.53)$$

Considering only the first two terms (up to $d\omega/dk$) and substituting Equation 4.53 into Equation 4.52, we have

$$E(z, t) \approx \exp(i(k_m z - \omega_m t)) \int_{-\infty}^{\infty} E_0(k) \exp\left(i\,(k - k_m)\left(z - \frac{d\omega}{dk}t\right)\right) dk. \quad (4.54)$$

Equation 4.54 represents a wave of frequency ω_m and wavevector k_m in a pulse with an amplitude varying with the values given by the integral. The value of the integral is constant in space z and time t if $z - (d\omega/dk)t$ is constant. Differentiating this constant value with respect to time gives

$$\frac{dz}{dt} = \frac{d\omega}{dk}. \qquad (4.55)$$

The quantity $d\omega/dk$ is consequently the velocity u_g of the amplitude variation of the pulse of electromagnetic wave and is known as the group velocity. The group velocity $u_g = d\omega/dk$ is the speed of propagation of the pulse of radiation and is the velocity that a signal can propagate and the velocity of radiation energy transfer. The small $d^2\omega/dk^2$ term in Equation 4.53 gives the change of the group velocity with wavenumber. The rate of change of the group velocity with frequency $du_g/d\omega = (1/u_g)(d^2\omega/dk^2)$ is known as the "group velocity dispersion."

We can determine a general relationship between the group velocity $u_g = d\omega/dk$ and the phase velocity $u = \omega/k = c/\eta$ of a pulse of light. Differentiating $\omega = ck/\eta$ gives

$$\frac{d\omega}{dk} = \frac{c}{\eta}\left(1 - \frac{k}{\eta}\frac{d\eta}{dk}\right) = \frac{c}{\eta}\left(1 - \frac{k}{\eta}\frac{d\eta}{d\omega}\frac{d\omega}{dk}\right)$$

so that

$$u_g = u\left(1 - \frac{k}{\eta}\frac{d\eta}{dk}\right) \tag{4.56}$$

and after rearranging $u_g = d\omega/dk$ terms, we have

$$u_g = \frac{u}{1 + (\omega/\eta)(d\eta/d\omega)}. \tag{4.57}$$

For a plasma the group velocity is given by

$$u_g = \frac{d\omega}{dk} = c\left(1 - \frac{\omega_p^2}{\omega^2}\right)^{1/2} = c\left(1 - \frac{n_e}{n_{crit}}\right)^{1/2} = c\,\eta. \tag{4.58}$$

Multiplying the group $c\,\eta$ and phase c/η velocities of light in a plasma gives c^2 as found with the multiplication of the group and phase velocities for a traveling wavefunction (see Equation 2.95).

An intense pulse of laser light propagating into a plasma initially has a low intensity and hence low critical density n_{crit} which increases towards the temporal peak of the pulse (see Equation 4.49). The plasma refractive index η can increase during the pulse up to the peak of the pulse. From Equation 4.58, the peak intensity region of the laser pulse has a higher group velocity so starts to "overtake" the leading edge of the pulse producing a sharpening of the laser pulse risetime as it propagates into plasma.

4.5 Cherenkov Radiation

Cherenkov radiation occurs when a charged particle travels through a transparent medium with a velocity greater than the phase velocity of light in the medium. As charged particles can travel only up to speeds equal to the vacuum speed of light, Cherenkov radiation is only produced in solids, liquids, and gases with a refractive index greater than one. With a refractive index above one, the phase velocity for the radiation in the medium is below the vacuum speed of light. Plasmas have refractive indices below one which means that the phase velocity of light in a plasma is greater than the vacuum speed of light, so Cherenkov radiation is not produced in a plasma.

Cherenkov radiation (sometimes written in Roman script as Cerenkov radiation) is named after the Soviet scientist Pavel Cherenkov (1904–1990). Cherenkov was awarded the Nobel prize in 1958 for his observation in 1934 of bluish light associated with radioactive material in water arising from particles moving faster than the speed of light in the water medium. Cherenkov shared the Nobel prize with colleagues Igor Tamm (1895–1971) and Ilya Frank (1908–1990) who developed the theory of Cherenkov radiation.

When a charged particle passes through a medium with a velocity greater than the phase velocity of light, perturbations in the medium produce a cone shaped radiation emission which propagates at a fixed angle to the particle velocity much like a shock wave propagates in air around an aircraft traveling faster than the sound speed. Many solids, liquids, and gases have higher refractive indices and hence lower phase velocities for light at shorter wavelengths, so charged particles from radioactive sources are more likely to have velocities exceeding the phase velocity at shorter wavelengths. Cherenkov radiation often appears as blue light (rather than a longer visible wavelength light) because of the refractive index increase of many materials at shorter visible light wavelengths. As originally observed by Pavel Cherenkov, nuclear reactor and other radioactive material immersed in water can emit an eerie blue Cherenkov light as high velocity particles from the radioactive material travel at speeds greater than the phase velocity of blue light in water.

For a charge moving uniformly through a medium, the charge itself does not radiate, but the medium is affected by the transiting charge. The background medium can oscillate as a charge passes so that charges in the medium are accelerated and radiate electromagnetic waves. For a charged particle moving at a velocity less than the speed of light in the medium, radiating waves from the background medium propagate radially from the source in the medium, with intensity falling as the inverse of distance squared. The electric fields of the electromagnetic waves created in the medium from different spatial position do not coherently overlap if the velocity of the charge is less than the phase velocity of light in the medium, so the intensities of light created in the medium are small and fall off with distance squared from each point along the charged particle trajectory.

The phase velocity of light in a medium is given by $\omega/k = c/\eta$ (see Section 4.4). Plasmas have refractive indices $\eta < 1$ and were discussed in Section 4.4. However, if a medium has a refractive index $\eta > 1$, we see that the velocity of a charged body can exceed the phase velocity of light in that medium. Most solids, liquids, and gases have refractive indices $\eta > 1$ for visible light. For example, for visible wavelengths, air has a refractive index of 1.000293, water has a refractive index of 1.33, window glass is 1.52, and diamond is 2.42. When the velocity of a charged

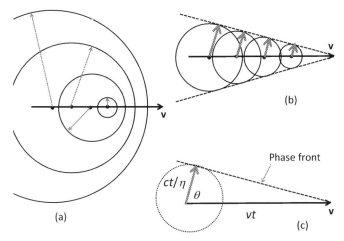

Figure 4.2   Schematics for the propagation of wavefronts arising from a charged particle of velocity **v** moving through a medium with refractive index η. For (a) the charged particle velocity $v < c/\eta$, for (b) and (c) the charge particle velocity $v > c/\eta$. Cherenkov radiation is produced along a phase front moving with a **k** wavevector at angle θ to the particle velocity **v** in (b) and (c) with $\cos \theta = c/(\eta v)$.

particle in a medium exceeds the phase velocity of light in that medium, electromagnetic fields created by media oscillations from different spatial positions along the path of the charged particle can coherently overlap, producing much higher light intensities as the fields add, with the intensity proportional to the square of the resultant electric and magnetic fields. This enhanced intensity of light is known as Cherenkov radiation.

The coherent overlap of oscillating electric fields of electromagnetic radiation as a charged particle with velocity greater than the phase velocity of light passes through a medium occurs at a particular cone angle around the charged particle velocity. In Figure 4.2, we show spheres of constant phase front emanating from sample discrete points along the trajectory of a charged particle. When the velocity v of the charged particle is less than the speed of light c/η in the medium, the phase fronts from different positions do not overlap. When the velocity v of the charged particles is greater than the speed of light c/η in the medium, phase fronts overlap for electromagnetic waves propagating at an angle θ to the particle velocity.

We show in Figure 4.2b spheres of constant phase from different spatial positions along the charged particle trajectory overlapping and coherently adding to produce a larger electric field. The overlap occurs for phase fronts moving at angle θ to the particle velocity **v**. In reality, emission occurs continuously along the charged particle track and not just at the selected intervals drawn in Figure 4.2.

Cherenkov light created at a point in a medium will travel a distance $(c/\eta)t$ in a time t, while the particle travels a distance vt (see Figure 4.2c). Angles between the particle velocity $\mathbf{v}$ and the direction of emitted Cerenkov radiation from that point are given by

$$\cos\theta = \frac{(c/\eta)t}{vt} = \frac{c}{\eta v}.$$ (4.59)

Cherenkov radiation with a phase velocity of c/η moves outwards with a wavefront parallel to the surface of a cone of half-angle $\pi/2 - \theta$ with θ given by Equation 4.59 (see Figures 4.2b and 4.2c). For Cherenkov radiation to appear for a particle velocity of v, we have that $\cos\theta < 1$, hence, using Equation 4.59 and as already discussed, we require that

$$v > \frac{c}{\eta}.$$

There is a similarity between Cherenkov radiation and a shock wave created by an aircraft flying faster than the speed of sound. In both, phase fronts augment to produce a high intensity at a particular angle. In both, the initial "seed" for the augmentation is the oscillation of the background medium.

A charged particle entering a plasma or other medium can be deaccelerated and radiate energy due to the deacceleration. In the presence of a magnetic field, synchrotron radiation can be produced as the particle rotates around the magnetic field lines with acceleration directed towards the center of rotation (see Chapter 5). For relativistic particles, there is a "beaming" effect on the emitted radiation such that the radiation is concentrated within a half-angle $1/\gamma_\eta$ to the velocity $\mathbf{v}$ of the particle, where γ_η is an appropriate Lorentz factor (see Section 3.5). When the refractive index η of the medium is such that $\eta < 1$, as it is in a plasma, Cherenkov radiation from fast charged particles cannot occur. Instead the radiation associated with the slowing down of fast relativistic particles in the medium dominates and the emission is concentrated within angles $1/\gamma_\eta$ to the velocity.

The Lorentz factor associated with the beaming effect on radiation emission angles is modified in a medium of refractive index $\eta > 1$ as the speed of light becomes c/η rather than c. The beaming angles θ around the particle velocity $\mathbf{v}$ where detected light is concentrated in the rest frame become

$$\theta \approx \frac{1}{\gamma_\eta} = \left(1 - \frac{v^2}{(c/\eta)^2}\right)^{1/2} = \left(1 - \eta^2\frac{v^2}{c^2}\right)^{1/2}.$$ (4.60)

For particles moving with a velocity v close to the speed of light c, the beaming angle $\theta \approx (1 - \eta^2)^{1/2}$. If the refractive index η is sufficiently close to unity, the beaming angles θ are dominated by the ratio v/c as for beaming in vacuum.

For a plasma with refractive index $\eta = (1 - (\omega_p/\omega)^2)^{1/2}$, if the refractive index effect dominates the beaming angle of emission in the rest frame and v approaches the speed of light c, we have from Equation 4.60 that the beaming angle of light emission in the rest frame is given by

$$\theta \approx \left(1 - \eta^2\right)^{1/2} = \frac{\omega_p}{\omega}. \tag{4.61}$$

Here ω_p is the plasma frequency (see Equation 4.46) and ω is the frequency of the radiated light. Equation 4.61 shows that for plasmas, the effect of the medium on the Lorentz factor and beaming angles is more important at lower frequencies ω. As the frequency ω of the emitted light increases, the beaming angle reduces $\propto 1/\omega$ until it reaches the value determined for vacuum propagation.

Equation 4.61 shows that there is often a cut-off in detected emission from relativistic particles in a plasma medium associated with a refractive index $\eta < 1$ if radiation is expected in a narrow beam angle. The medium produces a greater beaming angle $\theta \approx \omega_p/\omega$ than occurs in vacuum $\theta \approx 1/\gamma$ when

$$\omega < \gamma \omega_p.$$

Here $\gamma = 1/(1 - v^2/c^2)^{1/2}$ is the standard Lorentz parameter. The suppression of beaming due to the plasma refractive index is known as the Razin effect. The Razin effect can have a significant effect on emission from a plasma as radiation of frequency below the higher frequency $\gamma \omega_p$ is suppressed rather than the ordinary plasma frequency ω_p cut-off.

4.6 Thomson Scatter

Electromagnetic radiation incident on an electron causes the electron to accelerate (see Sections 4.1, 4.2, and 4.4). The electron acceleration, in turn, produces electromagnetic radiation which is then known as "scatter." In this section, we consider Thomson scatter where the scatter is from a single electron that is sufficiently isolated that scattering from any other nearby electrons does not add (or subtract) coherently to the scattered light intensity, where the electron is not constrained in a potential field, and where the photon energy is well below the rest mass energy of the electron.

Thomson scatter is named after the British Nobel prize winner Joseph John Thomson (1856–1940). The Nobel prize was awarded in 1906 to J. J. Thomson (as he was universally known) for work on what is now regarded as plasma physics, namely the electrical conductivity of gases. The scattering of light by isolated electrons is defined as Thomson scatter if the photon energies are not too high. The transition from low photon energy Thomson scatter to high photon energy scatter occurs when the photon energies approach the rest mass energy of the electron. Higher photon energy scatter (known as Compton scatter) is considered in Section 4.8.

In the presence of an oscillating electric field E associated with an electromagnetic wave, an electron experiences an acceleration $-eE/m_0$ if the electron velocities remain well below the speed of light. Using Equation 1.14, the electric field $E(\theta_A)$ created by the acceleration of an electron oscillating due to the electromagnetic wave at a distance r from the electron is then given by

$$E(\theta_A) = \frac{e}{4\pi \epsilon_0 c^2 r} \left(\frac{eE}{m_0} \right) \sin \theta_A, \tag{4.62}$$

where θ_A is the angle to the acceleration direction (in the direction of the electromagnetic electric field). Using Equation 1.15, the acceleration produces a radiated power $dP/d\Omega$ per unit solid angle such that

$$\frac{dP}{d\Omega} = \frac{e^2}{4\pi \epsilon_0} \frac{1}{4\pi c} \left(\frac{eE}{m_0}/c \right)^2 \sin^2 \theta_A. \tag{4.63}$$

The classical electron radius r_e (also known as the Compton or Lorentz radius or the Thomson scattering length) is the radius where the electrostatic potential energy of a sphere $e^2/(4\pi \epsilon_0 r_e)$ is equal to the electron rest energy $m_0 c^2$. It is a measure of the "size" of the electron. We have

$$r_e = \frac{e^2}{4\pi \epsilon_0} \frac{1}{m_0 c^2} = 2.818 \times 10^{-15} \text{ m}. \tag{4.64}$$

Equation 4.63 for the radiated power of an electron accelerated by an electromagnetic wave can be simplified using the classical electron radius. We have

$$\frac{dP}{d\Omega} = r_e^2 c \epsilon_0 E^2 \sin^2 \theta_A. \tag{4.65}$$

The instantaneous incident power per unit area (P_i/A) of electromagnetic radiation is given by

$$\frac{P_i}{A} = c \epsilon_0 E^2. \tag{4.66}$$

Consequently, the cross-section $d\sigma_S$ associated with a solid angle $d\Omega$ at angle θ_A to the electric field of the light is given by

$$d\sigma_S = \frac{dP}{P_i/A} = r_e^2 \sin^2 \theta_A d\Omega.$$

We can write for the differential scattering cross-section of a free electron

$$\frac{d\sigma_S}{d\Omega} = r_e^2 \sin^2 \theta_A, \tag{4.67}$$

where θ_A is the angle between the electric field of incoming radiation and the direction of the scattered radiation. Thomson scatter has an angular variation to the electric field direction of the incoming light as shown in Figure 1.5. Experimentally, the angular variation of Thomson scatter given by Equation 4.67 is observed in the scattering of laser light in gases (see e.g. Fruhling et al. [43]). The total scattering cross-section of a free electron into any angle is found by integrating over solid angle. We can write

$$\sigma_S = \int r_e^2 \sin^2 \theta_A d\Omega = 2\pi \int_{\theta_A=0}^{\pi} r_e^2 \sin^3 \theta_A a \, d\theta_A = \frac{8\pi}{3} r_e^2 \qquad (4.68)$$

using $d\Omega = 2\pi \sin \theta_A d\theta_A$ for a geometry with a symmetry in θ_A.

The classical electron radius r_e is numerically equal to 2.818×10^{-15} m which is much smaller than the dimensions of an atom (e.g. the Bohr radius $a_0 = 0.53 \times 10^{-10}$ m) or the wavelength of most electromagnetic radiation (gamma rays of e.g. 1 MeV photon energy have a wavelength of 1.24×10^{-12} m). The total scattering cross-section σ_S is 6.524×10^{-29} m^2. For moderate intensities of light, the small numerical values of the Compton radius and total Thomson scattering cross-section of the electron immediately indicate that the scattering of light by isolated electrons is usually a small effect and that little electromagnetic energy is usually lost by electron scatter. Another feature of Thomson scatter is that the scattering cross-sections given by Equations 4.74 and 4.68 are constant with the frequency of the radiation. The fraction of the power radiated in Thomson scatter is independent of the spectral content of the radiation.

Unpolarized light can be regarded as the independent supposition of two polarized beams with polarization direction at angle $\pi/2$ to each other. For scattering with unpolarized light, we can set one of the polarizations to be in the plane formed by the incident $\mathbf{k}_i$-vector and scattered light $\mathbf{k}$-vector (see Figure 4.3) and the other polarization to be at angle $\pi/2$ ("out of" the plane formed by the incident $\mathbf{k}_i$-vector and scattered light $\mathbf{k}$-vector). For the in-plane polarization, the angle of scattering θ_A relative to the electric field is related to the change in angle during scattering θ_γ (the angle between $\mathbf{k}_i$ and $\mathbf{k}$) by $\theta_\gamma = \pi/2 - \theta_A$ (see Figure 4.3). For the out-of-plane polarization, the angle of the $\mathbf{k}$-vector of the scatter light to the electric field is $\pi/2$. The total differential cross-section for unpolarized light is the average of the cross-section for the two linear polarizations. Using Equation 4.74, we can write

$$\left(\frac{d\sigma_S}{d\Omega}\right)_{unpol} = \frac{1}{2}\left[\left(\frac{d\sigma_S(\theta_\gamma)}{d\Omega}\right)_{pol} + \left(\frac{d\sigma_S(\pi/2)}{d\Omega}\right)_{pol}\right]$$

$$= \frac{r_e^2}{2}(\cos^2 \theta_\gamma + 1). \qquad (4.69)$$

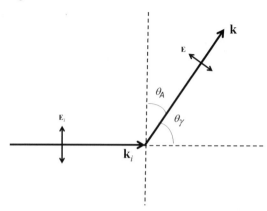

Figure 4.3   Scattering in the plane of the incident $\mathbf{k}_i$ and scattered $\mathbf{k}$ wavevectors showing the angle θ_A between the scattered wavevector and the electric field $\mathbf{E}_i$ of the incident wave in the plane of the scattering and the angle of scattering θ_γ (change of wavevector direction). The angle of the scattered wavevector $\mathbf{k}$ to the electric field component in the plane perpendicular to the illustrated plane is $\pi/2$.

Elliptically polarized light (see Section 4.2) exhibits an angular variation of Thomson scatter from a similar pattern as for Equation 4.67 at low ellipticity (close to linear polarization), changing to close to spherical symmetry as the degree of polarization ellipticity approaches circular polarization (see Fruhling et al. [43]). The total scattering cross-section $\sigma_S = (8\pi/3)r_e^2$ for unpolarized and elliptically polarized light integrated over all scattering angles θ_γ is the same as for linearly polarized light (see Equation 4.68).

In the frame of reference of the electron, Thomson scatter is not associated with a frequency shift of the light. However, Doppler frequency shifts observed in the laboratory frame of reference occur. The shift in frequency is given by an expression found in our treatment of Doppler effects in Section 2.6. From Equation 2.49, the Thomson scattered frequency ω_s observed in the laboratory frame is related to the incident light frequency ω by

$$\omega_s = \omega \gamma^2 \left(1 + (\hat{\mathbf{j}} - \hat{\mathbf{i}}) \cdot \mathbf{v}/c\right), \tag{4.70}$$

where $\mathbf{v}$ is the electron velocity and $\hat{\mathbf{i}}$ and $\hat{\mathbf{j}}$ are unit vectors in the direction of, respectively, the incident and scattered light propagation directions. For Thomson scatter from electrons moving with relativistic velocities, there is an upshift in the scattered light frequency of order γ^2. Thomson scatter of laser light from fast electrons is a useful mechanism for increasing the frequency of the laser light. It is possible to produce coherent X-rays from optical laser light due to the γ^2 scaling of the scatter [4].

Scattering from an ensemble of electrons with different velocities produces a shift in scattered frequency proportional to the velocity along the direction $\hat{\mathbf{j}} - \hat{\mathbf{i}}$ (see Equation 4.70). The scattered frequency shifts proportional to the electron velocity enable Thomson scatter to be used to measure electron velocity distributions in laboratory plasmas. For thermalized electrons, the electron temperature can be obtained from the width of a Thomson scattered spectrum. Doppler broadening for nonrelativistic plasmas is discussed in Section 7.3. Intense light from lasers is needed for Thomson scatter experiments because the Thomson scatter cross-section σ_S is small [54], [105].

If the electric field of the light scattering from an electron is sufficiently intense, electrons can be accelerated to relativistic velocities by the electromagnetic field itself. Laser light intensities typically above 10^{18} Wcm^{-2} are required for electron velocities to approach the speed of light (see Section 4.4). The electron acceleration $a_{||}$ in the direction of the electron velocity (parallel to the electric field $\mathbf{E}$ of the light) in the rest frame is such that

$$a_{||} = -\frac{eE}{\gamma m_0}.$$

Equation 3.44 gives an expression for the rest frame radiated power integrated over all angles for accelerations parallel and perpendicular to the moving frame velocity. Adapting Equation 3.44, the radiated power P in the rest frame due to Thomson scatter is

$$P = \frac{e^2}{4\pi\epsilon_0}\frac{2}{3c}\gamma^6\left(\frac{a_{||}^2}{c^2}\right). \tag{4.71}$$

The total scattering cross-section of a free electron is found by dividing the radiated power P by the instantaneous power of the light per unit area $c\epsilon_0 E^2$ (see Equation 4.66). Substituting in Equation 4.71 for $a_{||} = -eE/(\gamma m_0)$ and using Equation 4.64 for the classical electron radius r_e, we obtain

$$\sigma_S = \frac{8\pi}{3}r_e^2 <\gamma^2>^2 = \frac{8\pi}{3}r_e^2(1 + a_o^2/2)^2. \tag{4.72}$$

The cycle-averaged square of the relativistic Lorentz factor $<\gamma^2>$ is related to the reduced vector potential a_0 by Equation 4.30. The reduced vector potential a_o is related to laser intensities and wavelength as shown by Equation 4.15. The Thomson scattering cross-section σ_S increases proportionally to the square of the light intensity for intense light (intensity $>> 10^{18}$ Wcm^{-2}).

The angular variation of Thomson scattered light for high laser light intensities can be evaluated using Equation 3.58 for the power radiated by a charge accelerated in the direction of a relativistically moving charge. Substituting for the electron acceleration $a_{||} = -eE/(\gamma m_0)$ in an electromagnetic field, the power radiated per unit solid angle is given by

$$\frac{dP}{d\Omega} = \frac{e^2}{4\pi\epsilon_0}\frac{1}{4\pi c^3}\left(\frac{eE}{\gamma m_0}\right)^2\left(\frac{1}{1 - v\cos\theta/c}\right)^6 \sin^2\theta, \qquad (4.73)$$

where θ is the angle to the electron acceleration, the velocity, and the electric field of the light. The differential cross-section $d\sigma_S/d\Omega$ is found by dividing by the instantaneous power of the light per unit area $c\epsilon_0 E^2$ (see Equation 4.66). Using Equation 4.64 for the classical electron radius r_e, we obtain

$$\frac{d\sigma_S}{d\Omega} = \frac{r_e^2}{<\gamma^2>^2}\left(\frac{1}{1 - v\cos\theta/c}\right)^6 \sin^2\theta. \qquad (4.74)$$

Using the approximation given in Equation 3.54 for $1/(1 - v\cos\theta)$ at high γ, we can find an approximate expression for the differential cross-section for Thomson scatter at high laser intensity:

$$\frac{d\sigma_S}{d\Omega} \approx r_e^2(<\gamma^2>)^4\left(\frac{2}{1 + <\gamma^2>\theta^2}\right)^6 \theta^2$$

$$\approx r_e^2(1 + a_o^2/2)^4\left(\frac{2}{1 + (1 + a_o^2/2)\theta^2}\right)^6 \theta^2 \qquad (4.75)$$

At high intensities where the reduced electric potential a_0 is large and there are large Lorentz factors γ for the accelerated electrons, Thomson scatter will preferentially occur at angles θ close to the direction of the electric field of the light, though as for low intensity Thomson scatter, the cross-section still drops to zero in the direction $(\theta = 0)$ of the electric field of the incident light.

At high intensities, we have seen (see Section 4.1) that electrons execute a figure-8 motion oscillating in the electric field direction with a frequency equal to the frequency of the light and also oscillating at a frequency twice the frequency of the light in the direction of light propagation (see Figure 4.1). The oscillation in the direction of light propagation produces "second harmonic" radiation at twice the incident light frequency. The electron acceleration in the direction of light propagation (the z-direction) is given by Equation 4.9. Including the Lorentz increase in mass:

$$\frac{d^2z}{dt^2} = \frac{e^2E_0^2}{\gamma^2 m_0^2\omega c}\sin(\omega\tau)\cos(\omega\tau) = \frac{e^2E_0^2}{2\gamma^2 m_0^2\omega c}\sin(2\omega\tau)$$

An electron oscillates at a frequency 2ω in the direction of light propagation (the z-direction) and hence produces second harmonic radiation at frequency 2ω. We use Equation 3.61 for the power radiated by an accelerated charge with the acceleration perpendicular to the velocity of the charge. A value for the cross-section of second harmonic emission can be found by dividing the average radiated power by the incoming average power of the light $((1/2)c\epsilon_0 E_0^2)$. For high irradiances (high a_0 values), we obtain a differential cross-section $d\sigma_{SH}/d\Omega$ for the second harmonic emission

$$
\begin{aligned}
\frac{d\sigma_{SH}}{d\Omega} &\approx \frac{1}{2}r_e^2 \, (<\gamma^2>)^2 \left(\frac{eE_0}{m_0 c\omega}\right)^2 \left(\frac{1 - 2(<\gamma^2>)\theta^2 \cos(2\phi) + (<\gamma^2>)^2\theta^4}{(1 + (<\gamma^2>)\theta^2)^6}\right) \\
&\approx \frac{1}{2}r_e^2 \, (1 + a_o^2/2)^2 \; a_0^2 \left(\frac{1 - 2(1 + a_o^2/2)\theta^2 \cos(2\phi) + (1 + a_o^2/2)^2\theta^4}{(1 + (1 + a_o^2/2)\theta^2)^6}\right),
\end{aligned}
$$
(4.76)

where ϕ is an azimuthal angle around the electric field direction (see Figure 3.1) and θ is the angle to the electric field direction. The angular variation within the large brackets shows that emission is in the direction of the electric field (θ approaching zero). It is also apparent that the second harmonic cross-section is small at low intensities (small values of the reduced electric potential a_0) and only becomes significant relative to fundamental Thomson scatter for $a_0 >> 1$ (intensities $>> 10^{18}$ Wcm^{-2}).

Other harmonics of the incident light can also be scattered due to the nonlinear nature of the electron acceleration at ultra-high intensities ($a_0 >> 1$). In order to treat the dominant Thomson scatter, we have considered cycle-averaged values of the Lorentz factor $(<\gamma^2>)^{1/2}$. At ultra-high intensities, the change of the instantaneous Lorentz factor γ during a cycle of electromagnetic oscillation from unity when the electric field is zero (ignoring the drift velocity affect on γ) up to some maximum value at the peak of the electric field of the light becomes important. Due to the instantaneous electron mass increase γm_0, the electron oscillation becomes nonsinusoidal at the fundamental frequency in the electric field direction or nonsinusoidal at twice the fundamental frequency in the light propagation direction which means that higher harmonics are produced. Second and higher harmonic Thomson scatter of laser light off free electrons has been experimentally observed [4], [16], [43].

4.7 Coherent Scatter

If a high density of electrons are present in a plasma, Thomson scatter of the light can occur from the charge distribution of electrons with many electrons effectively radiating together. The electric field of the light scattered from each electron needs to be added to get the total electric field and the resulting intensity of scattered light is proportional to the square of the total electric field. A much larger scattered intensity occurs when the electric fields from many electrons add coherently.

Thomson scatter of coherent light from an electron charge distribution rather than single electrons is known as "coherent scatter." Coherent laser light, radio waves, and X-rays can undergo coherent scatter in appropriate media. A tutorial treatment of coherent scatter from plasmas illustrating the major features is presented in this section. More detailed discussions and examples of the use of coherent scatter for the diagnosis of laboratory plasmas are available (see e.g. Chapter 7 in Hutchinson [55] and Sheffield et al. [105]).

At a position $\mathbf{r}$ in a scattering plasma, the electric field amplitude per unit volume of scattered light of incident frequency ω and wavevector $\mathbf{k}_i$ can be characterized by

$$E_s(\mathbf{r}) = E_e n_e(\mathbf{r}) \exp\left(i(\mathbf{k}_i \cdot \mathbf{r} - \omega t)\right), \tag{4.77}$$

where $n_e(\mathbf{r})$ is the number density of electrons at position $\mathbf{r}$ and E_e is the scattered light electric field amplitude from a single electron. Periodic variations of electron density $n_e(\mathbf{r})$ are produced by waves in a plasma, where electrons alone or electrons tied to ions oscillate with a longitudinal motion parallel to the direction of wave propagation. Waves with frequency ω_e and wavevector $\mathbf{k}_e$ produce a periodic variation of electron density which we can represent as a perturbation of amplitude δn_e on a background uniform electron density n_0. We have the electron density varying in space and time as

$$n_e(\mathbf{r}) = n_0 + \delta n_e \cos(\mathbf{k}_e \cdot \mathbf{r} - \omega_e t). \tag{4.78}$$

We can expand the cosine variation of a wave in terms of exponentials:

$$\cos(\mathbf{k}_e \cdot \mathbf{r} - \omega_e t) = \frac{1}{2}\left(\exp(i(\mathbf{k}_e \cdot \mathbf{r} - \omega_e t)) + \exp(-i(\mathbf{k}_e \cdot \mathbf{r} - \omega_e t))\right)$$

Using this cosine expansion when substituting the expression (Equation 4.78) for the variation of electron density due to a wave in the plasma into Equation 4.77 gives an expression for the resulting electric field per unit volume due to scattering from a position $\mathbf{r}$ with three terms of different frequency and wavevector:

$$E_s(\mathbf{r}) = E_e n_0 \exp\left[i(\mathbf{k}_i \cdot \mathbf{r} - \omega t)\right]$$

$$+ \frac{E_e \delta n_e}{2}\left[\exp\left[i((\mathbf{k}_i + \mathbf{k}_e) \cdot \mathbf{r} - (\omega + \omega_e)t)\right] + \exp\left[-i((\mathbf{k}_i - \mathbf{k}_e) \cdot \mathbf{r} - (\omega - \omega_e)t)\right]\right]$$

$$\tag{4.79}$$

The scattered light can have frequencies and wavevectors of $(\omega + \omega_e, \mathbf{k}_i + \mathbf{k}_e)$, $(\omega - \omega_e, \mathbf{k}_i - \mathbf{k}_e)$, and $(\omega, \mathbf{k}_i)$.

An electron density wave can be associated with oscillation of the electrons relative to the ions or a much slower oscillation of the ions which drag along the scattering electrons (an ion sound wave). Electron plasma and ion sound waves are treated in books on plasma physics (see e.g. page 36 of Pert [88]). Like electromagnetic waves, plasma waves have "dispersion relationships" between the frequency ω_e and wavevector k_e of the wave. The kinetic motion associated with the temperature of the electrons (electron plasma wave) and ions (ion sound wave) acts as a restoring force on the oscillation of electrons or ions and adds a temperature-dependent frequency variation to the "zero-temperature" plasma frequency ω_p oscillation. An electron plasma wave has a dispersion relationship given by

$$\omega_e^2 = \omega_p^2 + \frac{3k_B T_e}{m_0} k_e^2, \tag{4.80}$$

where T_e is the electron temperature and m_0 is the electron rest mass. The ion sound wave has a dispersion relationship related to the speed ω_e / k_e of sound in a plasma with

$$\omega_e^2 = \frac{3k_B T_i + Z_i k_B T_e}{M} k_e^2, \tag{4.81}$$

where T_i is the ion temperature, M is the mass of the ions, and Z_i is the average charge of the ions (and hence the number of electrons per ion).[1]

The frequency shift $\pm \omega_e$ of coherent scattered light depends on the plasma frequency ω_p and electron temperature (for an electron plasma wave) and mainly the ion temperature (for the ion sound wave). The plasma frequency ω_p varies with the electron density n_e (see Equation 4.46), so the frequency shift due to the electron plasma wave can be used to deduce the electron density as well as the electron temperature. Comparing Equations 4.80 and 4.81, we see that due to the much greater ion mass M compared to the electron mass m_0, the frequency shift ω_e of scattered light is much less when scatter is off an ion sound wave rather than an electron plasma wave. A higher spectral resolution of the scattered light spectrum is needed to observe coherent scatter off an ion sound wave.

A plane wave of incident light of wavevector $\mathbf{k}_i$ at different positions $\mathbf{r}$ in the medium has a phase difference to the wave at the origin of the vector $\mathbf{r}$ equal to $+\mathbf{k}_i \cdot \mathbf{r}$. For a particular scatter direction defined by the wavevector $\mathbf{k}$, the phase difference of scattered light traveling towards some observer a large distance away is $-\mathbf{k} \cdot \mathbf{r}$ (see Figure 4.4). If the incoming light has a frequency ω, the scattered electric field E_s^* per unit volume from a position $\mathbf{r}$ in the plasma has a phase difference to the scattered light at the $\mathbf{r} = 0$ origin such that

$$E_s^*(\mathbf{r}) = E_e n_e(\mathbf{r}) \exp\left(i(\mathbf{k}_i \cdot \mathbf{r} - \mathbf{k} \cdot \mathbf{r} - \omega t)\right) = E_e n_e(\mathbf{r}) \exp(i(\mathbf{k}_i - \mathbf{k}) \cdot \mathbf{r} - \omega t).$$

---

[1] The speed ω_e / k_e of sound in a plasma is given by $(\omega_e / k_e)^2 = (\gamma_I k_B T_I + Z_i \gamma_E k_B T_e)/M$, where γ_I and γ_E are appropriate ratios of specific heats equal to $1 + 2/n_d$, where n_d is the number of degrees of freedom of the ions or electrons. For a one-dimensional sound wave, $n_d = 1$ for the ions, while $n_d \to \infty$ for the large number of electrons moving with the ions.

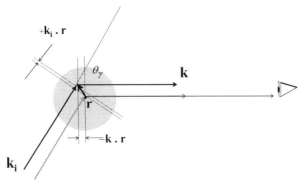

Figure 4.4   A schematic diagram showing the scattering of light from a medium with incident wave vector $\mathbf{k}_i$ and scattered wave vector $\mathbf{k}$. The phase differences in scattering from an electron at position $\mathbf{r}$ relative to notional scattering at a scattering center is shown. It is assumed that the observer is a long distance from the medium.

Introducing the vector difference $\Delta\mathbf{k}$ between the incoming wavevector $\mathbf{k}_i$ and scattered wavevector $\mathbf{k}$ such that $\Delta\mathbf{k} = \mathbf{k}_i - \mathbf{k}$, we have scattered electric fields (relative to $\mathbf{r} = 0$) given by

$$E_s^*(\mathbf{r}) = E_e n_e(\mathbf{r}) \exp(i(\Delta\mathbf{k} \cdot \mathbf{r} - \omega t)). \tag{4.82}$$

Simple geometry can be used to determine the magnitude of the wavevector difference $\Delta\mathbf{k}$ in terms of the angle θ_γ between the incident light (represented by the direction of the wavevector $\mathbf{k}_i$) and the scattered light direction (represented by wavevector $\mathbf{k}$). Figure 4.5 illustrates a sample orientation of wavevectors that can be used to determine the amplitude of $\Delta\mathbf{k}$. Assuming that the amplitudes of the incoming wavevector $|\mathbf{k}_i|$ and the scattered wavevector $|\mathbf{k}|$ are equal, we have

$$|\Delta\mathbf{k}| = \Delta k = 2k_i \sin\left(\theta_\gamma/2\right). \tag{4.83}$$

The incoming wavevector $|\mathbf{k}_i|$ and the scattered wavevector $|\mathbf{k}|$ are closer to equal when the incident wavevector is large with small wavelength, for example, the scattering radiation is an X-ray beam. With small changes of wavevector from the incident value $\mathbf{k}_i$ to $\mathbf{k} = \mathbf{k}_i \pm \mathbf{k}_e$ on scattering, Equation 4.83 is approximately accurate, but a more accurate formula is readily obtained (see Exercise 4.8). Significant differences for the incoming wavevector $|\mathbf{k}_i|$ and the scattered wavevector $|\mathbf{k}|$ are more likely with longer wavelength light, for example, visible and infrared scattering. With visible and infrared light, the plasma wavevector $|\mathbf{k}_e|$ has a value closer to the wavevector $|\mathbf{k}_i|$ of the scattering light.

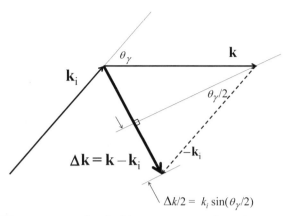

Figure 4.5   The geometry of an incident wavevector $\mathbf{k}_i$ and a scattered wavevector $\mathbf{k}$ for light undergoing scattering by an angle of θ_γ to the original light direction. The geometry illustrates that $\Delta k = 2k_i \sin\left(\theta_\gamma/2\right)$.

The differential scattering cross-section of a single free electron is given by Equation 4.74:

$$\frac{d\sigma_S}{d\Omega} = r_e^2 \sin^2 \theta_A,$$

where θ_A is the angle between the electric field of incoming radiation and the direction of the scattered radiation and r_e is the classical electron radius (given by Equation 4.64). The classical electron radius for single electron Thomson scatter is proportional to the square of the electron charge, so with coherent scatter, the differential cross-section for scatter from a large number of electrons allowing for phase differences in the scattered electric field becomes

$$\frac{d\sigma_S}{d\Omega} = r_e^2 |S(\Delta\mathbf{k})|^2 \sin^2 \theta_A, \tag{4.84}$$

where $S(\Delta\mathbf{k})$ is known as the form factor. Following Equation 4.82, the form factor is given by an integration of the number of electron scatterers while allowing for the effect of electric field additions (or subtractions) on the total scattered electric field. We have

$$S(\Delta\mathbf{k}) = \int n_e(\mathbf{r}) \exp\left(i\Delta\mathbf{k} \cdot \mathbf{r}\right) d\mathbf{r}, \tag{4.85}$$

where the integration extends from $\mathbf{r} = 0$ to a large distance $|\mathbf{r}| = \lambda_D$. An appropriate distance λ_D is the Debye length which is a distance in a plasma over which electron and ion charges average to be equal (see Equation 4.38).

The amount and nature of coherent scatter is dependent on the value of the form factor $S(\Delta \mathbf{k})$ as expressed in Equation 4.85. Substituting the electron density variation associated with a density wave of frequency ω_e and wavevector $\mathbf{k}_e$ as given by Equation 4.78 allows an integration of Equation 4.85. The absolute value of the form factor integrates to show that for coherent scatter, the form factor $|S(\Delta \mathbf{k})|$ is close to zero at all values of plasma wavevector $\mathbf{k}_e$ except where $\mathbf{k}_e = \Delta \mathbf{k}$ (see Exercise 4.10). Consequently, the direction and magnitude of $\mathbf{k}_e = \Delta \mathbf{k}$ is determined by the experimentally variable angle θ_γ between the incident light and the scattered light. The scattering geometry selects the magnitude of the wavevector $|\mathbf{k}_e| = k_e$ of the scattering density wave in the dispersion relationships for the electron plasma wave and ion sound wave (Equations 4.80 and 4.81). Usefully for plasma measurements, the relationship $\Delta \mathbf{k} = \mathbf{k}_e$ also means that coherent scattering only samples variations of electron density in the $\Delta \mathbf{k}$ direction.

For unpolarized light, the differential cross-section for coherent scatter is found following Equation 4.69. Unpolarized light can be said to be composed of two linearly polarized components with electric field directions, respectively, at an angle of $\pi/2$ to the light scatter direction and at an angle θ_A to the light scatter direction. We can write

$$\left(\frac{d\sigma_S}{d\Omega} \right)_{unpol} = \frac{r_e^2}{2} |S(\Delta \mathbf{k})|^2 (\cos^2 \theta_\gamma + 1), \tag{4.86}$$

where θ_γ is the angle of the scattered light wavevector $\mathbf{k}$ to the incident light wavevector $\mathbf{k}_i$. For light polarized in the plane of the light vectors, the angle θ_A to the electric field of the light is related to the angle of scatter θ_γ by $\theta_\gamma = \pi/2 - \theta_A$ (see Figure 4.3).

The forward scatter of incident coherent light when $\theta_\gamma \approx 0$ is more easily coherent as the scattering electrons are aligned closer to the same phase points for the incident and scattered light. A large wavelength λ_i for the incident light also enhances coherent scatter as the electron positions can be more readily in phase with the light. A large wavelength corresponds to small k_i value ($\lambda_i = 2\pi/k_i$). As $\Delta k = 2k_i \sin(\theta_\gamma/2)$, these two arguments suggest that a small value of Δk favors coherent scatter. The size of the coherent scatter integration in Equation 4.85 as given by the Debye length λ_D is also important. A smaller volume with smaller λ_D containing fewer electrons is more likely to produce coherent scatter. A parameter α_s known as the scattering parameter is used to determine if scatter is coherent. The scattering parameter is given by

$$\alpha_s = \frac{1}{\Delta k \, \lambda_D}. \tag{4.87}$$

When $\alpha_s \gg 1$, light scatter is probably coherent. When $\alpha_s \ll 1$, scatter is likely dominated by the contribution from single isolated electrons with a simple addition of the individual electron scatter intensities determining the magnitude of the scatter (as discussed in Section 4.6).

4.8 Compton Scatter

At high photon energies, the model of Thomson scatter of light with an electron oscillating in the electric field of the light and the electron scattering electromagnetic radiation at the same frequency in the electron frame as the incident radiation breaks down. Scattering of electromagnetic radiation with high photon energy results in a change of photon energy between the incident and scattered photons. The scattering process at high photon energies where there is a significant change of radiation frequency in the frame of the electron during scattering is known as Compton scatter. Compton scatter was discovered by Arthur Compton (1892–1962). An American physicist, Compton was awarded the Nobel prize in 1927 for his 1923 discovery of the particle-like behavior of light scatter at high photon energy.

We first examine the shift in radiation frequency occurring in Compton scatter. To find the shift in radiation frequency during Compton scatter, we need to recognize that a photon possesses momentum $p_\gamma = \hbar\omega/c$ as well as energy $\hbar\omega$ (see the discussion in Section 2.8). The scattering by an electron of a high energy photon with a high momentum can be regarded as a "collisional" process with energy and momentum conserved. The geometry of Compton scatter in the initial frame of reference of a scattering electron is shown in Figure 4.6. As the sum of the photon and electron energies before and after the scattering collision are equal, we can write

$$\hbar\omega + W_e = \hbar\omega_s + W_{e*}, \tag{4.88}$$

where ω is the photon frequency before the scattering, ω_s is the photon frequency after the scattering, and the electron energy changes from W_e before the collision to W_{e*} after the scattering collision. In the rest frame of reference of the electron before scattering, the conservation of momentum during photon scattering requires a balance between the momentum $\mathbf{p}_\gamma$ of the photon before scattering and the vector addition of the scattered photon and the electron momentum after scattering. We have

$$\mathbf{p}_\gamma = \mathbf{p}_{\gamma*} + \mathbf{p}_{e*}, \tag{4.89}$$

where $\mathbf{p}_{\gamma*}$ is the momentum of the scattered photon and $\mathbf{p}_{e*}$ is the momentum of the electron after scattering. As the electron is initially at rest in its frame of reference, the electron energy before the scattering collision is simply the rest mass energy $W_e = m_0 c^2$. The relativistic energy for the electron after scattering is given by Equation 2.73. We can write

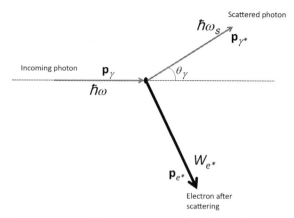

Figure 4.6   The geometry of Compton scatter of a photon by an electron initially at rest. Symbols used in the text for the photon energies ($\hbar\omega$, $\hbar\omega_s$), momentum ($\mathbf{p}_\gamma$, $\mathbf{p}_{\gamma*}$), and the angle θ_γ of photon deflection are shown.

$$W_{e*} = \sqrt{(p_{e*}c)^2 + (m_0c^2)^2}.$$

Substituting into the expression for the conservation of energy during the scattering (Equation 4.88) gives

$$\hbar\omega + m_0c^2 = \hbar\omega_s + \sqrt{(p_{e*}c)^2 + (m_0c^2)^2}.$$

Rearranging, the magnitude of the scattered electron momentum is given by

$$(p_{e*}c)^2 = (\hbar\omega - \hbar\omega_s + m_0c^2)^2 - (m_0c^2)^2. \tag{4.90}$$

The vector addition of the momenta of the photon and electron after the collision is equal to the momentum of the photon before the collision:

$$\mathbf{p}_{e*} = \mathbf{p}_\gamma - \mathbf{p}_{\gamma*}$$

Using the scalar product of the electron momentum after the collision yields the magnitude of the electron momentum after the scattering event. We can write that

$$p_{e*}^2 = \mathbf{p}_{e*} \cdot \mathbf{p}_{e*} = (\mathbf{p}_\gamma - \mathbf{p}_{\gamma*}) \cdot (\mathbf{p}_\gamma - \mathbf{p}_{\gamma*}) = p_\gamma^2 + p_{\gamma*}^2 - 2p_\gamma p_{\gamma*} \cos\theta_\gamma,$$

where θ_γ is the angle between the photon trajectory before the scattering event and after the scattering. Multiplying throughout by c^2, we obtain

$$p_{e*}^2 c^2 = p_\gamma^2 c^2 + p_{\gamma*}^2 c^2 - 2c^2 p_\gamma p_{\gamma*} \cos\theta_\gamma.$$

Substituting the momenta of the photons $p_\gamma = \hbar\omega/c$ and $p_{\gamma*} = \hbar\omega_s/c$ gives

$$p_{e*}^2 c^2 = (\hbar\omega)^2 + (\hbar\omega_s)^2 - 2(\hbar\omega)(\hbar\omega_s) \cos\theta_\gamma.$$

Using Equation 4.90 for the electron momentum after scattering,

$$(\hbar\omega - \hbar\omega_s + m_0 c^2)^2 - (m_0 c^2)^2 = (\hbar\omega)^2 + (\hbar\omega_s)^2 - 2(\hbar\omega)(\hbar\omega_s)\cos\theta_\gamma.$$

Evaluating the square, canceling, and rearranging terms yields

$$(\hbar\omega - \hbar\omega_s)m_0 c^2 = (\hbar\omega)(\hbar\omega_s)(1 - \cos\theta_\gamma).$$

Dividing both sides by $\hbar\omega\omega_s$ gives the relationship in the electron frame of the incident radiation frequency ω and the scattered radiation frequency ω_s. We have

$$\frac{c}{\omega_s} - \frac{c}{\omega} = \frac{\hbar}{m_0 c}(1 - \cos\theta_\gamma). \tag{4.91}$$

This equation can be rearranged to give an explicit expression for the scattered radiation frequency ω_s in terms of the incident radiation frequency ω:

$$\omega_s = \frac{\omega}{1 + (1 - \cos\theta_\gamma)\hbar\omega/(m_0 c^2)} \tag{4.92}$$

Equation 4.91 is often expressed in the wavelength of the incident λ and scattered λ_s light. In the frame of reference of the electron, we have different wavelengths before (λ) and after (λ_s) scattering given by

$$\lambda_s - \lambda = \lambda_C(1 - \cos\theta_\gamma), \tag{4.93}$$

where λ_C is the Compton wavelength defined by

$$\lambda_C = \frac{h}{m_0 c}. \tag{4.94}$$

The Compton wavelength λ_C has a numerical value of 2.426×10^{-12} m.

For wavelengths much longer than the Compton wavelength λ_C, there is little change in the scattered wavelength λ_s compared to the incident wavelength λ. If $\lambda \gg \lambda_C$, the photon energies are much less than the electron rest mass energy (i.e. $\hbar\omega \ll m_0 c^2$) and the photon-electron collision can be regarded as elastic.

The differential Compton cross-section for unpolarized radiation is derived using quantum electrodynamics and is known as the Klein-Nishina formula [62]. The differential cross-section for the scattering of radiation of frequency ω varies with the frequency ω_s of the scattered radiation as

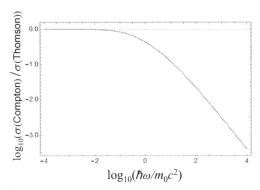

Figure 4.7   The cross-section for Compton scatter relative to the Thomson scatter cross-section as a function of photon energy in units of the electron rest mass energy $m_0c^2 = 511$ keV.

$$\frac{d\sigma_C}{d\Omega} = \frac{r_e^2}{2}\left(\frac{\omega}{\omega_s} + \frac{\omega_s}{\omega} - \sin^2\theta_\gamma\right), \tag{4.95}$$

where r_e is the Thomson scatter radius and θ_γ is the angle between the incident photon and the scattered photon. The relationship between the incident and scattered frequencies is given by Equation 4.93. For low frequency radiation where $\omega_s \approx \omega$, the differential cross-section shown in Equation 4.95 reduces to the Thomson scatter cross-section $(r_e^2/2)(1 + \cos^2\theta_\gamma)$ for unpolarized light (Equation 4.69). Using Equations 4.91 and 4.95, it is possible to integrate the differential cross-section over all solid angles to obtain the total Compton scatter cross-section. The total Compton scatter cross-section considering scatter into all solid angles is given by

$$\sigma_C = \sigma_S\frac{4}{3}\left[\frac{1+x}{x^2}\left(\frac{2x(1+x)}{1+2x} - \ln(1+2x)\right) + \frac{\ln(1+2x)}{2x} - \frac{1+3x}{(1+2x)^2}\right], \tag{4.96}$$

where $x = \hbar\omega/(m_0c^2)$ and σ_S is the total Thomson scatter cross-section. The total Compton scattering cross-section is plotted using Equation 4.96 in Figure 4.7. The Compton scatter cross-sections decrease with photon energies $> 0.1m_0c^2$ (>50 keV).

As well as the frequency shift in the frame of the electron due to the scattering process, Compton scatter exhibits frequency shifts in the laboratory frame due to the Doppler effect if the scattering electron has a significant velocity $\mathbf{v}$. The frequency ω' of incident radiation in the rest frame of the electron is given by Equation 2.46:

$$\omega' = \omega\gamma\left(1 - \frac{v}{c}\cos\theta_i\right), \tag{4.97}$$

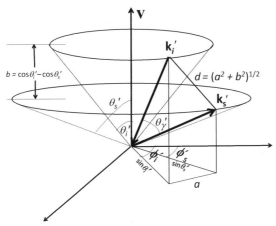

Figure 4.8   Relevant angles related to electron scattering in the frame of the electron are shown. Unit vectors for an incident radiation wavevector $\mathbf{k}'_i$ and a scattered wavevector $\mathbf{k}'_s$ are labeled along with the angle θ'_γ between the incident and scattered wavevectors. The different angles (θ'_i and θ'_s) of the two unit wavevectors to the electron velocity $\mathbf{v}$ are shown. The distances a, b, and d used to find relationships between the different angles are labeled.

where ω is the incident frequency of radiation in the laboratory frame, v is the velocity of the electron, and θ_i is the angle of the incident radiation to the electron velocity direction in the laboratory frame. The frequency ω' determines the Compton scatter cross-section (Equation 4.95) and scattered frequencies ω'_s in the electron frame (Equation 4.91). The scattered frequencies ω'_s in the electron frame can be determined from the incident frequency ω' in the electron frame using Equation 4.91 or 4.92. Modifying Equation 4.92,

$$\omega'_s = \frac{\omega'}{1 + (1 - \cos \theta'_\gamma)\hbar\omega'/(m_0 c^2)}, \tag{4.98}$$

where θ'_γ is the angle of the scattered radiation to the incident radiation in the electron frame. The scattered frequencies ω'_s in the electron frame are again Doppler shifted to ω_s in the laboratory frame, such that

$$\omega_s = \omega'_s \gamma \left(1 + \frac{v}{c} \cos \theta'_s \right), \tag{4.99}$$

where θ'_s is the angle of the scattered radiation propagation direction to the electron velocity in the electron frame.

Three-dimensional geometry shows how the angle θ'_γ of the scattered radiation to the incident radiation in the electron frame used in Equation 4.98 is related to the angle θ'_i of the incident radiation to the electron velocity and the angle θ'_s of the scattered radiation to the electron velocity (see Figure 4.8). Unit vectors into and out of the electron position in the direction of the incident and scattered radiation project onto a plane perpendicular to the electron velocity **v** such that there is a distance a between the tail of the incident unit vector and the head of the scattered unit vector. The cosine rule for a triangle in the plane perpendicular to the velocity gives that

$$a^2 = \sin^2 \theta_i' + \sin^2 \theta_s' - 2 \sin \theta_i' \sin \theta_s' \cos(\phi_i' - \phi_s'),$$

where ϕ_i' and ϕ_s' are azimuthal angles around the velocity direction. The difference between the components of the unit vectors in the direction of the incident and scattered radiation along the electron velocity direction is given by

$$b = \cos \theta_i' - \cos \theta_s'.$$

A triangle comprising sides with the unit vectors for the incident and scattered radiation directions and a distance d can be formed. The value of d is given by

$$d^2 = a^2 + b^2 = 2 - 2 \cos \theta_i' \cos \theta_s' - 2 \sin \theta_i' \sin \theta_s' \cos(\phi_i' - \phi_s')$$

upon using $\sin^2 \theta_i' + \cos^2 \theta_i' = \sin^2 \theta_s' + \cos^2 \theta_s' = 1$. The cosine of the angle θ_γ' between the directions of the incident $\mathbf{k}_i'$ and scattered $\mathbf{k}_s'$ radiation is then readily evaluated using the cosine rule for the triangle formed by the unit vectors for $\mathbf{k}_i'$ and $\mathbf{k}_s'$. We have

$$d^2 = 2 - 2 \cos \theta_\gamma'$$

as the length of each unit vector is one. Substituting for d^2 and rearranging, we obtain an expression for the angle θ_γ' between the incoming and Compton scattered light in the frame of the electron:

$$\cos \theta_\gamma' = \cos \theta_i' \cos \theta_s' + \sin \theta_i' \sin \theta_s' \cos(\phi_i - \phi_s) \tag{4.100}$$

The azimuthal angles $\phi_i = \phi_i'$ and $\phi_s = \phi_s'$ representing angles around the velocity **v** are invariant under a Lorentz transformation (see Section 3.5). Equation 4.100 is needed, for example, in Equation 4.98 to determine the precise change of frequency during Compton scatter.

Angles to the velocity in the electron frame can be converted to angles to the velocity in the laboratory frame using Equation 2.43. For the incident radiation at angle θ_i to the electron velocity in the laboratory frame, Equation 2.43 can be modified to give the angle θ_i' of the incident radiation to the velocity in the electron frame. We obtain

$$\cos \theta_i' = \frac{\cos \theta_i - v/c}{1 - (v/c) \cos \theta_i}. \tag{4.101}$$

Similarly, the conversion of scattered angles θ'_s to the electron velocity in the electron frame to scattered angles θ_s to the velocity in the laboratory frame is given by

$$\cos \theta_s = \frac{\cos \theta'_s + v/c}{1 + (v/c) \cos \theta'_i}. \tag{4.102}$$

As found for Thomson scatter with relativistic electrons, a γ shift in frequency of radiation going from the laboratory to the electron frame and then another γ frequency shift in going from the scattered light frequency in the electron frame to the laboratory frame can occur. A γ^2 increase in scattered photon energy is possible from electrons of kinetic energy $(\gamma - 1)m_0c^2$. Energy conservation limits the maximum photon energy increase possible with Compton scatter. The scattered photon energy must be less than the combined electron energy and incident photon energy. We must have

$$\hbar\omega_s < \gamma m_0 c^2 + \hbar\omega.$$

For highly relativistic electrons with large γ, the maximum scattered photon energies $\approx \gamma m_0 c^2 = 511\gamma$ keV.

Exercises

4.1 Capillary discharge lasers output extreme ultraviolet light at wavelength 46.9 nm with focused intensities $< 10^{14}$ Wcm^{-2} (see e.g. Macchietto et al. [71]). Evaluate the critical density for capillary discharge laser light in a plasma. [4.55×10^{23} cm^{-3}]

4.2 A solid target of up to several 100 nm thickness is almost instantaneously ionized by a high-intensity laser pulse. Assuming an electron density of 10^{23} cm^{-3} and a laser wavelength of 1.06 μm, estimate the laser intensity required for intensity-induced transparency to allow the laser pulse to transmit through the target.

[$\approx 1.4 \times 10^{22}$ Wcm^{-2}. The laser intensity can then directly heat the back of the target and accelerate electrons. The acceleration of electrons and indirectly ions by this process is known in the laser-plasma literature by the exotic name of "break-out afterburner" (abbreviated as BOA); see e.g. Yin et al. [118], [119]].

4.3 Photons can be regarded as massless particles moving at the speed of light c with energy $\hbar\omega = pc$, where p is the momentum of a photon. A photon can also be regarded as being composed of elliptically polarized components with an electric field rotating clockwise or anticlockwise and returning to the original direction after a wavelength. Assuming a radius $r_p = c/\omega$ of rotation for the photon momentum, show that the angular momentum of a circularly polarized photon has a magnitude $\hbar$. [This is an argument for the existence of the photon spin of $\pm\hbar$.]

4.4   Fast ignition inertial fusion seeks to ignite fusion reactions in laser-compressed deuterium and tritium fuel (see Section 2.4) by focusing a high power laser beam into the compressed fuel in order to locally heat it and start the fusion process [109]. Given that a high irradiance laser beam can accelerate electrons to a velocity $c/3$ in the direction of the beam (Equation 4.37), calculate the energy of electrons traveling at velocity $c/3$. Discuss if such electrons have sufficient energy to heat deuterium/tritium and ignite fusion reactions.

[31 keV. Electrons also gain energy transverse to the laser beam with velocities approaching c which can provide sufficient energy to locally heat the fuel.]

4.5   An electromagnetic wave propagates with a magnetic field B such that $B = (k/\omega)E$, where E is the electric field of the wave. Given that $k^2 = (\omega^2 - \omega_p^2)/c^2$ for a plasma with plasma frequency ω_p, show that in a plasma, the magnetic field of an electromagnetic wave is given by

$$B = \frac{\eta E}{c},$$

where η is the refractive index of the plasma.

4.6   The difference between the phase and group velocity of radio pulses passing through the ionosphere can provide a measure of the "total electron content" of the ionosphere. The ionospheric total electron content is a vertical spatial integration of the electron density. Show that the difference Δu in the phase and group velocity of electromagnetic pulses in a plasma of electron density n_e is given by

$$\Delta u = \frac{c}{(1 - n_e/n_{crit})^{1/2}} \frac{n_e}{n_{crit}},$$

where n_{crit} is the critical density for the frequency of the electromagnetic wave.

4.7   In morning sunlight, what is the predominant polarization direction of light from the sky directly above?

4.8   In coherent scatter, the amplitude of the scattered wavevector $|\mathbf{k}|$ can vary from the amplitude of the incident wavevector $|\mathbf{k}_i|$ due to up- or down-scatter off electron plasma and ion sound waves. Draw an appropriate vector diagram and show that the amplitude of the wavevector difference $\Delta\mathbf{k}$ defined by $\Delta\mathbf{k} = \mathbf{k} - \mathbf{k}_i$ is given by

$$|\Delta\mathbf{k}^2| = k^2 + k_i^2 - 2kk_i \cos\theta_y,$$

where θ_γ is the scattering angle (the angular change of direction of **k** from $\mathbf{k}_i$).

4.9 Explain and show why X-ray scattering in the forward direction has a change Δk in the wavevectors of incident **k** and scattered $\mathbf{k}_i$ light given by

$$\Delta k \approx k\,\theta_\gamma,$$

where θ_γ is the scattering angle (the angular change of direction of **k** from $\mathbf{k}_i$).

4.10 The form factor for coherent scattering from plasma waves is given by

$$S(\Delta \mathbf{k}) = \int n_e(\mathbf{r}) \exp\left(i\Delta \mathbf{k} \cdot \mathbf{r}\right) d\mathbf{r},$$

where the integration is over distances equal to the Debye length λ_D. The electron density $n_e(\mathbf{r})$ due to a density wave in a plasma is given by

$$n_e(\mathbf{r}) = n_0 + \delta n_e \cos(\mathbf{k}_e \cdot \mathbf{r} - \omega_e t).$$

Show that the absolute value of the form factor is given by

$$|S(\Delta \mathbf{k})| = \lambda_D^2 \delta n_e \left(\frac{\sin\left((\Delta \mathbf{k} - \mathbf{k}_e) \cdot \hat{\mathbf{z}}\lambda_D/2\right)}{(\Delta \mathbf{k} - \mathbf{k}_e) \cdot \hat{\mathbf{z}}} + \frac{\sin\left((\Delta \mathbf{k} + \mathbf{k}_e) \cdot \hat{\mathbf{z}}\lambda_D/2\right)}{(\Delta \mathbf{k} + \mathbf{k}_e) \cdot \hat{\mathbf{z}}} \right)$$

$$+ 2\lambda_D^2 n_0 \left(\frac{\sin(\Delta \mathbf{k} \cdot \hat{\mathbf{z}}\lambda_D/2)}{\Delta \mathbf{k} \cdot \hat{\mathbf{z}}} \right),$$

where $\hat{\mathbf{z}}$ is a unit vector in the direction of $\Delta \mathbf{k}$. [The form factor $|S(\Delta \mathbf{k})|$ has a pronounced maximum when $\Delta \mathbf{k} \pm \mathbf{k}_e$ and a maximum for forward scatter when $\Delta \mathbf{k} = 0$.]

4.11 The Compton scattered photon energy W_s for incident photons of energy W_i scattering off electrons at rest with rest mass energy $m_0 c^2$ is readily obtained from Equation 4.92:

$$W_s = \frac{W_i}{1 + (1 - \cos\theta_\gamma)W_i/(m_0 c^2)},$$

where θ_γ is the angular change in direction of the scattered photon relative to the incident photon. A "Compton camera" consists of scattering material with electron energy detection and a detector measuring the scattered photon energy. A Compton camera works by using the above relationship to find the position of a source of gamma rays by determining the angle θ_γ of scatter from measurements of the energy W_s of the scattered photon and the energy

W_{e*} of the scattered electron. Show that

$$\cos\theta_\gamma = 1 - m_0 c^2 \left(\frac{W_{e*}}{W_s(W_s + W_{e*})} \right).$$

4.12 The Compton wavelength λ_C can be experimentally measured by scattering X-rays by 90 degrees from a solid target. The spectrum of scattered X-rays from a solid has a peak at a wavelength λ_C above the incoming X-ray wavelength as expected for an isolated electron. The scattered spectrum also features scattering close to the incoming X-ray wavelength due to scattering off tightly bound electrons in the solid. Using Equations 4.93 and 4.94, explain how these two peaks arise.

4.13 The electron spin of $\hbar/2$ due to zitterbewegung is discussed briefly in a footnote in Section 2.9 (see also the paper by Hestenes [50]). In the zitterbewegung explanation of electron spin, the spin arises due to rotation at a radius r_Z of the electron mass m_0 at a frequency $2m_0c^2/\hbar$ (due to the electron wavefunction frequency of $m_0c^2/\hbar$). In the electron frame of reference, the phase velocity of the electron wavefunction is c. Show that the necessary radius r_Z to produce the observed electron spin of $\hbar/2$ is related to the Compton wavelength λ_C by

$$r_Z = \frac{1}{4\pi}\lambda_C.$$

5

Particle Motion and Radiation in Electric and Magnetic Fields

When particles distribute their energy by, for example, collisions, a thermal distribution with an associated temperature can be used to describe the particle energy distribution (see Chapter 6). When particle collisions and other energy distribution processes are negligible, particle energies are determined by the mechanics of particle acceleration. Charged particles gain (or lose) energy by interacting with electric fields. In this chapter, we examine particle interactions with electric and magnetic fields with an initial treatment of a process thought to produce cosmic rays and follow this with treatments of the radiation produced by charged particle acceleration in magnetic fields.

Magnetic fields accelerate moving charged particles, but we show in Section 5.1 that static magnetic fields do not affect the kinetic energy of charged particles apart from a small energy loss due to radiation. Magnetic fields are consequently used for the production of radiation from electron beams in devices such as synchrotrons and free electron lasers as the radiation energy is the main loss mechanism for the kinetic energy of the accelerated electrons. Magnetic fields are also useful for confining plasma as needed for fusion energy as the confined particle energies are not significantly changed by a magnetic field.

Magnetic fields used for plasma confinement are typically large (e.g. 12 Tesla in the ITER tokamak [18], [77]). Astrophysical plasmas such as those near a neutron star have strong embedded magnetic fields and even in interstellar space, there is a weak magnetic field ($\approx 10^{-10}$ Tesla). Radiation from laboratory and astrophysical plasmas produced due to electron motion in a magnetic field can be used to determine conditions in the emitting plasmas.

This chapter investigates the radiation emitted by accelerating particles in magnetic fields as found in plasmas, but also as used for the laboratory production of radiation. We examine synchrotron radiation and free electron laser radiation production in some detail. Synchrotrons and free electron laser devices are built as large laboratory structures. A synchrotron typically occupies the land area of a soccer field with a circular electron trajectory, while free electron lasers are linear with electrons accelerated over kilometers. Large circular particle trajectories in "storage rings" and linear accelerators minimize radiation losses from high energy particles.

Advanced synchrotrons are known as third-generation light sources. Third-generation light sources use "undulators" (alternating magnetic fields) to increase the radiation intensity above that of second-generation synchrotrons where simpler magnetic structures ("wigglers") are employed. First-generation light sources were often synchrotrons run parasitically alongside particle physics experiments, where a simple single bending magnet was used to produce light output. Free electron lasers produce even brighter coherent light and are known as fourth-generation light sources [15]. Synchrotrons and free electron lasers can both provide radiation from the infrared to X-ray range and are used for a large variety of laboratory experiments in the biological sciences, materials sciences, chemistry, and physics. As well as providing a broad spectral range of radiation, the light from synchrotrons and free electron lasers is collimated, is well-polarized, has a small source size, and is pulsed over times down to femtoseconds with good stability from pulse to pulse [115], [116].

5.1 The Motion of a Point Charge in an Electric or Magnetic Field

It is useful to examine in a general way the effects of electric and magnetic fields on particle energies. Electric and magnetic fields are measured by the behavior of charged particles.[1] Measuring the forces exerted on charged particles in the presence of an electric or magnetic field allows a determination of the field direction and strength. More fundamentally, the electric and magnetic fields in a system can be regarded as being defined by the motion of infinitesimally small "test" charges. Test charges need to be small so as not to affect the measured field values.

The classical force **F** on a particle of charge q due to an electric field **E** and magnetic field **B** is known as the Lorentz force and for particles at any velocity **v** is given by the expression

$$\mathbf{F} = \frac{d\mathbf{p}}{dt} = q\,(\mathbf{E} + \mathbf{v} \times \mathbf{B}). \qquad (5.1)$$

This Lorentz force (Equation 5.1) is not an equation that needs to be "derived." The Lorentz force empirically expresses how the electric **E** and magnetic **B** fields as

---

[1] The forces arising due to subatomic interactions are less obvious examples of charged particle interactions with fields. For example, the "magnetic moment" of subatomic particles arises due to the angular momentum or spin of charged particles and is often the ultimate physical basis for the measurement of magnetic fields. A compass needle, for example, can be used to measure the direction of a magnetic field because the spin and angular momentum of electrons in the needle material are aligned to produce a bulk effect known as ferromagnetism.

conventionally defined affect the motion of a body of charge q via the force $\mathbf{F}$. We can generalize the particle momentum such that $\mathbf{p} = \gamma m_0 \mathbf{v}$. Including the Lorentz γ factor with $\gamma = (1 - v^2/c^2)^{-1/2}$ in the expression for the momentum allows for a relativistic increase in momentum measured in the laboratory frame for particles moving close to the speed of light c.

The first principle of relativity is that "the laws of physics are the same in any inertial frame of reference" (see Section 2.2), so a systematic theory of relativity requires that the Lorentz equation (Equation 5.1) is applicable for observers in all frames of reference. Equation 5.1 or an equation derived from it can be used to determine the electric and magnetic field directions and strengths. The electric and magnetic field values and directions change depending on the frame of reference (see Section 3.2), but Equation 5.1 does not change.

Following Equation 2.65, the rate at which a force $\mathbf{F}$ does work is the power P given by

$$P = \mathbf{F} \cdot \mathbf{v}. \tag{5.2}$$

The rate of change of the energy of a particle in an electric and magnetic field is consequently given by

$$P = \frac{d(\gamma m_0 c^2)}{dt} = q\mathbf{v} \cdot \mathbf{E} \tag{5.3}$$

as $\mathbf{v} \cdot (\mathbf{v} \times \mathbf{B}) \equiv 0$. A temporally constant magnetic field has no effect on the energy of the particle, but does change the direction of the particle velocity and momentum.

As only electric fields and not magnetic fields change the energy of charged particles (see Equation 5.3), magnetic fields, not electric fields, are used to contain plasma in machines such as tokamaks studied in fusion energy research [18], [77]. Research aimed at achieving the production of fusion energy seeks to contain the deuterium (D) and tritium (T) isotopes of hydrogen in a hot plasma so that D and T nuclei collide and fuse, producing energetic helium (kinetic energy 3.5 MeV) and neutrons (with a kinetic energy of 14.1 MeV). The kinetic energy of the helium and the neutron results from the conversion of mass to energy following Equation 2.71. The energetic helium can act to heat D and T ions in a plasma, causing further fusion. When the helium heating produces a self-sustained fusion reaction, a condition of "ignition" is achieved. The neutrons escape the plasma and are used to heat a surrounding blanket of, for example, lithium which in turn heats water to drive a steam turbine producing electricity. Magnetic fields are used to confine D and T ions and the associated free electrons. A magnetic field changes the direction of ions and electrons and can confine them, but does not cause ions and electrons to lose energy (apart from a small radiation loss).

Equation 5.3 exemplifies why magnetic fields and not electric fields are used to accelerate charged particles to produce radiation. The kinetic energy of a charged particle remains constant in the presence of a steady-state magnetic field, with just a small loss due to the electromagnetic energy radiated. Cyclotron, synchrotron, and undulator devices developed to produce electromagnetic radiation consequently accelerate electrons using magnetic fields. Acceleration using static magnetic fields changes the velocity direction of particles, but has a negligible effect on the particle energy.

As electric fields change the energy of charged particles, it is worthwhile to consider charged particle energies in electric fields in some more detail. An electric field can arise electrostatically due to an electric potential and can be produced by a changing magnetic field. The electrostatic contribution to an electric field is represented by a scalar potential V with the magnetic field contribution represented by a vector potential $\mathbf{A}$. The electric field $\mathbf{E}$ is related to the scalar and vector potentials by

$$\mathbf{E} = -\nabla V - \frac{\partial \mathbf{A}}{\partial t}. \tag{5.4}$$

The scalar and vector potentials were used in our treatment of the Lorentz transformation of the electric and magnetic field between frames of reference (see Section 3.2). We also used the vector potential in our discussion of electron acceleration in an electromagnetic wave in Section 4.1. Substituting the scalar and vector potential expression for the electric field (Equation 5.4) into our expression for the change in the energy of a charged particle in an electromagnetic field (Equation 5.3), we have

$$\frac{d(\gamma m_0 c^2)}{dt} = q\mathbf{v} \cdot \mathbf{E}$$

$$= -q \left(\mathbf{v} \cdot \nabla V + \mathbf{v} \cdot \left(\frac{\partial \mathbf{A}}{\partial t} \right) \right). \tag{5.5}$$

If the electric field is purely a steady-state electrostatic field and there is no magnetic field, the vector potential $\mathbf{A}$ is zero.[2] An initially static charged particle then has a velocity $\mathbf{v}$ which is parallel to the electric field and parallel to the gradient ∇V of the scalar electric potential. For a purely electrostatic electric field, Equation 5.5 reduces to

$$\frac{d(\gamma m_0 c^2)}{dt} = -qv\frac{dV}{dx} = -q\frac{dx}{dt}\frac{dV}{dx} = -q\frac{dV}{dt},$$

where the x-direction is the direction of the potential gradient and consequently also the direction of the particle velocity v. Gathering up terms, we have for an electrostatic field that

---

[2] In the Lorenz gauge, see the appropriate definitions in Section 3.2.

$$\frac{d(\gamma m_0 c^2 + qV)}{dt} = 0. \tag{5.6}$$

Equation 5.6 means that in a purely electrostatic potential V, the quantity $\gamma m_0 c^2 + qV$ for a body of mass m_0 and charge q is temporally constant.

Instead of using the Lorentz force (Equation 5.1) to calculate the motion of a charged particle, it is possible to use a "Lagrangian" or "Hamiltonian" approach. We do not use the Lagrangian or Hamiltonian method elsewhere in this book, but it is appropriate to introduce the Lagrangian and Hamiltonian for particle motion in electric and magnetic fields and to present their relativistic forms. It is possible to determine the important equations of relativistic kinetics as we determined, for example, in Section 2.7 in a different way using the Lagrangian approach (see Chapter 2 of Landau and Lifshitz [63]).

With the Lagrangian form for particle motion in electric and magnetic fields, a functional L is defined[3] by an appropriate expression for the difference in the kinetic and potential energy of a body. For an energy conserving system, the minimum of the functional L determines the trajectory of the body. The minimum of a functional is found by solving the Euler-Lagrange equation. For a single particle with motion in the x-direction, we can write the Euler-Lagrange equation as

$$\frac{d}{dt}\left(\frac{\partial L}{\partial v}\right) = \frac{\partial L}{\partial x}, \tag{5.7}$$

where v is the particle velocity. A relativistic charged particle in a purely electrostatic potential accelerates in a direction parallel to the scalar potential gradient and has a Lagrangian written as

$$L = -m_0 c^2 \left(1 - \frac{v^2}{c^2}\right)^{1/2} - qV. \tag{5.8}$$

Substituting into the Euler-Lagrange equation (Equation 5.7), the Euler-Lagrange equation reduces to

$$\frac{d}{dt}\left(\frac{m_0 v}{(1 - v^2/c^2)^{1/2}}\right) = -q\frac{\partial V}{\partial x}. \tag{5.9}$$

The gradient of the scalar potential on the right-hand side equals the negative of the electric field ($\partial V/\partial x = -E$; see Equation 5.4). Equation 5.9 becomes

$$\frac{d(\gamma m_0 v)}{dt} = qE \tag{5.10}$$

---

[3] A functional can be defined as a function where the independent variable is itself a function. Their use in the "calculus of variations" extends well beyond the discussion presented here (see e.g. [45]).

which is equivalent to the expression expected from the Lorentz force ($dp/dt = qE$; see Equation 5.1). The derivative with respect to time t on the left-hand side of Equation 5.9 simplifies as follows:

$$\frac{d}{dt}\left(\frac{m_0 v}{(1 - v^2/c^2)^{1/2}}\right) = \left(\frac{m_0}{(1 - v^2/c^2)^{1/2}}\right)\frac{dv}{dt}\left(1 + \frac{v^2/c^2}{1 - v^2/c^2}\right)$$

$$= \left(\frac{m_0}{(1 - v^2/c^2)^{1/2}}\right)\frac{dv}{dt}$$

The force F in the stationary frame associated with the acceleration dv/dt of a mass m_0 is found by substituting into Equation 5.9:

$$F = \left(\frac{m_0}{(1 - v^2/c^2)^{1/2}}\right)\frac{dv}{dt} = qE \qquad (5.11)$$

Equation 5.11 further illustrates that the Lagrangian (Equation 5.8) for a charged particle in an electrostatic electric field is consistent with the Lorentz force (Equation 5.1).

In an environment where the electric field has a scalar potential V and vector potential $\mathbf{A}$ contribution, the Lagrangian for a particle of charge q and mass m_0 becomes

$$L = -m_0 c^2 \left(1 - \frac{v^2}{c^2}\right)^{1/2} + q\mathbf{v} \cdot \mathbf{A} - qV. \qquad (5.12)$$

It can be shown that this Lagrangian reduces to Equation 5.5 on substitution into the Euler-Lagrange equation for the (x, y, z) coordinate axes:

$$\frac{d}{dt}\left(\frac{\partial L}{\partial v}\right) = \frac{\partial L}{\partial r_i}, \qquad (5.13)$$

where r_i represents each of the x, y, and z Cartesian coordinates.

The Hamiltonian H is similar to the Lagrangian except that instead of the velocity of a body, the momentum is used and the Hamiltonian form is defined by the addition of expressions for the kinetic energy and potential energy of the body so that $H = L + 2V$. The Euler-Lagrange equation (Equation 5.7) transposes to two equations using the Hamiltonian H. The Hamilton equations for a body with varying position vector $\mathbf{r}$ and momentum $\mathbf{p}$ are:

$$\frac{dr_i}{dt} = \frac{\partial H}{\partial p_i},$$

$$\frac{dp_i}{dt} = -\frac{\partial H}{\partial r_i}, \qquad (5.14)$$

where p_i represents the momentum in the direction of the ith coordinate direction (x, y, or z). The relativistic Hamiltonian for a body of charge q and mass m_0 in an electric field is given by

$$H = c \left(m_0^2 c^2 + p^2\right)^{1/2} + qV, \tag{5.15}$$

where the momentum p is such that

$$\mathbf{p} = \gamma m_0 \mathbf{v} = \frac{m_0 \mathbf{v}}{(1 - v^2/c^2)^{1/2}}.$$

5.2 Acceleration of Particles

High temperatures in a plasma can lead to a thermal distribution of particle energies with high velocities (see Chapter 6). A thermal distribution requires processes such as collisions between particles. At low densities in plasmas and in laboratory particle accelerators, collisions between particles can be neglected and the interaction with electric fields causes particles to gain (or lose) energy. At low particle densities, the processes by which individual particles gain (or lose) energy dominates the particle energy distribution. In this section we briefly consider how charged particles can be accelerated to high velocities in laboratory-based particle accelerators and laser-produced plasmas and then spend time discussing the origin of high energy cosmic rays.

5.2.1 Particle Accelerators

Particle accelerators used for fundamental physics research and for application of the particles or their radiation use electric fields to accelerate the particles. The energy of charged particles can be increased by acceleration in an electric field, but their energy remains constant in a static magnetic field (apart from a small radiative energy loss). In a traditional high energy particle accelerator, charged particles are accelerated in a straight line (a linear accelerator) or in a large diameter (typically $>> 1$ m) circle so as to minimize radiation losses caused by acceleration perpendicular to the particle velocity. The Large Hadron Collider (LHC) near Grenoble is the world's largest and highest energy particle accelerator. The LHC has protons of energy up to 6.7 TeV $= 6.7 \times 10^{12}$ eV orbiting in a circle of circumference 26.7 km. The large circumference is needed to minimize radiation losses and to enable acceleration to 6.7 TeV as the maximum accelerating electric field in a large particle accelerator is $\approx 10^8$ V m^{-1}.

In the laboratory particle accelerators developed for high energy collider experiments or the production of radiation, electric potentials spaced around the acceleration path switch from positive to negative at a given frequency, so that particles traveling along the acceleration path experience a "traveling wave" of accelerating electric field polarity. For example, in a circular geometry, the circle can be divided into two D-shaped potentials which oscillate in polarity so that particles are always accelerated from one "dee" to the other. "Cyclotrons" use two hollow D-shaped sheet metal electrodes (called "dees") placed between the poles of an electromagnet which applies a static magnetic field perpendicular to the electrode plane and causes the particles to move between the dees with a circular trajectory of increasing radius with particle energy. "Synchrotrons" are descended from cyclotrons, but employ a time-varying magnetic field to bend the particle beam into a fixed radius trajectory. With a synchrotron, the magnetic field increases in time during the particle accelerating process so that the magnetic field is synchronized to the increasing kinetic energy of the particles.

5.2.2 Laser-produced Plasmas

Plasmas irradiated by lasers or high energy particles can accelerate high fluxes of particles over short distances. Plasma accelerators can be small as they can generate electric fields $\approx 10^{11}$ V m^{-1} much greater than the fields ($\approx 10^8$ V m^{-1}) available in large laboratory accelerators [51]. Such electric fields associated with laser light or plasma gradients enable acceleration to high energies (e.g. GeV) over short distances (usually < 1 cm; see Joshi et al. [59]). Wakefield acceleration where a laser or electron pulse is incident into a plasma creates a traveling wave of electric field gradient moving with the laser or electron pulse which accelerates charged particles in the direction of the laser or electron pulse. With a laser pulse, accelerated electrons also oscillate transversely, generating Thomson scatter of the laser light with an increase in photon energy in the laboratory frame by a factor γ^2 (see Equation 2.49 for incident and scattered radiation in the same direction). The generated radiation is directed within an angular range of $1/\gamma$ of the electron velocities due to the beaming effect (see Section 3.5). Here γ is the Lorentz parameter for the accelerated electrons, with $\gamma > 8,000$ achieved with laser acceleration [64]. The γ^2 increase of photon energy from the laser photon energy and $1/\gamma$ decrease in angle of emission due to beaming imply an approximate increase in the radiation flux (power per unit solid angle) due to Thomson scatter from wakefield accelerated electrons of γ^4. A significant number of electrons can oscillate in phase, so laser wakefield accelerated electrons can create a bright source of X-rays as well as a beam of high energy particles [73], [74].

Focused lasers incident onto solid targets can produce high energy particles with energies significantly greater than the temperature (< 1 keV) of thermalized plasma making up the bulk of the plasma ions and electrons typically created by laser heating of targets. High energy electrons have a small cross-section for interaction with other electrons and ions (see Chapter 8) and after being accelerated often escape the high plasma density created by laser irradiation of solid targets to form a superthermal population with large energy, but low density.

The interaction of high power laser light with electrons is examined in Section 4.1 where it is shown that a reduced vector potential a_o determines the acceleration of electrons. Using Equation 4.15, we have that the reduced vector potential a_o is related to the irradiance I measured in Wcm^{-2} by

$$a_o = \sqrt{\frac{I\lambda^2}{1.37 \times 10^{18}}}, \tag{5.16}$$

where the laser wavelength λ of focused laser light is measured in units of microns. For $a_o < 1$, at the critical density where the laser frequency equals the plasma frequency (see Section 4.4), electrons are resonantly accelerated to moderately high energies with superthermal distributions of temperature $\approx a_o^{2/3}$ MeV [24]. At higher irradiances $a_o > 1$, electrons are accelerated to produce energetic superthermal electrons with measured Lorentz parameters $\gamma \approx (1 + a_o^2)^{1/2}$ [24], [85], [114]. This scaling is consistent with our deduction in Section 4.3 of the maximum Lorentz parameter for electrons accelerated in an electromagnetic field (see Equation 4.31).

High energy electrons created in laser plasmas can penetrate through a solid target of typical thickness $1 - 100$ μm. Such high energy electrons can create a steep potential gradient from the largely unperturbed back surface of a target irradiated at the front by a high intensity laser. Protons from the target are accelerated in this potential gradient to energies $\approx 500\, a_o^2$ keV [85]. This process of ion acceleration is known in the laser-plasma literature as "target normal sheaf acceleration." The physics of the acceleration of electrons and ions in laser-produced plasmas is examined in detail in more specialized books (see e.g. Gibbon [47]), but due to the complexity of different processes, empirical particle energy scalings obtained from experiments and simulations as given here are important.

5.2.3 Cosmic Rays

Cosmic rays are protons and atomic nuclei that move through space with velocities close to the speed of light. Cosmic ray particles originate and undergo acceleration in the Sun and from outside the solar system in both our own galaxy and distant galaxies. Upon impact with the Earth's atmosphere, a cosmic ray produces secondary showers of particles in a direction closely following the original trajectory of the cosmic ray.

The secondary particles detected at the Earth's surface due to cosmic ray impact in the atmosphere provided the first evidence for high energy, short half-life "exotic" particles existing in addition to the protons, neutrons, and electrons making up terrestrial matter [84]. The detection at the Earth's surface of muons of half-life 1.56 μs was early evidence for relativistic time dilation as the time of travel in the

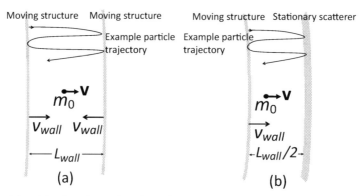

Figure 5.1   The Fermi acceleration of cosmic rays is illustrated with two scenarios: (a) two rapidly colliding structures such as colliding shock waves, (b) a fast moving structure such as a shock wave with a stationary scatterer. Symbols used in the text and illustrative trajectories for accelerated particles that can escape as cosmic rays are annotated.

Earth's frame of reference from the muon creation in the upper atmosphere to its detection at the Earth's surface is much longer than 1.56 μs (see the discussion in Section 1.2).

There is strong evidence that cosmic rays are generated by repeated particle reflections off rapidly moving structures or fields [6], [39]. In scenarios where a structure or a field is moving at high velocity, particles can be accelerated when reflected by the structure or field. A process known as first order Fermi acceleration [39] occurs when charged particles are repeatedly reflected by a rapidly moving shock wave[4] or a moving magnetic field[5] in a plasma. Possible geometries for cosmic ray acceleration are illustrated in Figure 5.1.

In order to quantitatively examine Fermi acceleration, consider a particle with velocity **v** and rest mass m_0 located between two particle reflecting structures (two "walls") which are moving with speeds v_{wall} towards the particle parallel (or antiparallel) to the particle velocity. Assume the reflection surfaces are separated at a

---

[4] A shock wave is a surface of higher density with a steep, abrupt density gradient moving through a fluid. Shock waves can form because the speed of a density oscillation wave in a fluid (known as a "sound wave") increases with density. A pulse of a sound wave can increase in amplitude and approach a step-like density increase as the peak of compression "overtakes" lower density, slower moving parts of the sound wave. For a detailed treatment of shock waves in fluids and plasmas, see Pert (Chapter 10) [88].

[5] A moving magnetic field generates an electric field that can change the kinetic energy of a charged particle, so our comments in Section 5.1 regarding the conservation of energy in a static magnetic field are not contradicted. In order for a generated electric field to increase the energy of a particle, the magnetic field needs to have a component out of the plane in which the particle is reflected, which usually means that the magnetic field is tangled and the plasma is turbulent [104].

particular time by a distance L_{wall}. The rate of energy increase dE/dt for the particle is determined by the number of particle-wall collisions per unit time ($\approx v/L_{wall}$) multiplied by the energy change ΔE per collision. The particle momentum Δp increase in a collision with a wall moving at speed v_{wall} is $\gamma m_0 v_{wall}$, where γ is the Lorentz factor for the particle. Assuming that the particle has a kinetic energy E much greater than the rest mass energy $m_0 c^2$ means that $E \approx pc$, where p is the particle momentum (see Equation 2.73). The energy increase ΔE per collision is then given by

$$\Delta E \approx \Delta p\, c = \gamma m_0 v_{wall}\, c.$$

The rate of energy increase of the particle with time is then

$$\frac{dE}{dt} \approx \frac{v}{L_{wall}} \gamma m_0 v_{wall}\, c \approx \frac{\gamma m_0 c^2\, v_{wall}}{L_{wall}} = E \frac{v_{wall}}{L_{wall}} \tag{5.17}$$

upon assuming $v \approx c$ and noting that the particle energy $E \approx \gamma m_0 c^2$. Equation 5.17 represents an equation of form

$$\frac{dE}{dt} \approx \frac{E}{\tau}, \tag{5.18}$$

where $\tau \approx L_{wall}/v_{wall}$ represents a timescale. The energy of particles accelerated by repeated reflection off rapidly moving surfaces grows exponentially at a rate proportional to the original energy of the particle. With a distribution of particle energies, higher energy particles are accelerated to much greater energy than lower energy particles. Some of the accelerated particles after several reflections will not be reflected further. These particles become cosmic rays traveling through space.

In order to deduce the energy distribution of particles accelerated by Fermi acceleration, we need to consider the number n of reflections, the energy gain ΔE per reflection, and the probability P_r of a particle remaining in the region of the reflecting structure. If the fractional change of energy on each reflection is β_r, then $\beta_r = \Delta E/E$ and the particle energy E increases from an initial value E_0 so that

$$E = E_0\, \beta_r^n. \tag{5.19}$$

After n reflections the number N of particles continuing to be reflected will drop from the original number N_0 trapped near the reflecting surfaces. We can write that

$$N = N_0\, P_r^n. \tag{5.20}$$

Combining Equations 5.19 and 5.20, the number n of reflections can be eliminated by considering the ratio of the logarithm of the particle number and energy:

$$\frac{\ln(N/N_0)}{\ln(E/E_0)} = \frac{\ln(P_r)}{\ln(\beta_r)}$$

Differentiating this expression, the number of particles per unit energy varies with particle energy E as

$$\frac{dN}{dE} = N_0 \left[\left(\frac{\ln P_r}{\ln \beta_r} \right) E_0^{\left(\frac{\ln P_r}{\ln \beta_r} \right)} \right] E^{\left(\frac{\ln P_r}{\ln \beta_r} - 1 \right)}. \tag{5.21}$$

The spectrum dN/dE of cosmic rays varies as $dN/dE \propto E^{-k}$ according to Equation 5.21. Cosmic ray numbers impinging on the Earth are measured to have a maximum at $\approx 10^9$ eV with the number of cosmic rays decreasing with increasing energy above 10^9 eV at a rate E^{-k}, where $k \approx 2.7 - 3$ [44]. The maximum energy of cosmic rays ($\approx 2 \times 10^{20}$ eV) is determined by cosmic ray scatter off the relativistically shifted cosmic microwave background as the cosmic rays travel over intergalactic distances (see Exercise 6.15). Comparing the observed E^{-k} variation of cosmic ray numbers to Equation 5.21 suggests that with the Fermi acceleration mechanism, the probability P_r of particles escaping in a reflection and the ratio of energy change β_r upon reflection are related by

$$\frac{\ln P_r}{\ln \beta_r} \approx 1 - k,$$

which means that

$$P_r \approx \frac{1}{\beta_r^{k-1}}. \tag{5.22}$$

A low particle energy increase on reflection (small β_r) is associated with a large probability P_r that a particle remains localized near the reflecting structure, while a large energy increase (large β_r) means a particle is likely to escape (there is a low probability P_r of remaining localized).

Simulations show that two colliding structures such as colliding shock waves are not necessary for cosmic ray acceleration. A single rapidly moving shock wave can accelerate particles as scattering processes cause particles to interact with the shock front many times [6]. The origin of cosmic rays in different energy ranges and from different sources is an interesting and continuing research topic [7], [44].

5.3 Emission from Particles in a Magnetic Field

Charged particles in a magnetic field with a velocity component perpendicular to the magnetic field direction undergo circular orbital trajectories around the magnetic field lines because of the Lorentz force (discussed in Section 5.1) directed towards the center of the orbit. For the remainder of this chapter we consider the kinetics of charged bodies in steady-state magnetic fields.

For particles with a charge q and a velocity component, $v_\perp$, perpendicular to a magnetic field $\mathbf{B}$ and, $v_\parallel$, parallel to the magnetic field, in the absence of an electric field, the acceleration due to the Lorentz force (Equation 5.1) in the laboratory frame is given by

$$\frac{dv_\perp}{dt} = \frac{q\,v_\perp B}{\gamma m_0}, \tag{5.23}$$

where the Lorentz parameter $\gamma = 1/(1 + (v_\parallel^2 + v_\perp^2)/c^2)^{1/2}$. We can balance this acceleration produced by the $\mathbf{v} \times \mathbf{B}$ Lorentz force with the centrifugal acceleration $v_\perp^2/r_L$, where r_L is the radius of the circular orbit around the magnetic field line (and is known as the Larmor or gyroradius):

$$\frac{q\,v_\perp B}{\gamma m_0} = \frac{v_\perp^2}{r_L}$$

Canceling a $v_\perp$ on both sides of the balance gives the angular frequency of the particle orbit in the laboratory frame:

$$\omega_c = \frac{v_\perp}{r_L} = \frac{qB}{\gamma m_0} \tag{5.24}$$

which is known as the cyclotron frequency, Larmor frequency, or gyrofrequency for the particle. The particles undergo circular cyclotron motion of angular frequency $\omega_c = qB/(\gamma m_0)$ in the frame of reference traveling at velocity $v_\parallel$, parallel to the magnetic field. For $v_\parallel \neq 0$ in the laboratory frame, the particle trajectory traces out a helical orbit. The Larmor or gyroradius in the laboratory frame is given by

$$r_L = \frac{\gamma m_0 v_\perp}{qB}. \tag{5.25}$$

If the particle velocity is at an angle, α, to the magnetic field, we can write that $v_\perp = v \sin \alpha$ and

$$\frac{dv_\perp}{dt} = \frac{q\,B\,v \sin \alpha}{\gamma m_0}, \tag{5.26}$$

where $dv_\perp/dt$ is directed perpendicular to the magnetic field. Using Larmor's formula (Equation 1.16) with acceleration measured in the rest frame (Equation 3.44), the radiated electromagnetic power associated with the acceleration is given by

$$P_S = \frac{q^2}{4\pi\varepsilon_0} \frac{2}{3c} \gamma^4 \frac{q^2 B^2}{(\gamma m_0)^2} \frac{v^2 \sin^2\alpha}{c^2}$$

$$= \frac{q^2}{4\pi\varepsilon_0} \frac{2}{3c} \gamma^4 \omega_c^2 \left(\frac{v}{c}\right)^2 \sin^2\alpha \qquad (5.27)$$

and is known as "cyclotron radiation" when $v \ll c$ or "synchrotron radiation" when v approaches the speed of light c. Here α is the angle of the particle velocity to the magnetic field, $\omega_c = qB/(\gamma m_0)$ is the cyclotron frequency, and the Lorentz parameter $\gamma = 1/(1 - v^2/c^2)^{1/2}$.

As a particle executes an orbit around a magnetic field, the particle undergoes an acceleration directed towards the center of the orbit and the electric field of the radiation due to the acceleration detected at a position far from the orbiting particle varies in time. For nonrelativistic particle motion with an observation angle perpendicular to the magnetic field lines, the radiated instantaneous power peaks when the acceleration is directed perpendicular to the line of sight ($\theta'_A = \pi/2$) and then oscillates in time as $\sin^2\theta'_A$ with $\theta'_A = \omega_c t$ (see Equation 1.15). The angular frequency of radiation from particles moving at speeds $v \ll c$ is consequently equal to the orbiting frequency $\omega_c = eB/m_0$ as $\gamma = 1$. If the angle of observation is not perpendicular to the magnetic field or the particle has a significant velocity $v_\parallel$ component along the magnetic field, when $v \ll c$, harmonics at frequencies $n\omega_c$ are produced, where n is an integer. The detected radiation is emitted as a harmonic for nonrelativistic particles as the electric field variation in time in the direction of observation departs from a single frequency sinusoidal oscillation, but has zero electric field values at the same time as the fundamental frequency ω_c. A range of harmonic sine waves of frequency $n\omega_c$ then sum to produce the electric field variation in time.

Particles moving at relativistic velocities have the angular variation of emission in the rest frame enhanced in the direction of the particle velocity (see Section 3.5). This causes departures from a pure sinusoidal oscillation of radiated electric field as detected in the rest frame and again leads to the production of harmonic frequencies. The spectrum of emission from relativistic particles (synchrotron radiation) is considered in Section 5.4.

If a particle velocity distribution is isotropic, we can average the expression for $\sin\alpha$ over the surface of a sphere and obtain the radiated power P_{av} per particle of cyclotron radiation for an ensemble of particles in a gas or plasma:

$$P_{av} = \int_0^\pi P_S \, 2\pi \sin\alpha \, d\alpha \qquad (5.28)$$

$$= \frac{q^2}{4\pi\varepsilon_0} \frac{2}{3c} \gamma^4 \omega_c^2 \left(\frac{v}{c}\right)^2 \int_0^\pi 2\pi \sin^3\alpha \, d\alpha$$

$$= \frac{q^2}{4\pi\varepsilon_0} \frac{2}{3c} \gamma^4 \omega_c^2 \left(\frac{v}{c}\right)^2 \frac{8\pi}{3} \qquad (5.29)$$

Considering electrons of charge $q = -e$, we can substitute $\omega_c = eB/(\gamma m_0)$ for the electron cyclotron frequency, use the expression for the classical electron radius (Equation 4.64) considered in our treatment of Thomson scatter (see Section 4.6), and obtain an expression for the cyclotron power radiated by an isotropic distribution of electrons in terms of the magnetic field B. The radiated power per electron of cyclotron radiation with an isotropic distribution of electron velocities becomes

$$P_{av} = \left(\frac{2}{3}\right)^2 r_e^2 c \left(\frac{v}{c}\right)^2 \gamma^2 B^2. \tag{5.30}$$

The energy density[6] of a magnetic field B is $B^2/(2\mu_0)$, so Equation 5.30 can be written in terms of an energy density U_m associated with the magnetic field causing cyclotron radiation. The radiated cyclotron power P_{av} for an isotropic distribution of electron velocities can then be presented in terms of the total Thomson cross-section $\sigma_S = (8\pi/3)r_e^2$ (see Equation 4.68) and the magnetic field energy density U_m:

$$P_{av} = \left(\frac{4}{3}\right) \sigma_S c \left(\frac{v}{c}\right)^2 \gamma^2 U_m \tag{5.31}$$

upon using $\mu_0 = 1/(c^2\epsilon_0)$ (see Section 1.8). As the total scattering cross-section σ_S is 6.524×10^{-29} m^2, we see that the power radiated by cyclotron emission in a plasma is small and needs to proceed for a long time before the energy radiated approaches the magnetic field energy density U_m.

It is shown in Section 7.9 that the power transfer from electrons to photons during inverse Compton scatter of light has an identical form as Equation 5.31 if the magnetic field energy density U is replaced by the energy density U_p of the radiation field. Having a similar expression for the power associated with the scatter of photons and the emission of cyclotron radiation is possibly unexpected, but arises fundamentally because the energy density of electromagnetic radiation is proportional to the magnetic field energy density of the radiation (see the discussion for Equation 1.12).

In plasmas, electrons are accelerated at greater rates and have larger velocities than ions due to their low mass. Synchrotrons and plasmas produce light more efficiently due to electron or sometimes positron motion rather than a heavier particle as their low rest mass causes the frequency ω_c of emission to be higher and for similar energy content, the velocity v is higher. The higher velocity v and higher frequency ω_c increase the radiated power proportionally to $\omega_c^2 v^2$. The increased velocity can also result in the relativistic factor $\gamma = 1/\sqrt{1 - v^2/c^2}$ enhancing light emission due to the γ^2 scaling (see Equations 5.30 and 5.31). As well as increases

---

[6] The energy density $B^2/(2\mu_0)$ of a magnetic field can be verified by considering the energy required to produce a magnetic field B inside a solenoid; see Exercise 5.2.

in the angle integrated emission, the maximum intensity of synchrotron radiation is enhanced compared to nonrelativistic cyclotron radiation, as synchrotron light is concentrated within a half-angle $1/\gamma$ around the direction of the velocity of the charged particles (see the discussion in Section 3.5).

5.4 Synchrotron Radiation

Synchrotron light is the electromagnetic radiation emitted when particles, moving at velocities close to the speed of light, change direction under the action of a magnetic field. A "synchrotron" is typically a large machine (about the size of a soccer field) that accelerates electrons to almost the speed of light. In a typical synchrotron, a vertical magnetic field causes the electrons to move in a large horizontal quasi-circular orbit resulting in bright light emitted in the direction of the electron velocity due to the "beaming" effect discussed in Section 3.5. Synchrotron light is also emitted when particles in plasmas moving at velocities approaching the speed of light are accelerated by magnetic fields (see Equations 5.27 and 5.29).

At relativistic particle energies, enhanced emission in the direction of the particle velocity ensures that synchrotron emission detected in the rest or laboratory frame comprises electric field oscillations significantly distorted from the sinusoidal variation of electric field associated with cyclotron radiation where the particles velocities are nonrelativistic (see Section 5.3). Synchrotron radiation comprises many closely spaced high harmonics of the relativistically increased cyclotron frequency $\omega_c = qB/(\gamma m_0)$. Such closely spaced harmonics typically merge into a quasi-continuum output due to small variations of particle velocities and the magnetic field strength experienced by accelerated particles.

The total radiated power detected in the rest frame from a charged particle in a synchrotron can be evaluated from Equation 3.44. With synchrotron radiation the acceleration of the charged particle is perpendicular to the particle velocity, so the acceleration $\mathbf{a}_{\parallel}$ parallel to the particle velocity is zero. For the orbiting particle acceleration $\mathbf{a}_{\perp}$ perpendicular to the velocity, we substitute the acceleration of a rotating body, namely $a_{\perp} = v_{\perp}^2/r_L$, where r_L is the Larmor radius. The total radiated power P for a particles of charge q is given by

$$P = \frac{q^2}{4\pi\epsilon_0}\frac{2}{3c^3}\gamma^4\frac{v_{\perp}^4}{r_L^2} \approx \frac{q^2}{4\pi\epsilon_0}\frac{2}{3}\gamma^4\frac{c}{r_L^2} \tag{5.32}$$

as the particle velocity $v_{\perp}$ with synchrotron emission is close to the speed of light c. Synchrotron machines uses electrons of charge $q = -e$ moving close to the speed of light. The γ factor for a synchrotron machine is consequently $\gamma \approx T_k/(m_0c^2)$, where T_k is the kinetic energy of the electrons and $m_0c^2 = 511$ keV is the electron rest mass energy (see Equation 2.72). The power radiated by a synchrotron is often expressed using the orbiting electron kinetic energy T_k. The power radiated by each electron is given by

$$P \approx \frac{e^2}{4\pi\epsilon_0} \frac{2c}{3r_L^2} \left(\frac{T_k}{m_0 c^2} \right)^4 . \tag{5.33}$$

The observed spectrum of synchrotron radiation is determined by the variation in time of the electric field arising from the acceleration of charged particles as seen in a rest frame of reference. Beaming effects cause the radiation field observed in the rest frame to be concentrated in a narrow range of angles close to the direction of the particle. With synchrotron radiation, an observer sees a pulse of light of duration much shorter than the period $2\pi/\omega_c$ of gyration of the charged particles around the magnetic field. The spectrum of observed light consequently extends over a much broader spectral region than $\omega_c/(2\pi)$.

In order to estimate the frequency range of synchrotron emission, we need to estimate the duration of a pulse of synchrotron emission from a single particle as detected by an observer. The angular range $\Delta\theta$ of an orbiting particle where the detected emission occurs is given by $\Delta\theta = 2/\gamma$. The geometry of synchrotron emission is schematically illustrated in Figure 5.2. By simple geometry, the angular range is also related to the Larmor radius r_L of the particle orbit by $\Delta\theta = \Delta s/r_L$, where Δs is the distance along the particle orbit where the detected emission occurs. The velocity change $|\Delta\mathbf{v}|$ of the orbiting particle over the distance Δs is given by

$$|\Delta\mathbf{v}| = 2v_\perp \sin(\Delta\theta/2) \approx v_\perp \, \Delta\theta$$

as the angular range $\Delta\theta$ is small (see the inset in Figure 5.2). Here $v_\perp$ is the velocity perpendicular to the magnetic field. The quantity $v_\perp$ is equal to the speed of the particle as it rotates in the Larmor orbit, so that $\Delta s = v_\perp \Delta t$, where Δt is the time taken by the particle to move the distance Δs around the Larmor orbit. Replacing $dv_\perp/dt$ by $|\Delta\mathbf{v}|/\Delta t$ in Equation 5.23, we have

$$\frac{|\Delta\mathbf{v}|}{\Delta t} = \frac{v_\perp \Delta\theta}{\Delta s/v_\perp} = \frac{q \, v_\perp B}{\gamma m_0}.$$

Simplifying, gives

$$\frac{\Delta\theta}{\Delta s} = \frac{q B}{\gamma m_0 v_\perp} = \frac{\omega_c}{v_\perp}, \tag{5.34}$$

where $\omega_c = qB/(\gamma m_0)$ is the cyclotron frequency for the particle. The distance Δs around the particle orbit where emitted light reaches the observer can now be determined using the angular range $\Delta\theta = 2/\gamma$ for emission. Using Equation 5.34:

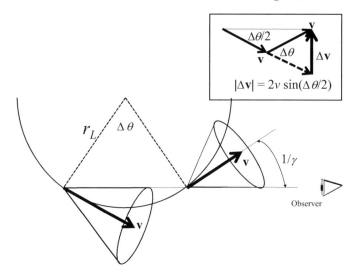

Figure 5.2   The geometry of synchrotron emission. Emission only occurs over a narrow range of angles $\Delta\theta = 2/\gamma$ due to the beaming effect associated with the relativistic velocity $\mathbf{v}$ of the emitting particles. The inset shows the geometry needed to work out the change in velocity $|\Delta\mathbf{v}| = 2v\sin(\Delta\theta/2)$ as a particle orbit rotates by $\Delta\theta$.

$$\Delta s = \frac{2v_\perp}{\gamma\omega_c} \tag{5.35}$$

Equation 5.35 means that the time for emission $\Delta t = 2/(\gamma\omega_c)$.

The arrival time of light at the observer is affected by the changing distance of the emitting particle from the observer. The first light emission takes longer to reach the observer, so the time duration of the observed light pulse is reduced compared to the duration of emission. The time duration Δt^* of the observed pulse of light is given by

$$\Delta t^* = \Delta t - \frac{\Delta s}{c}, \tag{5.36}$$

where $\Delta s/c$ is the time for light to travel the distance Δs over which emission occurs. As $\Delta t = \Delta s/v_\perp$, using Equation 5.35 the observed time duration of the synchrotron pulse of light is given by

$$\Delta t^* = \frac{2}{\gamma\omega_c}\left(1 - \frac{v_\perp}{c}\right). \tag{5.37}$$

The frequency of emission is consequently broad, extending to a maximum frequency $\approx 1/\Delta t^*$. When the particle velocity approaches the speed of light, this maximum frequency is much higher than the cyclotron frequency ω_c.

When particle velocities $v_\perp$ are close to the speed of light c, we can approximate Equation 5.37 using

$$1 - \frac{v_\perp}{c} = \frac{(1 - v_\perp/c)(1 + v_\perp/c)}{1 + v_\perp/c} = \frac{1 - v_\perp^2/c^2}{1 + v_\perp/c} \approx \frac{1}{2\gamma^2}. \tag{5.38}$$

The observed time duration of the synchrotron pulse is then simply related to the relativistic factor $\gamma = 1/\sqrt{1 - v_\perp^2/c^2}$. We have

$$\Delta t^* \approx \frac{1}{\gamma^3 \omega_c}. \tag{5.39}$$

The spectrum of radiation for synchrotron emission from highly relativistic particles extends up to a maximum critical frequency ω_{crit} defined by the inverse of Δt^*. Following standard convention, we set

$$\omega_{crit} = \frac{3}{2}\gamma^3 \omega_c = \frac{3qB}{2m_0}\gamma^2. \tag{5.40}$$

Converting the frequency in radians per second to cycles per second (hertz), the value of the synchrotron critical frequency is

$$v_{crit} = \frac{3}{4\pi}\gamma^3 \omega_c = \frac{3qB}{4\pi m_0}\gamma^2. \tag{5.41}$$

Strong synchrotron emission occurs up to the critical frequency v_{crit} which is a frequency higher by a factor $(3/2)\gamma^2$ compared to the cyclotron frequency for a nonrelativistic particle where $v_\perp << c$.

In Section 4.5, we discussed the Razin effect which causes a reduction in the beaming of light from relativistic particles in a plasma. Equation 4.61 shows that there is a cut-off in the beaming effect for relativistic particles in a plasma medium associated with a refractive index $\eta < 1$. The medium produces a much greater beaming angle $\Delta\theta \approx 2\omega_p/\omega$ than occurs in vacuum where $\Delta\theta \approx 2/\gamma$ when

$$\omega << \gamma\omega_p,$$

where ω_p is the plasma frequency (see Equation 4.46). The Razin effect for synchrotron radiation from a plasma stops the strong emission of lower frequency light where $\omega << \gamma\omega_p$. As the Lorentz parameter γ can be large for particles such as electrons with velocities close to the speed of light, the effective cut-off in low frequency emission moves to a greater frequency, well above the lower-limit plasma frequency ω_p applicable for nonrelativistic electrons.

The polarization of synchrotron light is in the plane formed by the particle acceleration vector and the direction of emission (see Section 1.3). In Section 9.4 we show that the polarization of light is not changed when moving between frames of reference. Linearly polarized light detected in astronomy is often associated with synchrotron radiation, so can indicate the presence of a regular magnetic field embedded in a plasma [23]. The predominant polarization of a large synchrotron device with electrons orbiting in a horizontal circle is horizontally linear as the electron acceleration and direction of observation are horizontal. Small deviations from a pure horizontal polarization can occur at small angles away from the horizontal axis of a synchrotron as electrons then have a small nonhorizontal acceleration component.

5.5 The Spectrum of Synchrotron Radiation

To obtain the spectrum of synchrotron radiation, the electric field variation in time associated with the pulsed output needs to be determined. The temporal electric field variation of synchrotron radiation can be calculated using the expression (Equation 3.59) for the radiated power detected in the rest frame as a function of angle θ to the velocity. Angles θ are related to distances s around the circumference of the orbiting charge by

$$\theta = \frac{s}{r_L},$$

where r_L is the Larmor radius (see Figure 5.2). The relative time t in the rest frame when synchrotron light reaches an observer needs to take account of the time of flight of photons from the emission distance s. The distance s over which synchrotron emission occurs is along a line of sight intersecting the synchrotron orbit (see Figure 5.2), so that using Equation 5.38 we have that

$$t = \frac{s}{v_\perp}\left(1 - \frac{v_\perp}{c}\right) \approx \frac{s}{v_\perp}\frac{1}{2\gamma^2}$$

and hence

$$s \approx 2v_\perp\gamma^2 t.$$

We can now relate the angle θ of a particle in the synchrotron orbit to the time t when emission is detected along a line of sight:

$$\theta = \frac{v_\perp}{r_L} = 2\gamma^2\frac{v_\perp}{r_L}t = 2\gamma^2\omega_c t, \tag{5.42}$$

where $\omega_c = v_\perp/r_L = qB/(\gamma m_0)$ is the cyclotron frequency. We have defined a critical frequency ω_{crit} related to the spectrum of synchroton emission with $\omega_{crit} = (3/2)\gamma^3 \omega_c$ (see Equation 5.40). In terms of the critical frequency

$$\theta = \left(\frac{4\omega_{crit}}{3\gamma}\right) t = \omega_{crit}^* t, \tag{5.43}$$

where we introduce

$$\omega_{crit}^* = \frac{4\omega_{crit}}{3\gamma} = 2\gamma^2 \omega_c.$$

As the particle rotates, we can set the angle θ to a line of sight using $\theta = \omega_{crit}^* t$. Using Equation 3.59, the power radiated as a function of time t is given by

$$\frac{dP}{d\Omega} = \frac{q^2}{4\pi\epsilon_0} \frac{1}{4\pi c^3} \frac{v_\perp^4}{r_L^2} \left(\frac{1}{1 - v_\perp \cos\theta/c}\right)^4 \left(1 - \left(\frac{\sin\theta \, \cos\phi}{\gamma(1 - v_\perp \cos\theta/c)}\right)^2\right)$$

$$\approx \frac{q^2}{4\pi\epsilon_0} \frac{c}{4\pi r_L^2} \left(\frac{1}{1 - v_\perp \cos(\omega_{crit}^* t)/c}\right)^4 \left(1 - \left(\frac{\sin(\omega_{crit}^* t) \, \cos\phi}{\gamma(1 - v_\perp \cos(\omega_{crit}^* t)/c)}\right)^2\right). \tag{5.44}$$

The variation of electric field amplitude $E(t)$ with time is proportional to the square root of $dP/d\Omega$ given by Equation 5.44. We can write

$$E(t) = E_0 \left(\frac{1}{1 - v_\perp \cos(\omega_{crit}^* t)/c}\right)^2 \sqrt{1 - \left(\frac{\sin(\omega_{crit}^* t) \, \cos\phi}{\gamma(1 - v_\perp \cos(\omega_{crit}^* t)/c)}\right)^2}, \tag{5.45}$$

where

$$E_0 = \frac{1}{\epsilon_0 c} \frac{q}{\sqrt{4\pi\epsilon_0}} \frac{1}{\sqrt{4\pi c^3}} \frac{v_\perp^2}{r_L}.$$

An example of the time dependence of the electric field in a pulse of synchrotron light is shown in Figure 5.3.

An electric field temporal pulse is composed of many sinusoidal electric field components with a large range of frequencies. The electric field $E(\omega)$ variation with frequency ω for synchrotron radiation is obtained from Equation 5.45 by taking the Fourier transform of $E(t)$:

$$E(\omega) = \int_{-\infty}^{\infty} E(t) \exp(i\omega t) dt \tag{5.46}$$

In the extreme relativistic limit when $\gamma \gg 1$, it is more accurate to evaluate the Fourier transform of the electric field using the approximations $(1 - v_\perp \cos\theta/c) \approx (1 + \gamma^2\theta^2)/(2\gamma^2)$ (Equation 3.53) and $\sin\theta \approx \theta$. Using $\theta = \omega_{crit}^* t$ and Equation 5.45, the Fourier transform of the electric field variation in time becomes

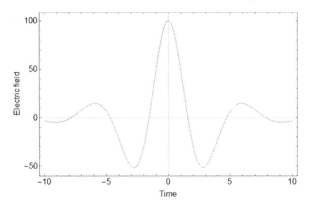

Figure 5.3   The electric field component for a pulse of synchrotron light at a
single frequency $\omega = 10\omega^*_{crit}$ (see the text). The electric field variation with time in
units of ω^{-1} is shown for $v_\perp/c = 0.9$ ($\gamma = 2.29$) and an azimuthal angle $\phi = \pi/2$.

$$E(\omega) \approx 8E_0\gamma^4 \int_0^\infty \frac{\sqrt{(1 - 2\gamma^2(\omega^*_{crit}t)^2 \cos(2\phi) + \gamma^4(\omega^*_{crit}t)^4)}}{(1 + \gamma^2(\omega^*_{crit}t)^2)^3} \cos(\omega t)dt. \quad (5.47)$$

Synchrotron radiation power $P(\omega)$ from large synchrotron facilities is commonly
quoted as the power per a constant 0.1% frequency bandwidth rather than the
power per a constant unit of frequency. The power $P(\omega)$ in units of power per
constant bandwidth is related to the power per unit frequency $dP(\omega)/d\omega$ such that
$P(\omega) \propto (dP(\omega)/d\omega) \, \omega$. The total power $P(\omega)$ output with synchrotron radiation is
proportional to the square of the electric field variation with frequency (see Equa-
tion 1.12) so the synchrotron radiation power from a single electron is given by

$$P(\omega) = \frac{1}{2}\epsilon_0 c |E(\omega)|^2 (10^{-3}\omega)dA, \quad (5.48)$$

where dA is the area of the beam. Using Equation 5.47 to deduce the Fourier transform of the electric field and then evaluating the synchrotron power using Equation
5.48, the spectrum of synchrotron radiation is shown in Figure 5.4. The output
power $P(\omega)$ increases with frequency up to the critical frequency ω_{crit} as defined
by Equation 5.40. The synchrotron power output then drops rapidly as ω increases
above ω_{crit}. The relative variation in synchrotron power shown in Figure 5.4 applies for all values of $\gamma \gg 1$, though the total output power (i.e. the absolute value
of the vertical power axis) varies strongly with γ (the power integrated over all
frequencies $\propto \gamma^4$; see Equation 5.32).

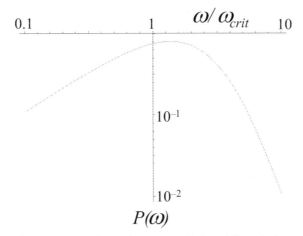

Figure 5.4 The spectrum of synchrotron radiation. The relative output power $P(\omega)$ is shown as a function of relative frequency ω/ω_{crit}, where ω_{crit} is the critical frequency. This evaluation is for azimuthal angle $\phi = \pi/2$.

It is useful to examine the development of a typical synchrotron spectrum as the electron energy increases from nonrelativistic values (γ close to 1) up to highly relativistic regimes ($\gamma >> 1$). For low energies, the electric field varies sinusoidally with a frequency equal to the cyclotron frequency ω_c. In synchrotron and cyclotron devices, radiation is detected at angles close to $\pi/2$ to the acceleration associated with the orbit around the magnetic field. In synchrotron and cyclotron devices, the spectrum of radiation at low energies is consequentially a single spectral line at the cyclotron frequency ω_c. In plasmas, where observations can be at some angle other than $\pi/2$ to the acceleration, cyclotron harmonics at frequency $n\omega_c$ (with n an integer) can occur as the electric field is always zero at the same node points as the fundamental frequency ω_c, but along an arbitrary line of sight, is no longer a simple sinusoid.

When the velocity of particles increases (v/c approaches 1), higher harmonics are produced even when the observation angle to the acceleration is close to $\pi/2$, as the electric field is no longer a simple sinusoid, but is distorted due to the beaming effect. Again the electric field is zero at the same node points as the fundamental frequency ω_c so that the contributions of the different frequencies to the total electric field are all harmonics $n\omega_c$.

For highly relativistic velocities (where $v \approx c$ and $\gamma >> 1$), the sinusoidal variation of electric field present at low velocities becomes a series of sharp pulses repeated at time intervals $2\pi/\omega_c$ when particles undergo a complete Larmor orbit (see Figure 5.3). The spectrum is a large number of harmonics, with the envelope of the power of each harmonic adding to give the spectrum shown in Figure 5.4. With synchrotron devices, there is typically a small spread of particle energies so that the cyclotron frequency $\omega_c = qB/(\gamma m_0)$ varies for different particles with slightly different energies and hence slightly different values of γ. The spread of energies and small variations of magnetic field B seen by different particles cause synchrotron emission at highly relativistic velocities to be effectively continuous once the spectral contribution from a large number of particles is summed.

5.6 Bending Magnets and Undulators

The output radiated power from charged particles moving at relativistic velocities in a large circular orbit synchrotron can be significantly enhanced by using magnetic fields to accelerate the particles into a motion with smaller radii of curvature. The power radiated by charged particles in a circular orbit is proportional to $1/r_L^2$, where r_L is the Larmor radius of curvature (Equation 5.33). Moving the Larmor radius for particles from values of tens of meters down to a centimeter scale can alone increase the power output by $> 10^6$. Further enhancement can be engineered by adding a sequence of bending magnets so that particles undergo a sine wave oscillation.

Synchrotron radiation sources and particle accelerator facilities often have particles rotating in large radii r_L "storage rings" to minimize radiation losses. Where radiation is needed, particles are injected into bending magnets producing a small Larmor radius. Employing alternating magnetic fields, particles injected into alternating magnetic pole pieces can undergo oscillatory half-orbital trajectories. If the radiation electric field produced by each pole piece is independent, the alternating magnetic field structure is known as a "wiggler" and the radiated power is proportional to the number of pole pieces. If the pole pieces are placed sufficiently close to each other, the electric fields from each pole piece can add coherently and the alternating magnetic field structure is known as an "undulator." For an undulator, the electric fields from each pole piece add, so that the radiated power is proportional to the square of the number of pole pieces because the power is proportional to the square of the electric field. With an undulator there is also a decrease of the angular variation of the radiation output in one plane, leading to an enhanced power per unit solid angle.

We can examine the motion of particles in an undulator by considering the Lorentz force due to the magnetic field (Equation 5.1). Radiated fields are relatively weak so that we ignore the radiation reaction force (see section 7.12) and any electric field. The force on a particle in an undulator is given by

$$\frac{d\mathbf{p}}{dt} = q(\mathbf{v} \times \mathbf{B}). \tag{5.49}$$

Assuming a Cartesian coordinate system for an undulator where the particle velocity $\mathbf{v}$ is in the x-direction and the magnetic field B_z is in the z-direction, the particles are accelerated in the y-direction. We have

$$\gamma m_0 \frac{dv_y}{dt} = q v_x B_z$$

upon using subscripts for the velocities v in different directions and assuming $|\mathbf{v}|$ is close in value to v_x.

In an undulator, there is a periodic oscillation along the x-direction of the magnetic field due to the alternate reversing of the polarity of pole pieces. If the spacing between pole pieces of the same polarity is λ_U, the magnetic field variation and particle oscillation has a regular structure with a "wavelength" of λ_U. We can write for the spatial variation of the magnetic field

$$B_z = B_0 \cos\left(\frac{2\pi x}{\lambda_U}\right). \tag{5.50}$$

If there are N_p pole pieces, this variation of magnetic field extends from say $x = 0$ up to $x = N_p \lambda_U$. The particle acceleration in the y-direction oscillates with distance x as

$$\gamma m_0 \frac{dv_y}{dt} = q \frac{dx}{dt} B_0 \cos\left(\frac{2\pi x}{\lambda_U}\right).$$

Multiplying both sides by dt, integrating the left-hand side with respect to dv_y and the right-hand side with respect to dx gives

$$\gamma m_0 v_y = q B_0 \int \cos\left(\frac{2\pi x}{\lambda_U}\right) dx = q B_0 \frac{\lambda_U}{2\pi} \int \cos\left(\frac{2\pi x}{\lambda_U}\right) d\left(\frac{2\pi x}{\lambda_U}\right).$$

The cosine integral is readily evaluated to give an expression for the particle velocity in the y-direction as a function of distance x along the undulator. We have

$$v_y = \frac{K_m c}{\gamma} \sin\left(\frac{2\pi x}{\lambda_U}\right), \tag{5.51}$$

where we introduce the dimensionless parameter K_m such that

$$K_m = \frac{q B_0 \lambda_U}{2\pi m_0 c}. \tag{5.52}$$

The parameter K_m is known as the "magnetic deflection parameter" for an undulator.

Particles in an undulator are deflected from paths parallel to the x-direction with a "deflection angle" θ_D given by

$$\theta_D = \frac{v_y}{v_x} \approx \frac{v_y}{c} = \frac{K_m}{\gamma} \sin\left(\frac{2\pi x}{\lambda_U}\right).$$

The maximum deflection angle of a particle is K_m/γ, so if $K_m < 1$, the relativistic cone of radiation emission with a half-angle $1/\gamma$ overlaps at all stages of the particle motion. The electric fields of the radiation emission at different stages of the particle motion can interfere and add. A value of $K_m < 1$ is consequently a criterion for an array of magnetic pole pieces to act as an undulator. If $K_m >> 1$, only a small angular range of the relativistic emission overlaps at different stages of the particle motion and so coherent addition of electric fields at different stages of the particle motion does not occur. The magnetic deflection parameter $K_m >> 1$ is the criterion for an array of pole pieces to act as a wiggler.

The angular radiation output from an undulator or wiggler in the direction of the particle oscillation (our y-direction) is over a range of angles K_m/γ, while in the direction of the magnetic field (our z-direction), the Lorentz beaming effect means the angular range of radiation is over $1/\gamma$. With an undulator, $K_m < 1$, and the angular range of emission is reduced in the direction of the particle oscillation. The net effect is a concentration of undulator power into a solid angle which is reduced by a factor K_m compared to the solid angle of emission from a simple bending magnet.

In the magnetic fields of an undulator, the Lorentz factor γ is constant as the particle energy remains constant apart from a small radiation energy loss. For particles with initial velocity **v** injected into an undulator

$$\gamma = \frac{1}{\sqrt{1 - v^2/c^2}} = \frac{1}{\sqrt{1 - (v_x^2 + v_y^2)/c^2}}.$$

Hence

$$\frac{v_x^2}{c^2} = 1 - \frac{1}{\gamma^2} - \frac{v_y^2}{c^2} = 1 - \frac{1}{\gamma^2} - \frac{K_m^2}{\gamma^2} \sin^2\left(\frac{2\pi x}{\lambda_U}\right).$$

Taking the square root of v_x^2/c^2, noting that $2\sin^2\alpha = 1 - \cos(2\alpha)$ for any angle α and $v_x/c \approx 1$, we have

$$\frac{v_x}{c} = 1 - \frac{1 + K_m^2/2}{2\gamma^2} + \frac{K_m^2}{4\gamma^2} \cos\left(2\frac{2\pi x}{\lambda_U}\right). \tag{5.53}$$

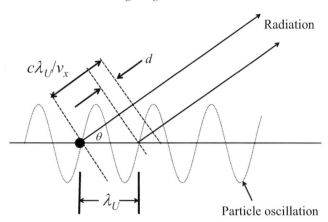

Figure 5.5 A schematic of an undulator showing the particle oscillation and radiation emission at angle θ. The geometry illustrates that the distance d between the particle motion and the electromagnetic wavefront is given by $d = c\lambda_U/v_x - \lambda_U \cos\theta$, where λ_U is the wavelength of the particle oscillation. For the two rays of emission illustrated, the radiation electric fields of wavelength λ are in phase if $d = n\lambda$, where n is an integer.

The average value of the particle velocity in the $x-$direction is simply

$$\frac{<v_x>}{c} = 1 - \frac{1 + K_m^2/2}{2\gamma^2}. \tag{5.54}$$

The time for a particle to travel a length λ_U along an undulator is $\lambda_U/<v_x>$. In this time, a wavefront of electromagnetic radiation travels a distance of $c\,\lambda_U/<v_x>$. After one period of particle motion through the undulator (distance λ_U), the spatial separation d of the particle and an electromagnetic wavefront propagating at an angle of θ to the x-axis along the direction of the radiation is given by

$$d = \frac{c\,\lambda_U}{<v_x>} - \lambda_U \cos\theta.$$

A schematic illustrating the derivation of the spatial separation d is shown in Figure 5.5.

For the coherent addition of electric fields, the wavelength λ of the radiation in the undulator must be such that $d = n\lambda$, where n is an integer. From Equation 5.54, for small K and large Lorentz factor γ:

$$\frac{c}{<v_x>} \approx 1 + \frac{1 + K_m^2/2}{2\gamma^2}$$

The wavelengths λ_r where coherent positive interference of emission from different pole pieces occurs are thus given by

$$n\lambda_r \approx \lambda_U \left(1 + \frac{1 + K_m^2/2}{2\gamma^2}\right) - \lambda_U \cos\theta.$$

For small angles θ, we have $\cos\theta = 1 - 2\sin^2(\theta/2) \approx 1 - \theta^2$ and obtain an equation for allowed undulator wavelengths

$$\lambda_r \approx \frac{\lambda_U}{2n\gamma^2}\left(1 + \frac{K_m^2}{2} + \gamma^2\theta^2\right). \tag{5.55}$$

Equation 5.55 is known as the "undulator equation." The value of λ_r with $n = 1$ is referred to as the resonant wavelength. Shorter output wavelengths with $n > 1$ are known as "harmonics."

The value of the magnetic deflection parameter K_m is proportional to the magnetic field B_0 and spacing $\lambda_U/2$ of the magnetic pole pieces (see Equation 5.52). Even with a fixed spacing of pole pieces (fixed $\lambda_U/2$), from Equation 5.55 we see that it is possible to tune the output wavelengths by changing the magnetic field value. Undulator wavelengths also vary with the angle θ to the axis of the undulator and so appropriate apertures are often employed to select the required wavelengths of output. The $\lambda \propto 1/\gamma^2$ dependence in the undulator equation ensures that undulator output can be at X-ray wavelengths ($\lambda < 1$ nm) as low mass particles such as electrons with energies > 0.5 GeV have $\gamma > 1,000$. If there are N_p magnetic pole pieces, the radiated power from an undulator varies as N_p^2, as the electric fields of the electromagnetic wave from each pair of pole pieces add coherently and the power is proportional to the square of the electric field.

5.7 Free Electron Lasers

Free electron lasers enhance radiation output from undulator structures by "microbunching" electrons so that the radiation emission from all electrons adds coherently. Microbunching occurs due to a ponderomotive force associated with an intense laser electromagnetic wave copropagating with the electron beam causing electrons to be spaced along the undulator with the same period as the magnetic pole pieces of the undulator. The output of a free electron laser is enhanced above that of an undulator by the number of electrons in a microbunch as the output is proportional to the square of the number of electrons, rather than being proportional to the number of electrons. For a review of free electron laser physics and operations, see Pellegrini et al. [87]. A schematic illustrating the methods of producing radiation from relativistic particles is shown in Figure 5.6.

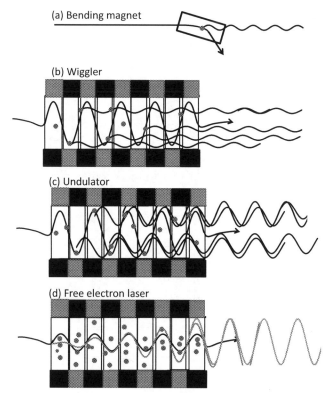

Figure 5.6 A schematic showing the production of radiation from relativistic electrons using (a) bending magnets (power $\propto n_e \gamma^4/r_L^2$), (b) wigglers (power $\propto n_e N_p \gamma^4/r_L^2$), (c) undulators (power $\propto n_e N_p^2 \gamma^4/r_L^2$), and (d) free electron lasers (power $\propto n_e^2 N_p^2 \gamma^4/r_L^2$). Here n_e is the number of electrons in the radiating volume, N_p is the number of magnetic pole pieces, γ is the relativistic Lorentz factor for the electrons, and r_L is the Larmor radius of the electron orbit in the magnetic field.

Free electron lasers operating in the visible and infra-red can utilize an optical "cavity" where mirrors reflect the light backward and forwards through an undulator. In the extreme ultraviolet and X-ray spectral regimes (wavelengths < 50 nm), mirrors and a cavity are not feasible due to the low reflectivity of mirrors. At shorter wavelengths, the laser electromagnetic radiation in a free electron laser can be "seeded" by injecting coherent light into an undulator or it can grow by amplification of spontaneous emission arising from the undulator output (this last method of operation is known as self-amplified spontaneous emission: SASE).

We represent the electric field of the laser light along the undulator length (*x*-direction) by

$$E_y = E_0 \cos(kx - \omega t + \psi), \tag{5.56}$$

where k is the radiation wavenumber, ω is the radiation frequency, and ψ represents the phase difference between the light and the electron oscillations in the undulator. The magnetic field variation along the undulator length is as shown for Equation 5.50 with

$$B_z = B_0 \cos (k_U x) . \tag{5.57}$$

We introduce here a "wavevector" k_U for the magnetic field oscillation with $k_U = 2\pi/\lambda_U$. The magnetic field oscillation with distance x along the undulator causes an electron velocity oscillation perpendicular to the undulator axis and the magnetic field (see Equation 5.58) such that

$$v_y = -\frac{K_m c}{\gamma} \sin (k_U x) . \tag{5.58}$$

The magnetic deflection parameter K_m in terms of k_U for electrons is given by

$$K_m = \frac{e B_0}{k_U m_0 c} . \tag{5.59}$$

Electrons in the undulator oscillate with a velocity v_y parallel to the electric field E_y of the laser light and are accelerated (or deaccelerated) by the laser electric field. The electrons move in a force $-eE_y$ so that the rate of energy transfer to an electron due to the force is $-eE_y v_y$. The kinetic energy of electrons oscillating due to the alternating electric field associated with an electromagnetic wave is termed the ponderomotive energy (see Chapter 4). The ponderomotive energy U of electrons in an undulator changes at a rate

$$\frac{dU}{dt} = -v_y e E_y = e E_0 \frac{K_m c}{\gamma} \sin (k_U x) \cos(kx - \omega t + \psi)$$

$$= e E_0 \frac{K_m c}{2\gamma} [\sin ((k + k_U)x - \omega t + \psi) + \sin ((k - k_U)x + \omega t - \psi)]$$

upon using the identity $\sin a \cos b = (\sin(a + b) + \sin(a - b))/2$. The Lorentz factor γ is constant with the electron oscillations due to the undulator magnetic field and we assume that the change of γ due to the laser electric field is small. Furthermore, we ignore the low spatial frequency variation of dU/dt (varying as $\sin ((k - k_U)x))$ and write that

$$\frac{dU}{dt} = eE_0 \frac{K_m c}{2\gamma} \sin\left((k + k_U)x - \omega t + \psi\right). \tag{5.60}$$

The velocity of the highly relativistic electrons used in undulators and free electron lasers is close to the speed of light c. The variation of the electron kinetic energy along the length of the undulator is, therefore, given by

$$\frac{dU}{dt} = \frac{dU}{dx}\frac{dx}{dt} \approx \frac{1}{c}\frac{dU}{dx}.$$

Within the phase term of Equation 5.60, a precise measurement of the electron velocity along the undulator axis (the x-direction) is needed to convert time t to distance x. The average value of the electron velocity $<v_x>$ in the x-direction is always slightly less than the speed of light c. In Section 5.6 we showed that particle velocities along an undulator are related to the magnetic deflection parameter K as given by Equation 5.54. We have

$$\frac{<v_x>}{c} = 1 - \frac{1 + K_m^2/2}{2\gamma^2} \approx \frac{1}{1 + (1 + K_m^2/2)/(2\gamma^2)} \tag{5.61}$$

as the deviation of the electron velocity $<v_x>$ from the speed of light c is small. The laser electromagnetic wave propagates at phase velocity $c = \omega/k$ (see Equation 1.7). We can equate time t to a value $x/ <v_x>$, where x is the distance along the undulator axis. The temporal phase variation ωt of the laser electromagnetic wave relates to the spatial phase variation with

$$\omega t = k\,x\,c/<v_x> = k\,x\left(1 + \frac{1 + K_m^2/2}{2\gamma^2}\right).$$

Rewriting the phase term of Equation 5.60 so that it represents the relative phase between the electromagnetic wave and the electron oscillation as a function of distance x along the undulator, we have

$$\frac{dU}{dx} = eE_0 \frac{K_m c^2}{2\gamma} \sin\Theta, \tag{5.62}$$

where the phase Θ of the electron oscillation relative to the laser electromagnetic wave is given by

$$\Theta = \left(k_U - \frac{k(1 + K_m^2/2)}{2\gamma^2}\right)x + \psi. \tag{5.63}$$

The ponderomotive force associated with gradients of ponderomotive energy was discussed in Chapter 4. The ponderomotive force $\mathbf{F}_p$ on an electron is given by $\mathbf{F}_p = -\nabla U$ (see the discussion leading to Equation 4.18). From Equation 5.62, there is a ponderomotive force $-dU/dx$ on each electron directed in the x-direction that seeks to reduce the phase Θ between the electron oscillation and the electro-magnetic wave to zero (or an integer multiple of π) by moving electrons to a zero value of dU/dx.

The phase Θ is reduced to zero when the phase ψ between the electron oscilla-tions and the electromagnetic wave oscillations reduces to zero. We saw in Section 5.6 that undulators produce enhanced light output when the electric fields of the electromagnetic wave produced from adjacent pole pieces of the same parity co-herently add. The undulator equation (Equation 5.55) gives the wavelengths λ_r of light which are resonant for an undulator. For angles parallel to the undulator axis with $\theta = 0$, the resonant ($n = 1$) wavelength is given by

$$\lambda_r \approx \frac{\lambda_U}{2\gamma^2} \left(1 + \frac{K_m^2}{2}\right). \tag{5.64}$$

The corresponding resonant wavevector $k_r = 2\pi/\lambda_r$ for a wavelength λ_r (see Equation 5.64) is given by

$$k_r \approx k_U \frac{2\gamma^2}{1 + K_m^2/2}. \tag{5.65}$$

Substituting $k = k_r$ from Equation 5.65 into the expression for the phase difference Θ between the electron oscillation and the laser electromagnetic wave (Equation 5.63), we see that Θ is zero when the phase difference ψ between the light and the electron oscillation is zero. The electrons and the electromagnetic wave are then in phase so that the ponderomotive force $dU/dx = 0$. The reduction to zero of the phase ψ between the electron oscillations and electromagnetic wave oscillation is the process that causes electrons to microbunch together in an undulator operating as a free electron laser.

In a free electron laser, electrons microbunch with separation in distance along the undulator axis equal to the wavelength of a resonant electromagnetic wave. The microbunched electrons then oscillate coherently in the transverse direction to the axis so that the oscillating charge q effectively becomes $q = -N_e e$, where N_e is the number of electrons in a microbunch. The output power from an accelerated charge is proportional to q^2 (see Equation 1.16), so the power output from a free electron laser is proportional to $N_e^2 N_p^2$, where N_p is the number of magnetic pole pieces in the undulator. The power output from a free electron laser is N_e times the power of an equivalent undulator.

A free electron laser can readily operate up to the critical frequency ω_{crit} for synchrotron radiation as defined by Equation 5.40. The critical frequency $\omega_{crit} = (3/2)\gamma^3\omega_c$, where ω_c is the cyclotron frequency of the undulator structure. Due to the γ^3 scaling, free electron lasing at photon energies up to 10 keV is readily achieved with a large electron accelerator.

Exercises

5.1 Consider a thin wire of short length dl with N_e conducting electrons per unit length along which flows a current I_C given by $I_C = N_e e v_e$, where v_e is the velocity of the electrons. A positive test charge q at a distance $\mathbf{r}$ at an angle of θ to $d\mathbf{l}$ is electrostatically attracted to the wire by the electrons and repelled by the positively charged ions in the wire. The charges per unit length N_e due to the electrons and ions are equal in magnitude, but opposite in sign so that the overall wire is neutrally charged. (a) If the charged particle is moving at a velocity $\mathbf{v}$ parallel to $d\mathbf{l}$ and allowing for relativistic effects on the electric fields produced by the electrons and ions in the wire, show that the respective electrostatic forces of attraction dF_- and repulsion dF_+ between the charge and the electrons and ions respectively are given by

$$dF_- = -\frac{1}{4\pi\epsilon}\frac{qN_e e dl}{r^2}\left[\left(1 - \frac{(v+v_e)^2}{c^2}\right)^{1/2}\sin\theta + \cos\theta\right],$$

$$dF_+ = \frac{1}{4\pi\epsilon}\frac{qN_e e dl}{r^2}\left[\left(1 - \frac{v^2}{c^2}\right)^{1/2}\sin\theta + \cos\theta\right].$$

(b) Hence, show that the net attractive force $dF = dF_+ + dF_-$ for electron velocities $v_e \ll c$ and a velocity v in the same direction as the current I is given by

$$dF = -\frac{\mu_0}{4\pi}\frac{I_C\,dl}{r^2}qv\sin\theta.$$

(c) Assuming the Lorentz equation $d\mathbf{F} = q\mathbf{v}\times d\mathbf{B}$ defines a magnetic field $d\mathbf{B}$ associated with the current carrying element, show that

$$d\mathbf{B} = \frac{\mu_0}{4\pi}\frac{I_C\,d\mathbf{l}\times\hat{\mathbf{r}}}{r^2},$$

where $\hat{\mathbf{r}}$ is a unit vector in the direction of $\mathbf{r}$. [For an answer to this Exercise, see Section 10.2.]

5.2 A cylindrical solenoid has wiring with N turns per unit length around a circular cross-sectional area of A and radius R. Use the Biot-Savart law to show that the magnetic field B produced inside such a solenoid of length l at the center of the solenoid is given by

$$B = \frac{\mu_0}{2}I_C N\int_{-l/2}^{l/2}\frac{R^2\,dx}{(R^2+x^2)^{3/2}},$$

where x is a distance along the length of the solenoid and I_C is the current in the wire. Hence show that for an infinitely long solenoid,

$$B = \mu_0 N I_C.$$

The "inductance" L_{ind} of a solenoid is defined by the voltage $L_{ind} dI_C/dt$. Show that the energy per unit length required to charge an infinitely long solenoid is $(1/2)L_{ind}I_C^2$. Faraday's law in an integral form shows that the voltage opposing the rise in current in a solenoid is $NA\, dB/dt$. Show that the inductance of the infinitely long solenoid per unit length is given by $\mu_0 N^2 A$ and that the energy per unit length required to achieve a magnetic field B in the solenoid is given by $A(1/2)B^2/\mu_0$. [This analysis confirms that the energy per unit volume contained in a magnetic field is $(1/2)B^2/\mu_0$.]

5.3   A particle of charge q and rest mass m_0 moves with a velocity $\mathbf{v}$ relative to the laboratory frame of reference and hence has a momentum $\mathbf{p}$ in the laboratory frame given by

$$\mathbf{p} = \frac{m_0 \mathbf{v}}{(1 - v^2/c^2)^{1/2}}.$$

An electrostatic potential accelerates the charge. The force $\mathbf{F}$ due to the potential must be such that

$$\mathbf{F} = \frac{d\mathbf{p}}{dt}.$$

Differentiate the above expression for the momentum $\mathbf{p}$ to show that in the laboratory frame

$$\mathbf{F} = \frac{m_0}{(1 - v^2/c^2)^{1/2}} \frac{d\mathbf{v}}{dt}.$$

5.4   Synchrotrons can accelerate particles to velocities close to the speed of light. Consider a synchrotron of circumference L_C with protons of kinetic energy T_k. Show that the magnetic field B needed to maintain the proton orbit within the synchrotron circumference is given by

$$B = \frac{2\pi T_k}{e\, c\, L_C}.$$

5.5   The Large Hadron Collider (LHC) near Geneva is the world's largest and highest energy particle accelerator. It is used for fundamental research in particle physics, notably producing the Higgs boson in 2012, a particle with a rest mass energy of 125 GeV. The LHC employs a circular "storage ring" of circumference 26.7 km containing protons of energy up to 6.7 TeV. Determine the magnetic field needed to keep these protons within the storage ring. [5.1 T]

5.6 Consider synchrotron radiation for particles with kinetic energy $T_k(t)$ measured relative to their rest mass energy $m_0 c^2$. Assuming that synchrotron radiation is the only process affecting the particle energy, show that the particle kinetic energy changes in time t from some initial time $t = 0$ according to the following relationship:

$$\ln\left(\frac{T_k(t)}{T_k(0)}\right) - \ln\left(\frac{T_k(t)+1}{T_k(0)+1}\right) + \frac{1}{T_k(t)+1} - \frac{1}{T_k(0)+1} = -\frac{8}{3}\sigma_S c\frac{U}{m_0 c^2},$$

where σ_S is the Thomson scatter cross-section and U is the energy density of the magnetic field.

5.7 If particle kinetic energies are significantly below the particle rest mass energy, show that the kinetic energy of particles varies due to cyclotron radiation such that

$$T_k(t) = T_k(0) \exp\left(-\frac{t}{t_0}\right),$$

where $t_0 \approx 4/B^2$ seconds for the magnetic field B measured in Tesla.

6

Photon and Particle Distribution Functions

To examine the emission or absorption of light from matter, we need to know how the numbers of interacting particles of matter change with particle energy. The distribution of particle numbers with energy is also needed in order to evaluate the total energy of an ensemble of particles and the way particles and energy move spatially and change temporally. In equilibrium, populations of quantum states are steady-state and the population variation with energy is characterized by a temperature. Equilibrium population distributions occur when there are sufficient interactions to enable a statistical distribution of particle energies.

All particles in equilibrium have energy distributions represented by a temperature, but the population distributions divide into two categories with different forms. Particles have Fermi-Dirac or Bose-Einstein distributions depending on whether the particles are respectively classed as fermions or bosons. The equilibrium population distributions are different as bosons can have any number of particles in a quantum state, while fermions are restricted to a maximum of one particle per quantum state (a restriction known as the Pauli exclusion principle). Particles such as electrons and protons are fermions and obey Fermi-Dirac statistic. Photons are an example of the boson class and obey Bose-Einstein statistics. We follow a treatment in Section 6.1 for the derivation of particle distribution functions that allows for the two types of distributions by allowing the number N_i of particles in a quantum state i to be either 0 or 1 for fermions, but to have any integer value for bosons.

The naming of the particle distributions as Fermi-Dirac or Bose-Einstein with particles labeled as fermions or bosons was instigated by the founder of relativistic quantum mechanics Paul Dirac (1902–1984). Later in life, Dirac preferred the expressions "Fermi" and "Bose" statistics, modestly dropping his own name and for symmetry, dropping Einstein's name. Dirac was born in Bristol and worked for most of his life at the University of Cambridge. Dirac shared the 1933 Nobel prize in physics with Erwin Schrodinger. The Nobel citation reads "for the discovery of new productive forms of atomic theory." Dirac's fundamental discoveries in quantum mechanics make him one of the most significant physicists of all time.[1]

Enrico Fermi (1901–1954) was born in Rome and worked in Italy and from 1938 in the United States, where he led the building and operation of the first nuclear reactor. The Chicago Pile-1 reactor sustained the first fission chain reaction in 1942. Fermi was awarded the Nobel prize in 1938 with the Nobel citation "for his demonstrations of the existence of new radioactive elements produced by neutron irradiation, and for his related discovery of nuclear reactions brought about by slow neutrons." Earlier Fermi developed the distribution function for particles where only one particle can occupy a quantum state. Fermi published a result in the same year as Dirac also solved the particle distribution problem [25], [38].

Satyendra Bose (1894–1974) was an Indian mathematician and physicist who contacted Einstein in order to publish his work on the distribution function of photons [98]. Particles such as photons which obey Bose-Einstein statistics were also termed "bosons" by Dirac after Bose. In the modern standard model of particle physics, it is recognized that other "force carrying" particles, namely gluons, and the W and Z bosons are also bosons (photons are the force carriers for electromagnetism). Whole atoms if the total spin is an integer also behave as bosons. When cooled bosonic atoms form Bose-Einstein condensates [91].

Wolfgang Pauli (1900–1958) was born in Austria, but worked in the United States from 1940 to 1946. Pauli developed several concepts in quantum mechanics including the principle bearing his name (the Pauli exclusion principle) that only one electron (or more generally one fermion) can occupy a quantum state. Pauli was awarded the Nobel prize in 1945 for developing the exclusion principle.

This chapter evaluates the equilibrium distribution of photon and particle energies. Expressions for particle energy distribution functions are derived using the idea of temperature and, for fermions, the chemical potential. The chemical potential is the energy needed to add another particle to an ensemble of particles. This leads to the Boltzmann ratio for the ratio of the population of two quantum states in an atom or ion and to the Saha-Boltzmann equation for the ratio of the populations of different ionization stages. The equilibrium distribution for radiation energy density and radiation intensity, the black-body radiation distribution is introduced.

---

[1] Paul Dirac was known to be extremely taciturn and precise in private, though he was an excellent public speaker and writer. Dirac became a professor at Florida State University in 1971 and undertook lecture tours. I heard Professor Dirac lecture on the fine structure constant during a visit to the Australian National University in 1975 [26]. I then attended a buffet lunch during his visit as a representative of the graduate students of the university. At the lunch I made myself innocuous by grabbing food and sitting at a corner table. To my surprise, the guest of honor Paul Dirac did the same and sat next to me without speaking. It was a typical Canberra summer day with a temperature of 35 °C, so I attempted to open a dialogue with "It's hot, isn't it?" Dirac replied, "Yes." I sensed that further conversation was not required, so we then both ate in silence.

Particle distribution functions for free electrons are presented: namely, the nonrelativistic Maxwellian distribution and the relativistic Maxwell-Juttner distribution.

6.1 Particle Distribution Functions

For a large number of particles, the probability that a particle is in an energy state E can be described by a quantity $f(E)dE$ proportional to the number of particles per unit volume with an energy between E and $E + dE$. The variation with particle energy of the number of particles for an equilibrium particle distribution can be characterized by two parameters: the temperature and the total number of particles. Instead of using the total number of particles, a quantity known as the chemical potential can be employed. The chemical potential is the energy required to add another particle to the group of particles.

For an equilibrium particle distribution function to exist, the different states of the system need to be accessible and there must be sufficient time without change of the system for the equilibrium distribution to be established. An equilibrium particle distribution function is the most probable distribution when the system is unchanged for some time. The length of time for equilibrium to be established depends on the rate of interaction, for example, the rate of collisions between particles.

In this section, we determine a general expression for equilibrium particle distribution functions leading to expressions for a collection of fermions (e.g. electrons), where only one fermion can occupy a quantum state, or a collection of bosons, where any number of bosons can occupy a quantum state. A relevant example of the boson class are photons. Any number of photons can occupy the same energy and spatial state. For photons, the same energy and spatial state is often referred to as a "mode."

We start our investigation of particle distribution functions by considering scenarios where temperatures may be low or modest, leading to the Fermi-Dirac distribution for fermions and the Bose-Einstein distribution for bosons. At the high temperatures where relativistic effects are important, the distinction between Fermi-Dirac and Bose-Einstein distributions largely disappears for laboratory plasmas as quantum state occupancies tend to be low and the "degeneracy" of fermions (number of quantum states for particles to occupy) has little effect. There are astrophysical situations where relativistic energies and particle degeneracy are both important (for e.g. in neutron stars where temperatures can approach 10^7–10^8 eV and densities approach 3×10^{17} Kg m^{-3}, comparable to the density of an atomic nucleus).

Consider the change ΔU of the total energy U of an assumed large number n_{tot} of particles distributed into a number of quantum states. For a constant temperature

T and constant chemical potential μ, any change in the total energy U of a large number of particles arises from a change in the entropy S of the system, plus any energy change due to a change Δn_{tot} of the number of particles. We have

$$\Delta U = T\Delta S + \mu \Delta n_{tot}. \tag{6.1}$$

In thermodynamics, entropy change ΔS is defined by the first term on the right and the chemical potential μ by the second term on the right. If the temperature is kept constant, there can be an energy change associated with an increase in the "disorder" represented by the entropy and an energy change when more particles are added (or removed).

The total energy U is determined by adding up the energy of each quantum state multiplied by the number of particles in the state, so that

$$U = \sum_i N_i n_i E_i, \tag{6.2}$$

where N_i is the number of particles allowed in the ith quantum state and n_i is the number of quantum states in the system with an energy E_i. For fermions, $N_i = 1$ or $N_i = 0$ particles per quantum state. For particles such as photons which are classed as bosons, N_i can range from 0, 1, 2, ... up to any integer.

The total number of particles is determined from a summation of the particle number over all the quantum states. We have

$$n_{tot} = \sum_i N_i n_i. \tag{6.3}$$

In statistical physics, the entropy S of a system is determined by the number of ways Q that particles can be arranged in the collection of quantum states of the system. We have that

$$S = k_B \ln Q. \tag{6.4}$$

We show below that this definition of entropy is consistent with the thermodynamic definition stated in terms of the change ΔU of total energy of the particles due to increasing entropy ΔS:

$$\Delta S = \frac{\Delta U}{T}$$

The logarithm in Equation 6.4 is useful in obtaining an approximate value for the entropy. Fortunately, the approximation we will use becomes very accurate when the number n_{tot} of particles is large. The number of distinct arrangements of n_{tot} indistinguishable particles into quantum states (which we are labelling as i) is given by probability theory. We have that the number of distinct arrangements Q is given by the factorial of the total number n_{tot} of particles divided by the product of factorials of the number n_i of particles in each quantum state i of the system[2]:

$$Q = \frac{n_{tot}!}{\prod_i n_i!} \tag{6.5}$$

The total number of ways that n_{tot} particles can be arranged is $n_{tot}!$. We are not concerned with the order of "placing" the particles in each of the quantum states i as we assume that they are indistinguishable, so Equation 6.5 divides the total number of arrangements by the number of ways $n_i!$ that n_i particles can be arranged in each quantum state i. This is done for each quantum state i giving rise to the product in the denominator. Taking the logarithm of Q gives

$$\ln Q = \ln(n_{tot}!) - \sum_i \ln n_i.$$

The Stirling approximation is a convenient simplification for the factorial of n_{tot}. For large x, we have

$$\ln(x!) \approx x \ln x - x.$$

The value of $\ln Q$ becomes

$$\ln Q = n_{tot} \ln n_{tot} - n_{tot} - \sum_i n_i \ln n_i + \sum_i n_i. \tag{6.6}$$

Returning to the energy balance equation (Equation 6.1), we can divide throughout by a small increment Δn_j of the population of n_j quantum states associated with a quantum state j. Taking the limits of small increments and using $S = k_B \ln Q$, we have a differential equation

$$\frac{\partial U}{\partial n_j} = k_b T \frac{\partial \ln Q}{\partial n_j} + \mu \frac{\partial n_{tot}}{\partial n_j}. \tag{6.7}$$

Each of these terms simplifies considerably. Using the summation of all quantum state energies given by Equation 6.2 we have

$$\frac{\partial U}{\partial n_j} = N_j E_j. \tag{6.8}$$

All values in the summation of U are zero in the partial derivative except when $i = j$. Similarly, using Equation 6.3 we have that

$$\mu \frac{\partial n_{tot}}{\partial n_j} = N_j \mu.$$

---

[2] Equation 6.5 assumes that particles have sufficient time and sufficient interaction for all the quantum states to be accessible. For fermions, collisions between particles often ensure equal accessibility, leading to equilibrium distributions. For photons and other bosons, the particles need to interact with electrons (or other fermions).

The partial derivative of $\ln Q$ also simplifies. We can differentiate $\ln Q$ using Equation 6.6. The total number of particles n_{tot} is constant when considering the partial derivative with respect to the population of the jth quantum state and again the summations are only nonzero upon partial differentiation when $i = j$. We have

$$\frac{\partial \ln Q}{\partial n_j} = -\ln n_j. \tag{6.9}$$

The result of these simplifications is

$$N_j E_j = -k_B T \ln n_j + \mu N_j.$$

Rearranging gives an expression for the occupancy of the jth quantum state

$$n_j = \exp\left(\frac{N_j(\mu - E_j)}{k_B T}\right). \tag{6.10}$$

Dropping the subscript j, we then have a probability for the occupation of a quantum state of energy E determined by

$$P(E) = \exp\left(\frac{N(\mu - E)}{k_B T}\right) / n_{tot}. \tag{6.11}$$

For bosons where any number N of particles can occupy the same quantum state, there is no energy involved in adding another particle to the ensemble of particles, so for bosons the chemical potential μ is zero. The average number of bosons in a quantum state (the Bose-Einstein particle distribution function) is determined in Section 6.4 (Equation 6.42), where we consider black-body radiation. In the derivation of the equilibrium radiation field (the black-body distribution), we need the average number of photons per quantum state. For photons, the quantum states are usually referred to as "modes."

For fermions, a quantum state can be either occupied or unoccupied. This criterion is known as the Pauli exclusion principle as it was first formulated by the physicist Wolfgang Pauli. The Pauli exclusion principle can be understood without detailed mathematics as follows. For fermions, the total wave function is antisymmetric, which means that changing the position vector $\mathbf{r}$ to $-\mathbf{r}$ changes the sign of the wavefunction and changes the sign of the particle spin. We can regard an antisymmetric total wavefunction as the criterion for defining a fermion. If two fermions are in the same state, consider a wavefunction representing both fermions. Interchanging the position vector changes nothing for the two-fermion wave function: The total wave function and spin are both unchanged. However, the total wave function and spin for fermions should change sign when the position vector $\mathbf{r}$ changes to $-\mathbf{r}$ as fermions (by definition) have an antisymmetric total wavefunction. The only way for the two-fermion wave function to change sign and remain the same is if the two fermion wavefunction is zero everywhere, which means that the two-fermion state cannot exist. Bosons have a symmetric total wavefunction and remain the same with position vector **r** to −**r** changes and hence any number of bosons can occupy a quantum state.

For fermions, the average occupancy $n(E)$ of a quantum state with energy E is determined by the ratio of the probability of occupation when a particle can be present in the quantum state ($N = 1$) to the addition of the two probabilities when a further occupancy is allowed ($N = 0$) and not allowed ($N = 1$). We obtain

$$n(E) = \frac{P(N=0)}{P(N=1) + P(N=0)}$$

$$= \frac{\exp((\mu - E)/k_B T)}{\exp((\mu - E)/k_B T) + 1}$$

$$= \frac{1}{1 + \exp((-\mu + E)/k_B T)}. \tag{6.12}$$

The particle energy distribution shown in Equation 6.12 is known as the Fermi-Dirac distribution.

We may have several quantum states with the same energy E or an energy within a small range E to $E + dE$. Equation 6.11 shows for fermions that if the energy E is the same for different quantum states then the probability $P(E)$ of occupation is the same. We can allow for the effect on the populations of different quantum states with the same energy by multiplying the average occupancy by a "degeneracy" g (also known as a statistical weight) which is the number of distinct quantum states with the same energy. The average occupancy of a discrete quantum state of energy E with a degeneracy g is given by

$$n(E) = \frac{g}{1 + \exp((-\mu + E)/k_B T)}. \tag{6.13}$$

At low densities and high temperatures when the average occupancy of free electron quantum states is much less than one, the chemical potential μ for free electrons is negative (see Appendix C). The exponential factor in the denominator of Equation 6.13 is then much greater than one and the occupancy of a quantum states simplifies to

$$n(E) = g \exp\left(\frac{\mu - E}{k_B T}\right). \tag{6.14}$$

Free electrons have many quantum states closely spaced in energy so that the quantum states form a "continuum" of energy states. For free electrons, we need to think about the density of quantum states as a function of energy. If the degeneracy or density of states in the energy range E to $E + dE$ is $g(E)$, the population of all quantum states with energy E to $E + dE$ is given by

$$f_{FD}(E)dE = \frac{g(E)dE}{1 + \exp((-\mu + E)/k_BT)}, \tag{6.15}$$

where $g(E)$ is the degeneracy density or density of quantum states at energy E. The degeneracy density $g(E)$ is the number of quantum states per unit energy per unit volume.

The thermodynamic definition of an entropy change ΔS defines entropy in terms of the change of total energy ΔU by specifying that $\Delta S = \Delta U/T$. This definition assumes that the chemical potential remains constant and the temperature remains constant. The statistical physics definition is given by Equation 6.4 with $S = k_B \ln Q$. We now show that the two definitions are consistent.

With constant chemical potential μ and constant temperature T, we can have an increase in entropy S associated with an increase of the total energy U by adding the entropy increase due to an increase in "disorder" (represented by $k_B \ln Q$) and the increase in entropy associated with the addition of particles. We have (see Equation 6.1) that the additional energy associated with an increase of the total particle number n_{tot} is given by $\mu \Delta n_{tot}$. The increase of entropy associated with the increase of particle number is consequently $(\mu/T)\Delta n_{tot}$. Adding the differential of the entropy increase due to disorder plus the entropy increase due to particle number with respect to the total energy U gives

$$\frac{dS}{dU} = \frac{\partial(k_B \ln Q)}{\partial U} + \frac{\mu}{T}\frac{\partial n_{tot}}{\partial U}.$$

We can write for any of the quantum states j that

$$\frac{\partial \ln Q}{\partial U} = \frac{\partial \ln Q}{\partial n_j}\frac{\partial n_j}{\partial U} = (-\ln n_j)\frac{1}{N_j E_j}$$

upon using Equation 6.9 for the differentiation of $\ln Q$ and Equation 6.8 for the differentiation of the total energy U. For the term involving the chemical potential μ, we write that

$$(\mu/T)\frac{\partial n_{tot}}{\partial U} = (\mu/T)\frac{\partial n_{tot}}{\partial n_j}\frac{\partial n_j}{\partial U} = \frac{\mu/T}{E_j}.$$

Using Equation 6.10 for the occupancy of the jth quantum state, namely $n_j = \exp(N_j(\mu - E_j)/k_BT)$, gives the required result that

$$\frac{dS}{dU} = \frac{\partial(k_B \ln Q)}{\partial U} + \frac{\mu}{T}\frac{\partial n_{tot}}{\partial U} = (-\ln n_j)\frac{k_B}{N_j E_j} + \frac{\mu/T}{E_j} = \frac{1}{T}.$$

This last result illustrates that the thermodynamic definition of entropy change $\Delta S = \Delta U/T$ is consistent with the statistical physics definition of entropy $S = k_B \ln Q$.

6.2 The Saha-Boltzmann Equation and Boltzmann Ratio

Two important equilibrium particle distributions are evaluated in this section: the Saha-Boltzmann equation and the Boltzmann ratio. The Saha-Boltzmann equation represents the ratio of the populations of differently charged ionic states, while the Boltzmann ratio gives the ratio of the populations of two quantum states within ions (or atoms) of the same charge.

Meghnad Saha (1893–1956) was an Indian astrophysicist who developed the Saha equation for the equilibrium ratio of the number of ions in different ionization stages in hot plasmas. He was elected a Fellow of the Royal Society in 1927 and after Indian independence from Britain became a member of the Indian parliament in 1952. The Saha equation generalizes to the Saha-Boltzmann equation if ratios between excited states of ions are considered. Ludwig Boltzmann (1844–1906) was an Austrian physicist who developed statistical mechanics and the definition of entropy (see Equation 6.4).

To determine expressions for the Saha-Boltzmann equation and Boltzmann ratio, we need the degeneracy $g(E)$ of the free electron quantum states and can then use Equation 6.15. In this section, we consider the nonrelativistic form of the Saha-Boltzmann equation. Relativistic effects on the Saha-Boltzmann equation are considered in Section 6.3.

The density $g(E)$ of quantum states of free particles like the electrons in a plasma can be evaluated using the idea of the density of states in a three-dimensional cubic cavity or box with infinitely steep potential sides and zero potential within the box. All wavefunctions for particles in a box with infinite potential walls have sinusoidal spatial variations with a node (zero wavefunction value) at the walls of the box (where the potential creating the box becomes infinite). The wavefunctions in a cubic cavity of sides L are often determined in undergraduate quantum mechanics course where it is shown that the volume in momentum p-space occupied by a single quantum state is $(h/(2L))^3$ provided L is much larger than the wavelength of the sinusoidal wavefunctions (see Exercise 6.2).

The degeneracy or number of quantum states of free particles with momentum between p and $p + dp$ per unit volume is given by

$$g(p)dp = 2\frac{(1/8)4\pi p^2 dp}{(h/(2L))^3 L^3} = \frac{8\pi p^2}{h^3}dp. \tag{6.16}$$

The volume in p-space of an annulus of radius p and thickness dp is $4\pi p^2 dp$. We divide here by the p-space volume of a single state $(h/(2L))^3$ and the volume L^3 of the potential box. The factor $1/8$ arises as all components of momentum p_x, p_y, and p_z are positive so only an octant of momentum space should be considered. The factor 2 allow for the two spin orientations for electrons. The degeneracy of quantum states of free particles given in Equation 6.16 is independent of the size L of the potential box. We can allow the potential box size L to approach infinity so as to have a truly "free" particle and still obtain the same density of states.

The number of particles with momentum p is determined by the degeneracy of free particles (Equation 6.16) multiplied by the average occupancy of the quantum states (Equation 6.12). We have for the number $n_p(p)\,dp$ of particles with momentum between p and $p + dp$ per unit volume:

$$n_p(p)dp = g(p)\,n(E)\,dp = \frac{8\pi p^2}{h^3}\frac{1}{1+\exp\left((-\mu + p^2/(2m_0))/k_BT\right)}\,dp \quad (6.17)$$

upon using $E = p^2/(2m_0)$.

The degeneracy $g(E)dE$ of quantum states with energy between E and $E + dE$ is obtained by equating $g(E)dE = g(p)dp$ and noting that $E = p^2/(2m_0)$ and $dE = p/m_0 dp$. The number of free electron quantum states with energies between E and $E + dE$ per unit volume is given by

$$g(E)dE = 4\pi \left(\frac{2m_0}{h^2}\right)^{3/2} E^{1/2}dE. \quad (6.18)$$

The number $n_p(E)dE$ of particles per unit volume with an energy between E and $E + dE$ is found by multiplying the number of states per unit volume $g(E)dE$ by the average occupancy $n(E)$ of the quantum states for a Fermi-Dirac distribution (Equation 6.12). We obtain for the number $n_p(E)$ of particles per unit volume per unit of energy:

$$n_p(E) = 4\pi \left(\frac{2m_0}{h^2}\right)^{3/2} E^{1/2}\frac{1}{1+\exp((-\mu + E)/k_BT)} \quad (6.19)$$

Integrating Equation 6.19 over all energies E, it is possible to determine a relationship between the chemical potential μ of free electrons and the total electron density n_e. We have

$$n_e = \int_0^\infty n_p(E)dE = 4\pi \left(\frac{2m_0}{h^2}\right)^{3/2} \int_0^\infty \frac{E^{1/2}}{1+\exp((-\mu + E)/k_BT)}dE. \quad (6.20)$$

Evaluation of Equation 6.20 involves numerical integration of so-called Fermi-Dirac integrals. Such evaluations show that the chemical potential μ is large and negative at high temperatures T and low densities n_e (see e.g. Figure 13.1 of Tallents [110] and for relativistic particles, see Appendix C). At high temperatures and low densities, the occupation of quantum states is low and does not approach the degeneracy limit. At low density and high temperatures, a particle added to an ensemble of free particles will likely go into a low energy free particle quantum state as the low energy states are not occupied. Adding a particle then reduces the average kinetic energy per particle which means the chemical potential is negative.

The degeneracy of free electrons becomes an important effect at much lower densities than for other particles (e.g. the nuclei of atoms or ions). The much smaller electron mass ensures that for similar temperatures, the electron momentum is much larger and the volume of space occupied by a single free particle quantum state is much larger (see Equation 6.18). It is, consequently, much easier to "fill up" the free electron quantum states.

In plasma physics, we want to relate the chemical potential for free electrons to the more readily measured electron number density (the number of electrons per unit volume). As the chemical potential μ is large and negative at all but the most dense of free electron densities (see Appendix C), the exponential in the denominator for Equation 6.20 is much greater than unity and the electron density n_e can be approximated by

$$
\begin{aligned}
n_e &= 4\pi \left(\frac{2m_0}{h^2}\right)^{3/2} \exp\left(\frac{\mu}{k_B T}\right) \int_0^\infty E^{1/2} \exp\left(-\frac{E}{k_B T}\right) dE \\
&= 2 \left(\frac{2\pi m_0 k_B T}{h^2}\right)^{3/2} \exp\left(\frac{\mu}{k_B T}\right).
\end{aligned}
\tag{6.21}
$$

Rearranging to obtain an expression for the chemical potential μ:

$$
\exp\left(\frac{\mu}{k_B T}\right) = \frac{n_e}{2} \left(\frac{h^2}{2\pi m_0 k_B T}\right)^{3/2}
$$

and hence

$$
\mu = k_B T \ln\left[\frac{n_e}{2} \left(\frac{h^2}{2\pi m_0 k_B T}\right)^{3/2}\right].
\tag{6.22}
$$

Using Equation 6.15 it is possible to determine a general relationship known as the Saha-Boltzmann equation for the ratio of the population of two discrete quantum states. We start by considering a quantum state in an ion of charge Z_i such that the quantum state is at energy ΔE_{ion} below the continuum; that is ΔE_{ion} is the ionization energy of the quantum state in the Z_i charged ion. To ionize the Z_i charged ion to an ion of charge $Z_i + 1$ involves removing an electron to the continuum of

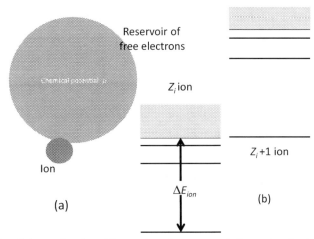

Figure 6.1   Schematic plots illustrating the energy balance in the ionization of a Z_i charged ion to a $Z_i + 1$ charged ion. (a) The chemical potential of the ion changes by $-\mu$ on ionization as an electron is added to the reservoir of free electrons. (b) The energy levels of a schematic Z_i charged ion and $Z_i + 1$ charged ion showing the ionization energy ΔE_{ion} of a quantum state of the Z_i charged ion.

free electron states and requires an energy input equal to ΔE_{ion}. The ionization energy ΔE_{ion} is defined as a positive quantity for our treatment.[3]

Adding an electron to the free electrons changes the free electron chemical potential by $+\mu$ and the ionic chemical potential by $-\mu$. The To obtain the relative abundance of the number density N_{Z_i+1} of $Z_i + 1$ charged ions and the number density N_{Z_i} of Z_i charged ions, we use Equation 6.13 noting that the chemical potential change for the ion is $-\mu$ and the energy change is $+\Delta E_{ion}$ (see Figure 6.1 for schematics illustrating the ionization energy and chemical potential changes on ionization). The ratio of the ion populations is such that

$$\frac{N_{Z_i+1}}{N_{Z_i}} = \frac{g_{Z_i+1}}{g_{Z_i}} \left(\exp\left(-\frac{\mu + \Delta E_{ion}}{k_B T} \right) + 1 \right) P_b, \tag{6.23}$$

where g_{Z_i+1} is the degeneracy of the $Z_i + 1$ charged ion quantum state and g_{Z_i} is the degeneracy of the Z_i charged ion quantum state. The factor P_b represents the fraction of "holes" (vacant quantum states) in the occupancy of the Z_i energy level and so gives a relative probability for the existence of the $Z_i + 1$ state formed by electrons vacating the Z_i energy level.

---

[3] We assume throughout that ionization energies written as ΔE_{ion} are positive quantities. This issue is highlighted because in atomic physics, ionization energies are sometimes quoted as being negative. For example, the ionization energy of the hydrogen atom is often written as $\Delta E_{ion} = -13.6/n^2$ eV, where n is the principal quantum number.

The probability P_b for the existence of the $Z_i + 1$ state is known as a Pauli blocking factor. With low degeneracy where the Z_i state populations have low occupancy, P_b approaches one, whereas if the lower energy quantum state is fully occupied, there are no $Z_i + 1$ states for an electron to be ionized into, so $P_b = 0$. Generally, the blocking factor P_b is one minus the occupancy of the Z_i energy level:

$$P_b = 1 - \left(\exp\left(-\frac{\mu + \Delta E_{ion}}{k_B T} \right) + 1 \right)^{-1} \tag{6.24}$$

Equation 6.23 then reduces to

$$\frac{N_{Z_i+1}}{N_{Z_i}} = \frac{g_{Z_i+1}}{g_{Z_i}} \exp\left(-\frac{\mu}{k_B T} \right) \exp\left(-\frac{\Delta E_{ion}}{k_B T} \right). \tag{6.25}$$

This equation is known as the Saha-Boltzmann equation. This version of the Saha-Boltzmann equation is valid, both when the average occupancy of free electron quantum states approaches unity and when the occupancy approaches zero [110]. The expression is universally valid independent of the occupancy of the quantum states. With an appropriate evaluation of the chemical potential μ, we see in Section 6.3 that Equation 6.25 is also valid when the electrons have relativistic velocities.

When electron occupancy of the free electron quantum states is low, we can use Equation 6.22 to obtain an expression for the Saha-Boltzmann equation using the electron density n_e rather than the chemical potential μ. We obtain the more commonly recognized form for the Saha-Boltzmann equation:

$$\frac{N_{Z_i+1}}{N_{Z_i}} = \frac{g_{Z_i+1}}{g_{Z_i}} \frac{2}{n_e} \left(\frac{2\pi m_0 k_B T}{h^2} \right)^{3/2} \exp\left(-\frac{\Delta E_{ion}}{k_B T} \right) \tag{6.26}$$

Equation 6.25 shows that the population of an excited quantum state within an atom or ion is proportional to $g \exp(-E/k_B T)$, where E is the energy of the quantum state relative to the ground state and g is the degeneracy of the excited quantum state. There is no change of the free electron chemical potential in exciting or de-exciting from one excited quantum state to another within the same ion charge. The ratio of two discrete quantum state populations n per unit volume which we label for the higher energy state with subscript "u" and for the lower energy state with subscript "l" is given by

$$\frac{n_u}{n_l} = \frac{g_u}{g_l} \exp\left(-\frac{\Delta E}{k_B T} \right), \tag{6.27}$$

where the energy difference of the two quantum states is ΔE and g_u and g_l are, respectively, the upper and lower quantum state degeneracies. Equation 6.27 is often referred to as the "Boltzmann ratio" for bound electron quantum states.

6.3 The Relativistic Form of the Saha-Boltzmann Equation

The Saha-Boltzmann equation for the population ratio of two quantum states in ions (or atoms) of different charge depends on the degeneracy of free electrons as a free electron is created when an ion or atom is ionized to a higher charge. In Section 6.2, we determined the degeneracy of free particles by considering a three-dimensional cavity or box with infinitely steep potential sides and zero potential within the box. The allowed quantum states within the box have discrete values of momentum and a degeneracy density per unit volume which is independent of the volume of the box.

The degeneracy of free particles in a three-dimensional cavity or box with infinitely steep potential sides and zero potential within the box are the same in momentum space for particles with relativistic momentum as given for non-relativistic particles (Equation 6.16) provided the box is sufficiently large. It is possible to show for a cavity or box where the box dimensions are much greater than the Compton wavelength $\lambda_C = \hbar/(m_0 c)$ of the particles, that particles with relativistic momentum have the same density of quantum states in momentum space (Equation 6.16) as nonrelativistic particles [1]. The density of states of relativistic particles in a box is examined in Exercise 6.11.

Equation 6.16 is valid for relativistic particles provided we consider the degeneracy density for a box or cavity with sides greater than the Compton wavelength. The Compton wavelength λ_C is 2.426×10^{-12} m for electrons. Considering a box associated with a single electron implies that plasmas with electron densities an order of magnitude or so less than $1/\lambda_C^3 \approx 10^{35}$ m^{-3} have free electron quantum states densities given by Equation 6.16. Such an upper density limit is extremely high. For example, the density of all the electrons in a typical solid is less than 10^{30} m^{-3}. The compressed plasmas required for inertial fusion ignition in the laboratory and found at the center of the Sun have electron densities less than 10^{33} m^{-3} [67], [68], [80].

The degeneracy $g(W)dW$ of relativistic quantum states with energy between W and $W + dW$ is obtained as for nonrelativistic particles by equating $g(W)dW = g(p)dp$. The total relativistic energy is given by Equation 2.73:

$$W = \sqrt{p^2 c^2 + m_0^2 c^4}$$

so that

$$p = \left(\frac{W^2}{c^2} - m_0^2 c^2 \right)^{1/2}.$$

Differentiating the total relativistic energy we find that

$$\frac{dW}{dp} = \frac{c^2 p}{(p^2 c^2 + m_0^2 c^4)^{1/2}} = \frac{c(W^2 - m_0^2 c^4)^{1/2}}{W}.$$

The number of relativistic free electron quantum states with energies between W and $W + dW$ is found by substituting for momentum p and dp/dW into Equation 6.18:

$$g(W)dW = \frac{8\pi p^2}{h^3} \frac{dp}{dW} dW = \frac{8\pi}{h^3} \frac{1}{c^3} \left(W^2 - m_0^2 c^4\right)^{1/2} W dW \qquad (6.28)$$

We convert this expression for the degeneracy of relativistic free electron quantum states to a variation with the Lorentz parameter γ using $W = \gamma m_0 c^2$. The number of relativistic free electron quantum states with energies corresponding to the Lorentz parameter between γ and $\gamma + d\gamma$ per unit volume is

$$g(\gamma)d\gamma = \frac{8\pi}{h^3} \frac{(m_0 c^2)^3}{c^3} \left(\gamma^2 - 1\right)^{1/2} \gamma d\gamma. \qquad (6.29)$$

The number density of relativistic electrons can be related to the chemical potential μ as was undertaken for nonrelativistic electrons (Equation 6.20). We integrate the Fermi-Dirac distribution (Equation 6.15) using the degeneracy given by Equation 6.29. We need to substitute the kinetic energy E by $E = (\gamma - 1)m_0 c^2$ to arrive at an integration with respect to the Lorentz parameter γ:

$$n_e = \frac{8\pi}{h^3} \frac{(m_0 c^2)^3}{c^3} \int_{\gamma=1}^{\infty} \frac{\left(\gamma^2 - 1\right)^{1/2} \gamma}{1 + \exp((-\mu + (\gamma - 1)m_0 c^2)/k_B T)} d\gamma \qquad (6.30)$$

This expression is valid where both relativistic effects and degeneracy effects due to the full occupation of quantum states are important. Plasmas where degeneracy and relativistic velocities are important can only be expected in the hottest and most dense bodies such as the early universe and in, say, dwarf and neutron stars. Equation 6.30 is evaluated in Appendix C.

Relativistic particles occupy high energy free electron quantum states and the number of the accessible states increases with particle energy. For electron densities below, say, the extreme value of $\approx 10^{35}$ m^{-3}, relativistic thermal distributions of particles have a large negative chemical potential (see Appendix C). The denominator in Equation 6.30 then has the exponential term much greater than one and we can write that

$$n_e = \frac{8\pi}{h^3} \frac{(m_0 c^2)^3}{c^3} \exp\left(\frac{\mu + m_0 c^2}{k_B T}\right) \int_{\gamma=1}^{\infty} \left(\gamma^2 - 1\right)^{1/2} \gamma \exp\left(-\frac{\gamma m_0 c^2}{k_B T}\right) d\gamma. \qquad (6.31)$$

The integration is soluble as a special function:

$$\int_{\gamma=1}^{\infty} \gamma \left(\gamma^2 - 1\right)^{1/2} \exp\left(-\frac{\gamma m_0 c^2}{k_B T}\right) d\gamma = \frac{k_B T}{m_0 c^2} K_2(m_0 c^2 / k_B T), \qquad (6.32)$$

where $K_2(m_0 c^2 / k_B T)$ is a modified Bessel function of the second kind and order two evaluated at an inverse temperature value of $m_0 c^2 / k_B T$. The electron density n_e is then related to the chemical potential μ by

$$n_e = \frac{8\pi}{h^3} \frac{(m_0 c^2)^2}{c^3} \exp\left(\frac{m_0 c^2}{k_B T}\right) \exp\left(\frac{\mu}{k_B T}\right) K_2(m_0 c^2 / k_B T) \, k_B T. \qquad (6.33)$$

The exponentiated factor containing the chemical potential μ is then given by

$$\exp\left(-\frac{\mu}{k_B T}\right) = \frac{8\pi}{h^3} \frac{(m_0 c^2)^2}{c^3} \exp\left(\frac{m_0 c^2}{k_B T}\right) K_2(m_0 c^2 / k_B T) \frac{k_B T}{n_e}. \qquad (6.34)$$

The chemical potential for a nondegenerate relativistic distribution of particles in an equilibrium state can be written as

$$\mu = -k_B T \, \ln\left[\frac{8\pi}{h^3} \frac{(m_0 c^2)^2}{c^3} \exp\left(\frac{m_0 c^2}{k_B T}\right) K_2(m_0 c^2 / k_B T) \frac{k_B T}{n_e}\right]. \qquad (6.35)$$

Using the general form of the Saha-Boltzmann equation (Equation 6.25) we can substitute Equation 6.34 for the term containing the chemical potential μ:

$$\frac{N_{Z_i+1}}{N_{Z_i}} = \frac{g_{Z_i+1}}{g_{Z_i}} \exp\left(-\frac{\mu}{k_B T}\right) \exp\left(-\frac{\Delta E_{ion}}{k_B T}\right)$$

$$= \frac{g_{Z_i+1}}{g_{Z_i}} \frac{8\pi}{h^3} \frac{(m_0 c^2)^2}{c^3} \exp\left(\frac{m_0 c^2}{k_B T}\right) K_2(m_0 c^2 / k_B T) \frac{k_B T}{n_e} \exp\left(-\frac{\Delta E_{ion}}{k_B T}\right)$$

$$\qquad (6.36)$$

Equation 6.36 is the relativistic form of the Saha-Boltzmann equation. Equation 6.36 enables an evaluation of the equilibrium ratio of the population of a $Z_i + 1$ charged ion relative to the population of a bound quantum state in a Z_i charged ion when electron temperatures $k_B T$ are sufficiently high that relativistic effects need to be taken into account.

We can evaluate the electron temperatures $k_B T$ needed for relativistic effects to be important in the relativistic Saha-Boltzmann equation by considering a large amplitude asymptotic expression for the modified Bessel function $K_2(m_0 c^2 / k_B T)$ (see Appendix B). When $k_B T < m_0 c^2$, we have that

$$K_2(m_0 c^2 / k_B T) \approx \frac{\sqrt{\pi}}{2} \left(\frac{2 k_B T}{m_0 c^2}\right)^{1/2} \exp\left(-\frac{m_0 c^2}{k_B T}\right) \left[1 + \frac{15}{8} \frac{k_B T}{m_0 c^2}\right]. \qquad (6.37)$$

Substituting this expression for the modified Bessel function into Equation 6.36 gives an approximate expression for the Saha-Boltzmann equation valid when relativistic effects are small, but not insignificant. We obtain the Saha-Boltzmann expression for nonrelativistic electrons (Equation 6.26) multiplied by a relativistic correction factor:

$$\frac{N_{Z_i+1}}{N_{Z_i}} \approx \frac{g_{Z_i+1}}{g_{Z_i}} \frac{2}{n_e} \left(\frac{2\pi m_0 k_B T}{h^2} \right)^{3/2} \exp\left(-\frac{\Delta E_{ion}}{k_B T} \right) \left[1 + \frac{15}{8} \frac{k_B T}{m_0 c^2} \right] \quad (6.38)$$

Here the relativistic correction factor is contained within the square bracket.

Using Equation 6.38, we can estimate that when $k_B T \approx 0.05\, m_0 c^2$, relativistic effects increases the Saha-Boltzmann ratio N_{Z_i+1}/N_{Z_i} from the non-relativistic limit by a factor of $\approx 10\%$. A temperature $k_B T = 0.05\, m_0 c^2$ corresponds approximately to a temperature 26 keV for electrons. Temperatures ≈ 26 keV are required for efficient production of fusion energy in inertial fusion [67], [68], [80]. In inertial fusion, the ionization is determined by the Saha-Boltzmann equation as densities are high (peaking at $n_e \approx 5 \times 10^{32}$ m^{-3} [67], [68], [80]). Our treatment here shows that relativistic effects should be included in some inertial fusion ionization calculations.

6.4 Black-body Radiation

Electrons, protons, the nuclei of atoms and matter in general exhibit wave-like properties (the quantum mechanical wavefunction), but for many applications, material bodies can be treated as particle-like point sources of mass and charge. In a kind of symmetry, light has a wave-like nature and a particle-like nature. The complementarity principle discussed as a footnote in our initial wave treatment of light (Section 1.1) states that both light and matter exhibit wave-like and particle-like properties depending on the measurement or interaction being considered.

Light particles are now known as photons. The particle nature of light is apparent in the photo-electric effect and was discussed originally by Einstein [34] who used the term "light quanta" (*lichtquant* in German).[4] In the photo-electric effect, light irradiated material only emits an electron when the energy of a photon of the light exceeds the binding energy of the material. Light is absorbed by matter in units of energy and if the unit of light energy is less than the quantum differences for the energies in the material irradiated, the light is not absorbed.

In a more theoretical approach, the particle-like nature of light is seen to arise from time-dependent solutions of the Schrodinger equation (see Section 2.9) and,

---

[4] The term "photon" was introduced in a paper published in 1926 [66], though the term "light quanta" and, particularly the idea of "virtual quanta" rather than "virtual photons" (see Section 7.7) continued to be used for many years (see e.g. Jackson [58]).

as we show in the section, from time-dependent solutions of the wave equation (Equation 1.5). In the wave equation approach, the wave equation for a radiation field can be shown to transform to a harmonic oscillator variation in time which has discrete energy eigenvalues that are equally spaced for each frequency. Each unit of the equally spaced energy eigenvalues represents a photon.

We now expand on this argument for the particle-like properties of light. The wave equation for light in vacuum (Equation 1.5) can be written as

$$\nabla^2 \mathbf{E} - \frac{1}{c^2}\frac{\partial^2 \mathbf{E}}{\partial t^2} = 0. \tag{6.39}$$

The electric field $\mathbf{E}$ for a light field varies spatially and temporally. The spatial variation in $\mathbf{r}$ is characterized by a wave vector $\mathbf{k}$ such that $\mathbf{E}(t) = \mathbf{E}_0(t)\exp(i\mathbf{k}\cdot\mathbf{r})$. Substituting into Equation 6.39, we have

$$-k^2\mathbf{E}_0(t) - \frac{1}{c^2}\frac{\partial^2 \mathbf{E}_0(t)}{\partial t^2} = 0.$$

Using $c = \omega/k$, where ω is the angular frequency of the light, the variation of the electric field amplitude in time reduces to

$$\frac{\partial^2 \mathbf{E}_0(t)}{\partial t^2} = -\omega^2 \mathbf{E}_0(t). \tag{6.40}$$

Equation 6.40 has the form of an equation for a harmonic oscillator of frequency ω. Classical harmonic oscillators include the oscillation of a pendulum and the oscillation of a vibrating spring. Quantum mechanical solutions are often solved in undergraduate quantum mechanics courses (see e.g. Appendix A3 in Tallents [110]). Like all quantum mechanical energies, the energy eigenvalues are quantized. The energy eigenvalue solutions for the harmonic oscillator equation are such that

$$E_{n_p} = \left(n_p + \frac{1}{2}\right)\hbar\omega = \left(n_p + \frac{1}{2}\right)h\nu, \tag{6.41}$$

where n_p is an integer, ν is the frequency of the light, h is Planck's constant, and $\hbar = h/(2\pi)$. Equation 6.41 illustrates that the energies of a radiation field are quantized in units of $h\nu$. Each quanta $h\nu$ of radiation is known as a photon. The integer n_p represents the number of photons of frequency ν in the light field.

The German physicist Max Planck (1858–1947) gives his name to the proportionality constant h between the frequency of radiation and the energy of the quanta of radiation. Planck was awarded the 1918 Nobel prize in physics for "... his discovery of energy quanta" (as cited by the Nobel committee).

The equilibrium radiation field for light is known as black-body radiation or as Planckian radiation (in honor of Max Planck). When a radiation field interacts strongly with particles of matter via emission and absorption processes, the

radiation field can acquire a distribution of the number of photons between modes
with a characteristic temperature equal to the temperature of the matter. The black-
body distribution of photon energies is obtained by first considering the density
of possible wavevectors **k** in a spatial cavity (known as the "density of modes").
The black-body radiation intensity is found by multiplying the density of modes
by the mean number of photons per mode and the energy of each photon. Photons
are classed as bosons, so it is possible to have any number of photons in a mode.
The actual number of photons in a mode for black-body radiation is determined by
the characteristic temperature of the radiation field. We undertake an evaluation of
the black-body radiation intensity in the frame of reference of the radiation field.

The probability of the presence of n_p photons with energy $h\nu$ for a radiation
temperature of T is needed in order to determine the equilibrium radiation field. As
photons are bosons, the chemical potential is zero. Adding a photon to a collection of photons does not release or require any energy. For bosons, the parameter
N in Equation 6.11 can represent any number of photons in the same mode. The
probability of having n_p photons per mode follows from Equation 6.11:

$$P_{n_p} = \frac{\exp(-n_p h\nu/k_B T)}{\sum_{n'=0}^{\infty} \exp(-n' h\nu/k_B T)} = \left(1 - \exp\left(\frac{-h\nu}{k_B T}\right)\right) \left[\exp\left(-\frac{h\nu}{k_B T}\right)\right]^{n_p}$$

The summation in the denominator represents the total occupancy of all quantum states and can be evaluated using the solution for an infinite series given by
$\sum_{n'=0}^{\infty} r^{n'} = 1/(1 - r)$. The average number of photons n_{av} per mode is then given
by

$$n_{av} = \sum_{n_p=0}^{\infty} n_p P_{n_p} = \left(1 - \exp\left(\frac{-h\nu}{k_B T}\right)\right) \sum_{n_p=0}^{\infty} n_p \left[\exp\left(-\frac{h\nu}{k_B T}\right)\right]^{n_p}.$$

The infinite series $\sum_{n=0}^{\infty} n r^n = r/(1-r)^2$ so

$$n_{av} = \frac{\exp\left(-h\nu/k_B T\right)}{1 - \exp(-h\nu/k_B T)}$$

which simplifies to

$$n_{av} = \frac{1}{\exp(h\nu/k_B T) - 1}. \tag{6.42}$$

This value of n_{av} can be generalized to represent the average occupancy of bosons
in a quantum state of energy E after replacing $h\nu$ by E. Equation 6.42 then becomes
the Bose-Einstein particle distribution function. There is a resemblance to the average occupancy of fermions (the Fermi-Dirac distribution given by Equation 6.12),

except the minus in the denominator for bosons becomes a plus for fermions and we can set the chemical potential μ to zero for bosons.

The density of modes for a radiation field can be obtained by considering the density of modes inside a defined space. Consider the space inside a conducting cube of sides L. A steady-state radiation field requires that the electromagnetic field has nodes so that the electric field is zero at the walls of the cube. With electric field nodes of the electromagnetic waves at the walls, the number of half-wavelengths of the light directed parallel to a wall is an integer. In any direction parallel to a wall, the wavevector of the electromagnetic field in that direction has a value of some integer times π/L. The volume of a single mode in k-space is thus $(\pi/L)^3$.

The number $\rho_k dk$ of modes per unit volume with a wavenumber $k = |\mathbf{k}|$ between k and $k + dk$ is given by

$$\rho_k dk = 2\frac{(1/8)4\pi k^2 dk}{(\pi/L)^3 L^3}.$$

The volume $4\pi k^2 dk$ in k-space between k and $k + dk$ is here divided by the volume $(\pi/L)^3$ of a mode in k-space and the volume L^3 of the conducting cube. The factor $1/8$ arises as only positive k_x, k_y, and k_z have meaning and need to be counted, while the factor 2 accounts for the possible two components of polarization for an electromagnetic wave. Simplifying and converting wavenumber $k = 2\pi v/c$ to frequency v, the density $\rho_v dv$ of modes per unit volume with a frequency between v and $v + dv$ becomes

$$\rho_v dv = \frac{8\pi v^2}{c^3} dv. \tag{6.43}$$

The density of modes given by Equation 6.43 is independent of the size L of the assumed cavity and so is universally applicable for all cavities. Allowing the size L of the cavity to approach infinity means that the density of modes applies to a freely propagating radiation field, such as the cosmic microwave background. The density of modes is also applicable to some small black-body cavities, such as the "hohlraums" used in inertial fusion research (see Exercise 6.4). More generally, the mode density (Equation 6.43) applies to any radiation field, though for a radiation field which is not a black-body, modes may have zero occupancy of photons.

The black-body radiation energy density $W_p(v)dv$ per unit volume in a frequency between v and $v + dv$ is given by a multiplication of the density of modes ρ_v, the energy hv per photon and the mean or average number n_{av} of photons per mode for an equilibrium distribution of photons. Using Equation 6.42 for the mean number of photons per mode, we have for the black-body radiation energy density that

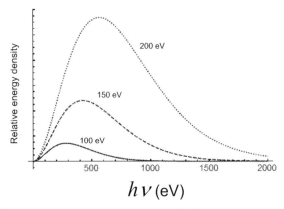

$$h\nu\,(\text{eV})$$

Figure 6.2   The relative energy density per unit photon energy per unit volume for black-body radiation as a function of photon energy. Three sample radiation temperatures (100 eV, 150 eV, and 200 eV) as indicated are plotted.

$$W_p(\nu)d\nu = \rho_\nu\,n_{av}\,h\nu\,d\nu = \frac{8\pi h\nu^3}{c^3}\frac{1}{\exp(h\nu/k_B T) - 1}d\nu. \qquad (6.44)$$

The radiation energy density for a black-body radiation field as a function of photon energy for some sample temperatures is illustrated in Figure 6.2.

In unit time, the black-body radiation energy density within a distance equal to c will pass through a surface. The radiation energy density $W_p(\nu)d\nu$ is consequently equal to $I_{bb}(\nu)\,d\nu/c$, where $I_{bb}(\nu)$ is the intensity of the black-body radiation. The black-body intensity $I_{bb}(\nu)$ calculated in this way represents the power per unit area per unit frequency over all possible angles (4π steradian). This black-body radiation intensity in the frame of the radiation is given by

$$I_{bb}(\nu) = \frac{8\pi h\nu^3}{c^2}\frac{1}{\exp(h\nu/k_B T) - 1}. \qquad (6.45)$$

Black-body radiation $I_{bb}(\nu)$ has a maximum at a photon energy of $h\nu = 2.8k_B T$. Integrating the black-body spectrum over all frequencies produces a spectrally integrated intensity varying as $(k_B T)^4$ (see Exercise 6.3). The spectrally integrated flux of radiation incident per unit area on the walls of a black-body is given by $\sigma_{SB} T^4$, where $\sigma_{SB} = 5.78 \times 10^{-8}$ W m^{-2} K^{-4} is known as the Stefan-Boltzmann constant. In a time δt, one quarter of the radiation energy from a black-body cavity volume of $A\,c\,\delta t$ impinges on the surface area A of black-body cavity (three quarters of the radiation is propagating at angles which do not strike the surface in a time δt, see Exercise 6.7). Using Equation 6.44, this means that the Stefan-Boltzmann constant is related to some of the constants determining the black-body radiation energy density by

$$\sigma_{SB} = \frac{8\pi h}{c^3} \frac{c}{4} \frac{k_B^4}{h^4} \frac{\pi^4}{15} = \frac{8\pi^5}{60} \frac{k_B^4}{h^3}. \tag{6.46}$$

The integration of the frequency v dependent quantities in Equation 6.44 over all frequency reduces to a value $k_B^4 \pi^4/(15h^4)$.

We show in Section 7.1 that I/v^3 is invariant under a Lorentz transformation. The black-body radiation intensity component $8\pi h v^3/c^2$ is consequently invariant under a Lorentz transformation as $I_{bb}(v) \propto v^3$. However, when the radiation is detected in other frames of reference, a Doppler shift in the observed frequencies v' occurs such that $v/v' = \gamma(1 + v\cos\theta'/c)$ (see Equation 2.47). Black-body radiation intensities in a frame of reference S' moving relative to a black-body source (frame S)[5] appear with a Doppler shift such that

$$I'_{bb}(v') = \frac{8\pi h v'^3}{c^2} \frac{1}{\exp\left[(hv'/k_B T)\ \gamma(1 + v\cos\theta'/c)\right] - 1}. \tag{6.47}$$

Equation 6.47 indicates that the temperatures T' observed in a frame moving relative to the black-body radiation source are such that

$$T' = \frac{T}{\gamma(1 + v\cos\theta'/c)}, \tag{6.48}$$

where T is the temperature of the black-body radiation in the frame of the radiation source, v is the relative velocity of the observer to the black-body emission and θ' is the angle of observation to the relative velocity in the moving observers frame of reference.

The cosmic microwave background (CMB) is an example of near-isotropic black-body radiation [41], [92], [108]. In the early universe, particle and radiation temperatures were in equilibrium due to frequent interactions via photon absorption, emission and scatter. As the universe expanded, the likelyhood of photon absorption and scatter reduced with Thomson scatter of photons with free electrons (see Section 4.6) acting as the last significant photon/particle interaction process. At a time 378,000 years after the Big Bang, the universe had cooled to a temperature $T \approx 3,000$ K and almost all electrons had recombined with protons to form either atomic hydrogen (76% abundance) or atomic helium (24% abundance). Further Thomson scatter off free electrons then ceased and the CMB photons of temperature $T \approx 3,000$ K with $k_B T \approx 0.26$ eV had insufficient energy to ionize hydrogen or helium as hydrogen has an ionization energy of 13.6 eV and helium an ionization

---

[5] The proper frame of reference is here frame S. Elsewhere, we usually have frame S' as the proper frame moving with the measured quantity.

energy of 24.59 eV. The CMB stopped interacting with particles and simply propagated freely in space with a constant temperature in the frame of reference of the
last Thomson scattering event.

We are now moving away from the last scattering events that created the CMB
at a velocity just below the speed of light. The angle θ' of the present observations
of the CMB relative to the CMB frame velocity in our frame of reference on planet
earth is close to zero radians. Equation 6.48 for the observed CMB temperature T'
then becomes

$$T' = \frac{T}{\gamma(1+v/c)} = T\frac{(1-(v/c)^2)^{1/2}}{1+v/c} = T\frac{((1-v/c)(1+v/c))^{1/2}}{1+v/c}$$

$$= T\left(\frac{1-v/c}{1+v/c}\right)^{1/2} = \frac{T}{1+z} \tag{6.49}$$

upon using the expression for the cosmological redshift z given by Equation 2.52,
namely

$$1+z = \left(\frac{1+v/c}{1-v/c}\right)^{1/2}.$$

The detected CMB temperature ($T' = 2.725$ K) varies from the emitted temperature
T by a large factor $1/(z+1)$ with $z \approx 1,100$ due to the expansion of the universe.

As well as a $1/(z+1) \approx 1/1,100$ drop in the measured CMB temperature due to
the expansion of the universe, there is also has a much smaller temperature variation associated with the velocity of the solar system relative to the CMB and even
smaller variations due to tiny temperature variations in the initial CMB. When
searching for small anisotropies in the CMB temperature due to anisotropies in
the early universe[6], a relativistic effect due to the solar system velocity relative
to the CMB at the last scattering event needs to be taken into account. The CMB
temperature is $+0.00335$ Kelvin viewed in directions towards the constellation Leo
and -0.00335 Kelvin viewed in directions π radians opposite to Leo towards the
constellation Aquarius. This angular variation of the CMB arises because the local group of galaxies move together with a velocity of ≈ 600 km s^{-1} relative to the
cosmological microwave background [92]. Exercise 6.14 considers an approximate
relationship between the velocity of the local group of galaxies relative to the CMB
and the CMB temperature anisotropy.

---

[6] Small angular anisotropy in the CMB arises due to small temperature fluctuations of matter at the time of the
last photon scattering event, now imprinted as small variations of the CMB temperature as a function of the
angle of observation [92]. The observed small anisotropies of the near-uniform early universe are believed to
be the precursors to the formation of galaxies and other nonuniformities in the present universe.

6.5 The Maxwellian Distribution

Free particles at moderate to high temperatures not bound in a potential have closely spaced quantized energies approaching a continuum of possible energy states. In this section, we develop the particle distribution function appropriate for free fermionic particles at modest temperatures. We consider the non-relativistic expression for the distribution of particle speeds and energy where all significant particle speeds are much less than the velocity of light. Such nonrelativistic particle distribution functions are known as Maxwellian or Maxwell-Boltzmann distributions. Later this treatment of free particle distribution function is generalized (in Section 6.6) to high temperatures where the particle velocities approach the speed of light.

This average occupancy $n(E)$ of a quantum state given by Equation 6.12 can be utilized to obtain an expression for the distribution of the speeds and energies of free fermions. Assuming a three-dimensional distribution of particle velocities, the number of particles per unit volume $f_v(v)dv$ with speed between v and $v + dv$ is given by the proportionality

$$f_v(v)dv \propto 4\pi v^2 n(E)dv,$$

where $E = (1/2)m_0 v^2$ is the particle energy for rest mass m_0. The factor $4\pi v^2 dv$ is the velocity space volume corresponding to the speed range v to $v + dv$ given by the volume of a shell of radius v and thickness dv. The expression for the particle distribution of speeds can then be written as

$$f_v(v)dv \propto 4\pi v^2 \frac{dv}{1 + \exp\left(((1/2)m_0 v^2 - \mu)/(k_B T)\right)}. \tag{6.50}$$

To convert the proportionality constant here to an absolute value of the distribution of speeds requires normalization. We choose to require that integrating over all possible speeds gives a value of unity, so that the particle distribution function represents the probability of finding a particle with speed in the range v to $v + dv$. We then have that

$$\int_0^\infty f_v(v)dv = 1.$$

The probability distribution function with this normalization will give the number of particles per unit volume with speeds between v and $v + dv$ if we multiple $f_v(v)dv$ by the number density of all particles (of any velocity). The integrations in the general case are not analytic, but at high temperatures, the chemical potential is large and negative,[7] so the exponential in the denominator of Equation 6.50 becomes much larger than one and we neglect the one. We write that

---

[7] The chemical potential μ for free particles is discussed in Section 6.2 and Appendix C.

$$f_v(v)dv \propto 4\pi v^2 \frac{dv}{1 + \exp\left(((1/2)m_0 v^2 - \mu)/(k_B T)\right)} \propto 4\pi v^2 \exp\left(-\frac{m_0 v^2}{2k_B T}\right) dv.$$

After normalization so that $\int_0^\infty f_v(v)dv = 1$, we obtain the Maxwellian distribution of speeds with

$$f_v(v)dv = \left(\frac{m_0}{2\pi k_B T}\right)^{\frac{3}{2}} 4\pi v^2 \exp\left(-\frac{m_0 v^2}{2k_B T}\right) dv. \tag{6.51}$$

By equating the number $f_E(E)dE$ of particles with energy between E and $E + dE$ to the Maxwellian distribution of speeds using $f_v(v)dv = f_E(E)dE$ (and noting that $E = (1/2)m_0 v^2$ and $dE = m_0 v dv$), we can obtain an expression for the Maxwellian distribution of energies:

$$f_E(E)dE = \frac{2}{\sqrt{\pi}} \left(\frac{1}{k_B T}\right)^{\frac{3}{2}} E^{1/2} \exp\left(-\frac{E}{k_B T}\right) dE \tag{6.52}$$

Many plasmas have equilibrium distributions given by the Maxwellian distribution. In equilibrium, the absolute values of the particle velocity and energy distributions are proportional to the total number density, but the probability distribution with velocity and energy varies only with the temperature.

6.6 Maxwell-Juttner Distributions for Relativistic Particles

At very high temperatures, the Maxwellian distribution (Equation 6.51) can predict particle velocities exceeding the speed of light. Clearly, an alternative expression allowing for relativistic effects as velocities approach the speed of light is needed. The defining temperature where this is necessary is usually regarded as being extremely high, with temperature $k_B T$ values approaching the rest mass energy $m_0 c^2$ of the particles.

Just as for nonrelativistic particles, relativistic particles have statistical population distributions.[8] Temperatures where particles have velocities approaching the speed of light have statistical distributions f_{MJ} varying as $f_{MJ} \propto \exp(-E/k_B T)$ with a kinetic energy $E = (\gamma - 1)m_0 c^2$. Rather than consider the distribution of speeds as we did for non-relativistic particles (Equation 6.51), we consider the distribution of particle momentum. Assuming a three-dimensional distribution of the particle momentum of free particles, the number of particles per unit volume $f_{MJ}(p)dp$ with momentum between p and $p + dp$ is given by the proportionality

---

[8] Thermodynamics and statistical mechanics are inherently nonlocal in space and time which is a particular conceptual problem when individual particle velocities approach the speed of light and have different frames of reference. Nevertheless, detailed treatments show that relativistic particles can be treated statistically (see the discussion by Dunkel et al. [27]).

$$f_{MJ}(p)dp \propto 4\pi p^2 \exp\left(-\frac{(\gamma-1)m_0 c^2}{k_B T}\right)dp.$$

We want to convert this distribution to an energy distribution. For relativistic particles, the energy is readily measured in terms of the value of the Lorentz parameter
γ so we write that

$$f_{MJ}(\gamma)d\gamma \propto 4\pi p^2 \exp\left(-\frac{(\gamma-1)m_0 c^2}{k_B T}\right)\frac{dp}{d\gamma}d\gamma.$$

The particle momentum $p = \gamma m_0 v = \gamma m_0 c \beta$, where $\beta = v/c$. Recalling that $\gamma = 1/\sqrt{1-v^2/c^2} = 1/\sqrt{1-\beta^2}$, a little manipulation shows that

$$\beta = \frac{1}{\gamma}\left(\gamma^2 - 1\right)^{1/2}.$$

The differentiation of the momentum yields

$$\frac{dp}{d\gamma} = m_0 c\frac{d(\gamma\beta)}{d\gamma} = m_0 c\gamma\left(\gamma^2 - 1\right)^{-1/2} = \frac{1}{\beta}.$$

The particle energy distribution function for relativistic particles then has the following proportionality:

$$f_{MJ}(\gamma) \propto (\gamma\beta)^2\frac{dp}{d\gamma}\exp\left(-\frac{(\gamma-1)m_0 c^2}{k_B T}\right) = \gamma^2\beta\exp\left(-\frac{(\gamma-1)m_0 c^2}{k_B T}\right)$$

$$= \gamma\left(\gamma^2 - 1\right)^{1/2}\exp\left(-\frac{(\gamma-1)m_0 c^2}{k_B T}\right)$$

$$= \gamma\left(\gamma^2 - 1\right)^{1/2}\exp\left(\frac{m_0 c^2}{k_B T}\right)\exp\left(-\frac{\gamma m_0 c^2}{k_B T}\right) \qquad (6.53)$$

The normalization of the particle energy distribution function removes the proportionality in Equation 6.53 and proceeds by requiring that

$$\int_{\gamma=1}^{\infty} f_{MJ}(\gamma)d\gamma = 1.$$

We employ a normalization term equal to $1/[\exp(m_0 c^2/k_B T)\,Z(T)]$ with $Z(T)$
given by

$$Z(T) = \int_{\gamma=1}^{\infty} \gamma\left(\gamma^2 - 1\right)^{1/2}\exp\left(-\frac{\gamma m_0 c^2}{k_B T}\right)d\gamma = \frac{k_B T}{m_0 c^2}K_2(m_0 c^2/k_B T), \quad (6.54)$$

where $K_2(m_0 c^2/k_B T)$ is a modified Bessel function of the second kind and order
two evaluated at an inverse temperature value of $m_0 c^2/k_B T$. The factor
$\exp(m_0 c^2/k_B T)$ in the denominator of our normalization term cancels the equivalent term in our proportionality expression for $f_{MJ}(\gamma)$ (Equation 6.53). The variation of the normalization parameter $Z(T)$ with temperature is shown in Figure 6.3.

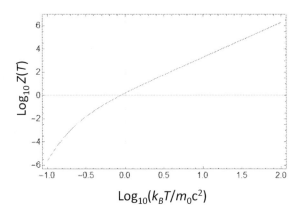

Figure 6.3   The parameter $Z(T)$ as a function of temperature in units of the rest mass energy. $Z(T)$ is used to normalize the Maxwell-Juttner particle distribution function. Both axes are plotted on logarithmic scales.

After multiplication by the normalization term, the Maxwell-Juttner particle distribution function for free relativistic particles with an isotropic (three-dimensional) statistical distribution is given by

$$f_{MJ}(\gamma) = \frac{1}{Z(T)}\gamma \left(\gamma^2 - 1\right)^{1/2} \exp\left(-\frac{\gamma m_0 c^2}{k_B T}\right). \tag{6.55}$$

The Maxwell-Juttner expression is given here as a function of the relativistic Lorentz factor γ. The distribution can be readily converted to a distribution function in terms of the kinetic energy E using $E = (\gamma - 1)m_o c^2$, where $m_o c^2$ is the rest mass energy of the particles.

In the limit of temperatures T where the particle velocities are well below the speed of light (where γ approaches unity), Equation 6.55 is equivalent to the Maxwellian distribution (Equation 6.52). Sample Maxwell-Juttner distribution functions are compared to Maxwellian distributions in Figure 6.4. Figure 6.4a shows that at temperatures of $0.1m_0 c^2$ (corresponding to a temperature of 51 keV for electrons), there is a small discrepancy between the Maxwellian and the Maxwell-Juttner distributions. From Figure 6.4b we see that at temperatures of $10m_0 c^2$ (corresponding to a temperature of 5.1 MeV for electrons), the difference between the Maxwellian and Maxwell-Juttner distributions is significant. The behavior of the Maxwell-Juttner distribution over a large dynamic range is shown in 6.5. There is a cut-off in the maximum kinetic energy of particles which increases linearly with temperature (maximum kinetic energy $\propto$ maximum $\gamma - 1 \propto$ temperature). The maximum kinetic energy $\approx 25k_B T$. The peak of the Maxwell-Juttner distribution for highly relativistic particles can be shown to occur when $\gamma = 2k_B T/m_0 c^2$ (see Exercise 6.12).

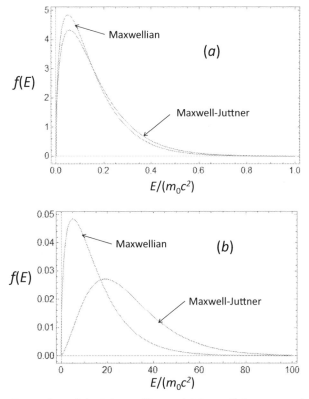

Figure 6.4   Examples of the Maxwellian and Maxwell-Juttner particle distribution functions for temperatures (a) $k_B T = 0.1 m_0 c^2$, and (b) $k_B T = 10 m_0 c^2$. The distribution functions are plotted as a function of the particle kinetic energy in units of the particle rest mass energy.

With relativistic particles, temperatures in different directions can sometimes be significantly different. A particle acceleration process may produce a relativistically high temperature in a single direction with a Maxwellian distribution at a more modest temperature perpendicular to the acceleration direction. Collision cross-sections between particles drop rapidly at high particle energies, so an acceleration process can dominate the particle distribution function. For example, in laser-produced plasmas, significant potential changes can arise on the surfaces of planar targets which accelerate ions and electrons perpendicular to the surface planes, but have little affect on motion parallel to the planes (see Section 5.2.2). Perpendicular to the target planes, ions and electrons can be accelerated to relativistic energies. It is useful to consider a one-dimensional statistical distribution of particles at temperatures when the particle velocities are affected by relativistic effects.

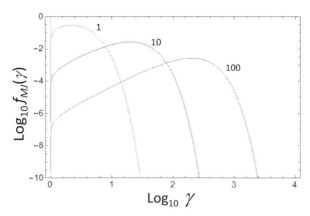

Figure 6.5   The Maxwell-Juttner particle distribution function for relativistic particles plotted as a function of the Lorentz γ parameter for the particles. A value of
$\gamma - 1$ is equal to the particle kinetic energy in units of the rest mass energy $m_0 c^2$.
Curves are shown for temperatures $k_B T = 1 \times m_0 c^2$, $10 \times m_0 c^2$ and $100 \times m_0 c^2$ as
labeled. Both axes are plotted on logarithmic scales.

For a one-dimensional distribution of particles, the number of particles $f_{MJ}(p)dp$
with momentum between p and $p + dp$ is given by the proportionality

$$f_{MJ}(p)dp \propto \exp\left(-\frac{(\gamma - 1)m_0 c^2}{k_B T}\right) dp.$$

For a one-dimensional distribution, the particle energy distribution function for
relativistic particles has the following proportionality Figure 6.5:

$$f_{MJ}(\gamma) \propto \frac{dp}{d\gamma} \exp\left(-\frac{(\gamma - 1)m_0 c^2}{k_B T}\right) = \frac{1}{\beta} \exp\left(-\frac{(\gamma - 1)m_0 c^2}{k_B T}\right)$$

$$= \gamma \left(\gamma^2 - 1\right)^{-1/2} \exp\left(-\frac{(\gamma - 1)m_0 c^2}{k_B T}\right)$$

$$= \gamma \left(\gamma^2 - 1\right)^{-1/2} \exp\left(\frac{m_0 c^2}{k_B T}\right) \exp\left(-\frac{\gamma m_0 c^2}{k_B T}\right) \qquad (6.56)$$

We employ a normalization term $1/[\exp(m_0 c^2/k_B T) Z_{1D}(T)]$ with a revised $Z_{1D}(T)$
given by

$$Z_{1D}(T) = \int_{\gamma=1}^{\infty} \gamma \left(\gamma^2 - 1\right)^{-1/2} \exp\left(-\frac{\gamma m_0 c^2}{k_B T}\right) d\gamma = K_1(m_0 c^2/k_B T), \qquad (6.57)$$

where $K_1(m_0 c^2/k_B T)$ is a modified Bessel function of the second kind and order
one evaluated at an inverse temperature value of $m_0 c^2/k_B T$. The Maxwell-Juttner
particle distribution function for free relativistic particles with a statistical distribution at a high temperature in only one dimension is given by

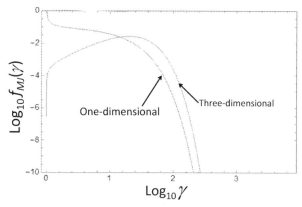

Figure 6.6 The Maxwell-Juttner particle distribution function for relativistic particles plotted as a function of the Lorentz γ parameter for the particles assuming a one-dimensional and three-dimensional distribution (as labeled). The value of $\gamma - 1$ is equal to the particle kinetic energy in units of the rest mass energy $m_0 c^2$. Curves are shown for temperature $k_B T = 10 \times m_0 c^2$. Both axes are plotted on logarithmic scales.

$$f_{MJ}(\gamma) = \frac{1}{Z_{1D}(T)} \gamma \left(\gamma^2 - 1\right)^{-1/2} \exp\left(-\frac{\gamma m_0 c^2}{k_B T}\right). \tag{6.58}$$

A comparison of a distribution functions with a one- and three-dimensional statistical distribution of particles is shown in Figure 6.6.

6.7 Relativistic Equations of State

An "equation of state" is a tabulation or mathematical expression that relates "state" parameters such as the energy density, pressure, material density and temperature of matter. At moderate temperatures (where particles have nonrelativistic velocities) and moderate densities (where degeneracy effects are negligible), plasma quantities are often related by ideal gas laws with an allowance if needed for energy expended in ionizing atoms or other processes taking up energy. We consider in this section equation of state departures from ideal gas laws where particles have sufficient density to have high occupancy of quantum states and where particles have relativistic velocities.

Plasmas can have separate equations of state for the electrons and ions with the ions (and unionized atoms) alone contributing to the material density, but with both electrons and ions contributing to the plasma pressure. In laboratory plasmas, the electrons and ions often have different temperatures with equilibrium population distributions as collisions between particles transfer energy at rates proportional to the ratio of the particle masses. Electrons equilibrate energy with ions on a timescale $\approx (1/2) \times 1836M_A$ longer than they equilibrate energy to other electrons, where M_A is the atomic mass of the ion species and 1,836 is the ratio of the mass of protons to electrons (see Exercise 8.2 and Spitzer [107]). Particle collisions are examined in Chapter 8.

There are many examples where knowledge of an equation of state is needed in the modeling and simulation of plasma fluids. In a fluid, the material density varies proportionally to pressure gradients in the fluid. The equation of state relationship between density and pressure is needed to solve for the material density as pressure gradients can be converted to density gradients using the equation of state. The equation of state relationship between energy density and temperature is needed to determine the temperature of a plasma for a known amount of energy input. The heat capacity is usually quoted here: It is the energy required to raise the temperature of a unit mass of material by a unit of temperature. In cosmology, the "equation of state" usually refers to the ratio of the pressure P_p to the energy density U_T of material in the universe with a parameter w representing this ratio:

$$w = \frac{P_p}{U_T} \tag{6.59}$$

In cosmology, the energy density U_T includes the rest mass energies of particles and their kinetic energies.

Gaseous and plasma changes of density ρ in response to a pressure gradient ∇P_p can be modeled using the fluid equations of continuity and an equation of motion. The equation of continuity represents conservation of mass. Material density changes in time $(d\rho/dt)$ are due to the velocity $\mathbf{u}$ of the fluid so that for fluids with fluid velocities $|\mathbf{u}| << c$, the equation of continuity can be written as

$$\frac{\partial \rho}{\partial t} + \nabla \cdot (\rho \mathbf{u}) = 0 \tag{6.60}$$

as $d\rho/dt = \partial \rho/\partial t + \nabla \cdot (\rho \mathbf{u})$. The equation of motion is a fluid version of Newton's law that the net force on a body produces an acceleration, where the force is equal to the body's mass multiplied by the acceleration. For the fluid equation of motion, the net force on the fluid is replaced by a pressure gradient (the net force per unit area) and the mass is replaced by density (the mass per unit volume). The fluid equation of motion for nonrelativistic fluid velocities $\mathbf{u}$ can be written as

$$\rho \left(\frac{\partial \mathbf{u}}{\partial t} + (\mathbf{u} \cdot \nabla)\mathbf{u} \right) = -\nabla P_p \tag{6.61}$$

as $d\mathbf{u}/dt = \partial \mathbf{u}/\partial t + (\mathbf{u} \cdot \nabla)\mathbf{u}$. For fluids with nonrelativistic velocities $\mathbf{u}$ in a rest frame, Equations 6.60 and 6.61 can be used to determine density variations in time and space using an equation of state relationship between pressure P_p and the material density ρ (mass per unit volume). Equations 6.60 and 6.61 are usually solved numerically.

A fluid may have bulk motion at velocity $\mathbf{u}$ which is nonrelativistic, but if the temperature of particles is sufficiently high, random particle velocities of particularly the much less massive electrons can approach the speed of light. We need an equation of state modification between pressure P_p, the material density ρ and the energy density U_k to take into account the relativistic nonlinear relationship of velocity and momentum at high velocities (v approaching c). The kinetic energy T_k of a particle is related to the momentum p by Equation 2.74:

$$T_K = \sqrt{p^2 c^2 + m_0^2 c^4} - m_0 c^2$$

Rearranging we can write for the kinetic energy

$$T_k = m_0 c^2 \left[\left(1 + \left(\frac{p}{m_0 c} \right)^2 \right)^{1/2} - 1 \right]. \tag{6.62}$$

The velocity v of a particle is generally related to the energy W by $v = dW/dp$ (see Equation 2.77). The rest mass energy $m_0 c^2$ does not change with the particle momentum p so that

$$v(p) = \frac{dW}{dp} = \frac{p}{m_0} \left[1 + \left(\frac{p}{m_0 c} \right)^2 \right]^{-1/2}. \tag{6.63}$$

In gases or plasmas, the mass density ρ is dominated by the mass of the atoms or ions. However, electrons contribute to the plasma pressure. The plasma pressure to be used in Equation 6.61 is the addition of the pressure due to the ions and electrons. Due to their light mass, electrons have relativistic velocities at much lower temperatures than atoms or ions.

We now consider a distribution of particles with the same mass m_0 in a gas or plasma. The material density ρ needed to solve Equation 6.60 and 6.61 is simply given by

$$\rho = m_0 n_{tot}, \tag{6.64}$$

where n_{tot} is the total number of particles per unit volume. The kinetic energy U_k per unit volume of all the particles can be found by integrating the kinetic energy $T_k(p)$ of the number of particles $n_p(p)\, dp$ per unit volume with momentum between p and $p + dp$:

$$U_k = \int_0^\infty T_k(p)\, n_p(p)\, dp \tag{6.65}$$

The scalar product of a particle velocity $\mathbf{v}$ and the particle momentum $\mathbf{p}$ can be expanded in Cartesian coordinates (x, y, z) as

$$\mathbf{v} \cdot \mathbf{p} = vp = v_x p_x + v_y p_y + v_z p_z.$$

The velocities multiplied by momentum can be averaged over the particle distribution with the average indicated by $<vp>$. For an isotropic distribution of particle velocities and momentum, these averages are the same in all directions so that

$$<v_x p_x> = <v_y p_y> = <v_z p_z> = \frac{1}{3} <vp> . \tag{6.66}$$

The pressure P_p due to the particles is the integration of all the particle momentum crossing a surface of unit area in unit time. We can choose a surface of unit area perpendicular to, say, the x-axis in Cartesian coordinates and write for the pressure

$$P_p = \int_0^\infty <v_x p_x> n_p(p) \, dp = \frac{1}{3} \int_0^\infty v(p) \, p \, n_p(p) \, dp, \tag{6.67}$$

where $v(p)$ is the particle velocity corresponding to momentum p.

Substituting $v(p)$ from Equation 6.63 into Equation 6.67 we obtain an expression for the pressure P_p of a distribution $n(p)$ of particles:

$$P_p = \frac{1}{3} \int_0^\infty \frac{p^2}{m_0} \left[1 + \left(\frac{p}{m_0 c} \right)^2 \right]^{-1/2} n_p(p) \, dp \tag{6.68}$$

The kinetic energy U_k per unit volume of all the particles is given by Equation 6.65 which can be rewritten as

$$U_k = \int_0^\infty m_0 c^2 \left[\left(1 + \left(\frac{p}{m_0 c} \right)^2 \right)^{1/2} - 1 \right] n_p(p) \, dp. \tag{6.69}$$

Consider a nonrelativistic distribution of particle velocities where $p \ll m_0 c$. The pressure P_p becomes for a non-relativistic distribution:

$$P_p \approx \frac{1}{3} \int_0^\infty \frac{p^2}{m_0} n(p) \, dp \tag{6.70}$$

The energy density U_k for a nonrelativistic distribution is given by

$$U_k = \int_0^\infty m_0 c^2 \left[\left(1 + \left(\frac{p}{m_0 c} \right)^2 \right)^{1/2} - 1 \right] n_p(p) \, dp$$

$$\approx \int_0^\infty m_0 c^2 \left[\frac{p^2}{2(m_0 c)^2} \right] n_p(p) \, dp$$

$$= \frac{3}{2} P_p \tag{6.71}$$

upon expanding the square root in the limit $p << m_0 c$ and noting the similarity of the second integral with Equation 6.70.

Equation 6.71 shows a relationship where the energy density $U_k = (3/2)P_p$, where P_p is the pressure. As well as assuming that the momentum of particles is small ($p << m_0 c$), to obtain Equation 6.71 we have simply assumed that the particle velocities and momentum are isotropic (directed equally in all directions). The form of the particle distribution function $n_p(p)$ does not affect the relationship $U_k = (3/2)P_p$ provided energy is distributed equally in the x, y, and z coordinate directions (as assumed for Equation 6.66). The three co-ordinate axes represents three degrees of freedom. In some gases, plasmas and other systems, there are more degrees of freedom. For example, molecules may have rotational and vibrational degrees of freedom. Energy is then distributed so that that $U_k = (n/2)P_p$, where n is the number of degrees of freedom.

A system may be composed of ultra-relativistic particles where $p >> m_0 c$, $v \approx c$ and $T_k \approx p\,c$. Using Equation 6.67 we have for the pressure of a highly relativistic particle distribution:

$$P_p \approx \frac{1}{3} \int_0^\infty c\,p\,n_p(p)\,dp \qquad (6.72)$$

Using Equation 6.65 for the energy density, we have for the energy density of a highly relativistic particle distribution:

$$U_k \approx \int_0^\infty c\,p\,n_p(p)\,dp = 3P_p \qquad (6.73)$$

The form of the particle distribution function $n_p(p)$ does not affect this relationship. If there are more than (or less than) three degrees of freedom, we have $U_k = n\,P_p$, where n is the number of degrees of freedom. Equation 6.73 shows a similar proportionality of the pressure and energy density for highly relativistic particle distributions as found for nonrelativistic distributions, but with a change of the $w = P_p/U_k$ parameter from $w = 2/3$ (non-relativistic) to $w = 1/3$ (highly relativistic).

To relate the plasma pressure P_p and the energy density U_k of a distribution of particles to the temperature $k_B T$ of the particles, we need to substitute in Equations 6.68 and 6.69 an expression for the particle distribution function $n_p(p)$. We found in Section 6.2 (Equation 6.17) that

$$n_p(p) = \frac{8\pi p^2}{h^3} \frac{1}{1 + \exp\left((-\mu + p^2/(2m_0))/k_B T\right)}, \qquad (6.74)$$

where μ is the chemical potential. The chemical potential μ is large and negative if the particle density n_{tot} is sufficiently small that quantum states are not close to being fully occupied. Such a particle distribution function is said to be "nondegenerate" as the "degeneracy" (number of particles occupying an energy level) does not affect the particle distribution. Expanding Equation 6.68, the pressure P_p of a distribution $n(p)$ of particles can be expressed as

$$P_p = \frac{8\pi}{3h^3 m_0} \int_0^\infty \left[1 + \left(\frac{p}{m_0 c} \right)^2 \right]^{-1/2} \frac{p^4}{1 + \exp\left((-\mu + p^2/(2m_0))/k_B T \right)} \, dp.$$

(6.75)

Expanding Equation 6.69, the kinetic energy U_k per unit volume of all the particles can be expressed as

$$U_k = \frac{8\pi m_0 c^2}{h^3} \int_0^\infty \left[\left(1 + \left(\frac{p}{m_0 c} \right)^2 \right)^{1/2} - 1 \right] \frac{p^2}{1 + \exp\left((-\mu + p^2/(2m_0))/k_B T \right)} \, dp.$$

(6.76)

In the limit of a nonrelativistic distribution of particles ($p << m_o c$) with a sufficiently low density so the chemical potential μ is large and negative (see Appendix C), Equation 6.75 for the pressure P_p of the particles reduces to the ideal gas relationship $n_{tot} k_B T$, where n_{tot} is the total number of particles per unit volume (see Exercise 6.16). The total energy U_k in the nonrelativistic, low density limit reduces to $(3/2) n_{tot} k_B T$.

With a relativistic particle distribution, the chemical potential μ for a low density particle distribution which is not degenerate is given by Equation 6.35. We can write for a nondegenerate relativistic particle distribution that

$$\exp\left(\frac{\mu}{k_B T} \right) = \frac{h^3}{8\pi} \frac{c^3}{(m_0 c^2)^2} \exp\left(-\frac{m_0 c^2}{k_B T} \right) \left[K_2(m_0 c^2/k_B T) \right]^{-1} \frac{n_{tot}}{k_B T},$$

where n_{tot} is the total number of particles per unit volume and $K_2(m_0 c^2/k_B T)$ is the modified Bessel function of the second kind (see Appendix B). We can substitute into Equation 6.75 for the particle pressure as follows:

$$P_p = \frac{8\pi}{3h^3 m_0} \int_0^\infty \left[1 + \left(\frac{p}{m_0 c} \right)^2 \right]^{-1/2} p^4 \exp\left(\frac{\mu - p^2/(2m_0)}{k_B T} \right) dp$$

$$= \frac{1}{3m_0} \frac{c^3}{(m_0 c^2)^2} \exp\left(-\frac{m_0 c^2}{k_B T} \right) \left[K_2(m_0 c^2/k_B T) \right]^{-1} \frac{n_{tot}}{k_B T} I_{Pint},$$

(6.77)

where we represent the integration over momentum by

$$I_{Pint} = \int_0^\infty \left[1 + \left(\frac{p}{m_0 c}\right)^2\right]^{-1/2} p^4 \exp\left(-\frac{p^2/(2m_0)}{k_B T}\right) dp$$

$$= (m_0 c)^5 \int_0^\infty \frac{x^4}{(1+x^2)^{1/2}} \exp\left(-\frac{x^2}{2\Theta}\right) dx. \tag{6.78}$$

Here $x = p/(m_0 c)$ and $\Theta = k_B T/(m_0 c^2)$ is a reduced temperature expressed in units of the particle rest mass energy.

The kinetic energy U_k given by Equation 6.76 can be treated in a similar manner to the pressure. For a nondegenerate relativistic particle distribution, we can write for the total kinetic energy of the particles:

$$U_k = \frac{8\pi m_0 c^2}{h^3} \int_0^\infty \left[\left(1 + \left(\frac{p}{m_0 c}\right)^2\right)^{1/2} - 1\right] p^2 \exp\left(\frac{\mu - p^2/(2m_0)}{k_B T}\right) dp$$

$$= \frac{c^3}{m_0 c^2} \exp\left(-\frac{m_0 c^2}{k_B T}\right) \left[K_2(m_0 c^2/k_B T)\right]^{-1} \frac{n_{tot}}{k_B T} I_{Uint}, \tag{6.79}$$

where we represent the integration over momentum by

$$I_{Uint} = \int_0^\infty \left[\left(1 + \left(\frac{p}{m_0 c}\right)^2\right)^{1/2} - 1\right] p^2 \exp\left(-\frac{p^2/(2m_0)}{k_B T}\right) dp$$

$$= (m_0 c)^3 \int_0^\infty [(1+x^2)^{1/2} - 1] x^2 \exp\left(-\frac{x^2}{2\Theta}\right) dx. \tag{6.80}$$

In the limit of $\Theta \ll 1$ ($k_B T \ll m_0 c^2$), expanding the square root to two terms, this integration simplifies to

$$I_{Uint} \approx (m_0 c)^3 \int_0^\infty (1/2) x^4 \exp\left(-\frac{x^2}{2\Theta}\right) dx.$$

$$= (m_0 c)^3 \frac{3\sqrt{\pi}}{2^{3/2}} \Theta^{5/2}. \tag{6.81}$$

Returning to Equation 6.79, we can expand $K_2(m_0 c^2/k_B T)$ as outlined in Appendix B and use Equation 6.81 to show that in the nonrelativistic limit $U_k = (3/2)n_{tot}k_B T$.

To obtain approximate expressions for the relativity "correction" when $\Theta = k_B T/m_0 c^2 < 1$, but Θ is not negligible, we can simplify the expression for the integration I_{Uint} as for Equation 6.81, but include an extra term in the expansion of the square root:

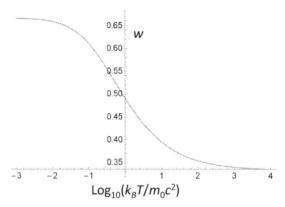

Figure 6.7   The equation of state parameter w for the ratio of the pressure P_p to energy per unit volume U_k of a distribution of particles as a function of the particle temperature k_BT (in units of the particle rest mass energy m_0c^2). The horizontal particle temperature axis is on a logarithmic scale.

$$I_{Uint} \approx (m_0c)^3 \int_0^\infty [(1/2)x^2 - (1/8)x^4] x^2 \exp\left(-\frac{x^2}{2\Theta}\right) dx.$$

$$= (m_0c)^3 \frac{3\sqrt{\pi}}{2^{3/2}} \Theta^{5/2} (1 - (5/4)\Theta)$$

Allowing for the mildly relativistic approximation of $K_2(m_0c^2/k_BT)$ (see Appendix B), we obtain an expression for the kinetic energy per unit volume including a "correction" valid when $k_BT < m_0c^2$:

$$U_k \approx \left(\frac{3}{2}\right) n_{tot} k_B T \left[\frac{1 - (5/4)k_BT/m_0c^2}{1 + (15/8)k_BT/m_0c^2}\right] \tag{6.82}$$

The multiplier in the square bracket is a correction to the expression for the total kinetic energy U_k for a distribution of n_{tot} particles of temperature k_BT allowing for relativistic particle velocities, but assuming $k_BT < m_0c^2$.

The ratio w of the particle pressure to the particle kinetic energy simplifies to a ratio of two integrations (based on Equations 6.78 and 6.80) with values dependent on the reduced temperature $\Theta = k_BT/(m_0c^2)$:

$$w = \frac{P_p}{U_k} = \frac{1}{3} \frac{\int_0^\infty x^4(1+x^2)^{-1/2} \exp\left(-x^2/(2\Theta)\right) dx}{\int_0^\infty [(1+x^2)^{1/2} - 1] x^2 \exp\left(-x^2/(2\Theta)\right) dx} \tag{6.83}$$

The variation of the ratio w of particle pressure to energy density evaluated using Equation 6.83 is shown in Figure 6.7. At nonrelativistic temperatures ($k_BT/m_0c^2 \ll 1$)), the ratio w approaches $w = 2/3$, while for highly relativistic temperatures ($k_BT/m_0c^2 \gg 1$)), w approaches $w = 1/3$. Using Figure 6.7 and the correction for relativistic effects enclosed in the square bracket of Equation 6.82, a relativistic correction for the pressure variation from the ideal gas law can be estimated for particles with temperatures $k_B T < m_0 c^2$.

Exercises

6.1 The Schrodinger equation for an electron in a potential V can be written as

$$\left(-\frac{\hbar^2}{2m_0}\nabla^2 + V\right)\psi = E\psi,$$

where ψ is the electron wavefunction; E is the energy state of the electron; and $\hbar = h/(2\pi)$, where h is Planck's constant. For a free electron with $V = 0$, show that $\psi = \exp(ikz)$ is a solution of Schrodinger's equation with $k = 2\pi/\lambda_{DB}$, where $\lambda_{DB} = h/p$ for p the electron momentum. [The wavelength λ_{DB} of the free electron wavefunction is known as the De Broglie wavelength.]

6.2 Consider a three-dimensional potential box or cavity of sides L with an infinite potential at the walls of the box and zero potential within the box. For a single electron confined within the box, show that the components of momentum in Cartesian coordinates p_x, p_y, and p_z parallel to the sides of the box must have values of $nh/(2L)$, where n is an integer. Hence show that the volume in momentum space occupied by a quantum state of the electron in the box is $h^3/(2L)^3$.

6.3 The energy of black-body radiation per unit of frequency per unit volume is given by

$$W_p(\nu) = \frac{8\pi h\nu^3}{c^3}\frac{1}{\exp(h\nu/k_B T) - 1}.$$

Show that the spectrally integrated energy density per unit volume W_{tot} is given by

$$W_{tot} = \frac{8\pi^5}{15h^3c^3}(k_B T)^4.$$

[You need the definite integral $\int_0^\infty x^3/(\exp(x) - 1)dx = \pi^4/15$.]

6.4 Inertial fusion research seeks to compress a plastic pellet containing deuterium and tritium inside a black-body cavity. The black-body radiation impinging on the pellet ablates some material which causes compression of the deuterium and tritium via Newton's third law. In inertial fusion research, black-body cavities are usually known by the German word *hohlraum*. Determine the radiation temperature inside a small hohlraum of cylindrical shape with a length of 6.44 mm and diameter of 3.55 mm used for research at the National Ignition Facility in California if the total radiation energy in the hohlraum is 1 MJ. [1 keV]

6.5 The solar surface is typically 5,800 K, while sunspots are regions appearing darker with lower temperatures. If a sunspot has a temperature of 4,200 K, determine the relative spectrally integrated intensity of the sunspot compared to other parts of the solar photosphere. [0.275]

6.6 Given the solar photosphere temperature of 5,800 K and that the radius of the Sun is 6.957×10^5 km, determine (a) the mass of solar material converted to energy per second. (b) The present mass of the Sun is 1.989×10^{30} kg and the Sun is 4.6×10^9 years old. Estimate the fraction of solar mass that has been converted to energy in the life of the Sun. [(a) 4.3×10^9 kg s^{-1}, (b) 3×10^{-4}]

6.7 Consider isotropic radiation of energy density W_{rad} at a distance of z from a surface of area A. In a time δt, show that radiation within a cone of half-angle α impinges on the surface, where

$$\cos \alpha = \frac{z}{c\delta t}.$$

A cone of half-angle α subtends a solid angle of $2\pi (1 - \cos \alpha)$ steradian. Use this to show that in a time δt, the radiation energy impinging on the surface from all possible distances z is $W_{rad}(1/4) A c \delta t$.

6.8 Show that the black-body intensity $I_{bb}(\lambda)$ in units of power per unit area per unit wavelength per unit steradian is given by

$$I_{bb}(\lambda) = \frac{2hc^2}{\lambda^5} \frac{1}{\exp(hc/(\lambda k_B T)) - 1},$$

where λ is the wavelength of the radiation. Hence, for long wavelength radiation show that

$$I_{bb}(\lambda) \approx \frac{2c}{\lambda^4} k_B T.$$

[The long wavelength expression is an example of the Rayleigh-Jeans approximation for black-body radiation intensity.]

6.9 A laser pulse is focused so that the intensity variation in radius r from the focal center and in time t is such that

$$I(r, t) = I_0 \exp\left(-\frac{r^2}{r_0^2}\right) \exp\left(-\frac{t^2}{t_0^2}\right).$$

Show that the energy E of the pulse is such that

$$E = 4\pi^{3/2} r_0^2 t_0.$$

Evaluate the number of photons in the laser pulse if the laser wavelength is one micron, the focal radius is one micron, the laser pulse duration is one picosecond and the peak laser intensity is 10^{16} W cm^{-2}. [$\approx 10^{16}$]

6.10 A particle with relativistic energy in a three-dimensional potential box of sides L with infinitely high potential walls has been shown [1] to have sinusoidally varying wavefunctions as found for Exercise 6.1 with components of $\mathbf{k}$ parallel to the sides of the box with values of k_x, k_y, and k_z determined by three equations of form

$$\tan(k_i L) = \frac{2(W + m_0 c^2)\hbar k_i}{\hbar^2 (k_i^2 - k^2)c - 2m_0 c(W + m_0 c^2)},$$

where i represents the components x, y or z and $k^2 = k_x^2 + k_y^2 + k_z^2$. In the nonrelativistic limit where the total energy $W \approx m_0 c^2$ and the momentum $\hbar k_i << m_0 c$, show that the components of momentum in Cartesian coordinates p_x, p_y, and p_z parallel to the sides of the box have values of $nh/(2L)$, where n is an integer.

6.11 Show that the expression for the appropriate relativistic k_x, k_y and k_z values in a three-dimensional potential box as listed in Exercise 6.10 reduces to $\tan(k_i L) \approx 0$ when the Compton wavelength $\lambda_C = h/(m_0 c)$ is such that $L >> \lambda_C$. Hence show that when $L >> \lambda_C$, the components of momentum are given by $nh/(2L)$, where n is an integer as was found for nonrelativistic particles.

6.12 Consider the Maxwell-Juttner distribution $f_{MJ}(\gamma)$ given by Equation 6.55. For a highly relativistic distribution where $k_B T >> m_0 c^2$, show that the peak of the particle distribution occurs when $\gamma = 2k_B T/m_0 c^2$.

6.13 Show that the redshift parameter z is related to the Lorentz γ parameter for an astronomical object by

$$z = \gamma - 1 + \sqrt{\gamma^2 - 1} \approx 2\gamma,$$

where the last relationship only applies for large redshift.

6.14 The cosmic microwave background temperature of $T' = 2.725$K is measured with a small angular anisotropy in temperature known as the dipole pattern with a "hot pole" and "cold pole" temperature variation of ± 0.00335K at angles π radians apart. The temperature difference $\Delta T' = 0.0067$K associated with the dipole is due to the velocity Δv of the local cluster of galaxies relative to the CMB. The "hot pole" is associated with a velocity decrement of $-\Delta v$ and the "cold pole" with a velocity increase of Δv above the velocity v of the frame of the last scattering event that produced the CMB. Using Equation 6.48, show that

$$\Delta T' \approx \frac{T}{\gamma} \left(\frac{1}{1 + (v + \Delta v)/c} - \frac{1}{1 + (v - \Delta v)/c} \right)$$

assuming that v has a value approaching the speed of light c. Hence show that

$$\frac{\Delta T'}{T'} \approx -\frac{\Delta v}{c}.$$

6.15 Extremely high energy cosmic rays collide with the cosmic microwave background (CMB) to produce new particles as the CMB spectral frequencies are Doppler shifted to sufficiently high photon energies in the particle frame to enable particle-photon interactions. Use the Doppler shift formula (Equation 2.46) to show that cosmic rays made up of protons of energy E_{max} and rest mass energy $m_p c^2$ can produce a pion of rest mass $m_\pi c^2$ from the most abundant CMB photons if

$$E_{max} > \frac{(m_\pi c^2)(m_p c^2)}{2 \times 2.8 k_B T},$$

where T is the CMB temperature. As protons lose their energy when interacting with the CMB in this way, evaluate the maximum proton energy you would expect to observe when the protons in cosmic rays interact with the present day CMB to produce a proton and a pion. [$\approx 10^{20}$ eV. This upper limit for cosmic ray energies is known as the Greisen-Zatsepin-Kuzmin limit [48]. In approximate agreement with our answer, cosmic rays with energy over 3×10^{20} eV have not been observed [9], [22].]

6.16 The pressure of an equilibrium distribution of particles with momenta p allowing for relativistic and degeneracy effects can be written as

$$P_p = \frac{8\pi}{3h^3 m_0} \int_0^\infty \left[1 + \left(\frac{p}{m_0 c} \right)^2 \right]^{-1/2} \frac{p^4}{1 + \exp\left((-\mu + p^2/(2m_0))/k_B T \right)} dp.$$

For a nondegenerate distribution of particles, the chemical potential μ is given by

$$\mu = k_B T \ln \left[\frac{n_{tot}}{2} \left(\frac{h^2}{2\pi m_0 k_B T} \right)^{3/2} \right],$$

where n_{tot} is the total number of particles per unit volume. Show that the pressure of particles where relativistic and degeneracy effects are small is given by the ideal gas relationship

$$P_p = n_{tot} k_B T.$$

[You need the integral $\int_0^\infty p^4 \exp(-p^2/A) dp = (3\sqrt{\pi}/8) A^{5/2}$.]

6.17 Plasmas studied for the production of fusion energy may require temperatures of 25 keV for an efficient fusion process. The electron pressure in such plasmas is usually evaluated assuming ideal gas laws. Estimate the percentage reduction in electron pressure from the ideal gas law value due to relativistic effects for an electron temperature of 25 keV. [14%]

7

Radiation Emission, Absorption, and Scatter in Plasmas

Special relativity is important in understanding radiation produced by plasmas and the interaction of radiation with plasma when the velocities of plasma particles are high. Electrons achieve relativistic velocities at lower energies due to their much smaller mass compared to ions, so this chapter concentrates on relativistic electron effects on the emission, absorption, and scattering of light in plasmas. Nonrelativistic or low velocity treatments are presented first in most cases. Our relativistic expression are compared to the nonrelativistic, so that relativistic "corrections" can be applied for small, but not insignificant relativistic changes.

Radiation emission, absorption and scatter are the processes whereby photons interact with matter. These radiative processes transfer energy spatially in a plasma, cause radiation fields to become thermalized with a black-body distribution and can ensure particles become thermalized with Boltzmann population ratios. Light emission, absorption, and scattering effects are also used for the measurement of plasma quantities.

In this chapter, we find some combinations of radiative quantities that remain Lorentz invariant in different frames of reference. Spectral line radiation emission and absorption is examined. An intrinsic relativistic effect – the Doppler broadening of spectral lines is considered in some detail. The rate of continuum emission associated with free electrons accelerating in the field of an ion is examined using a novel perspective, intrinsically linked to relativity concepts. The method of "virtual photons" is used to evaluate expressions for continuum emission. The idea is that in the frame of an electron, a "collision" with an ion appears to the electron as a pulse of electric field with an associated magnetic field, in other words an electromagnetic pulse. The continuum emission arises from the Thomson or Compton scatter of the electromagnetic pulse off the electron. The chapter presents a discussion of inverse Compton scatter where the radiation field can gather energy by Compton scatter in a plasma of relativistic electrons. An introduction to the radiation reaction force on electrons emitting light is presented.

7.1 Radiative Transfer

We now consider the emission and absorption from a volume of gas or plasma with a density of many emitters per unit volume. If the emission processes from individual particles in a gas or plasma are independent of each other, the total power of emission is given by a sum of the power emitted by each particle. Some radiative quantities of independent emitters are Lorentz invariant: that is, the quantities remain constant under a Lorentz transformation between frames of reference.

Charged particles radiate electromagnetic radiation when accelerated. A general expression for the electromagnetic power radiated by a single accelerated charged body or particle in the reference frame of the body or particle was considered in Section 1.3. The acceleration and radiation of particles due to magnetic fields was considered in Chapter 5. We deal with issues related to the radiation from accelerated electrons in a plasma and transitions between quantum states in this chapter. However, the detailed atomic physics relating to the acceleration and consequent radiation of electrons bound in atomic and ionic quantum states needs to be evaluated using time-dependent quantum mechanics and is not covered in this book (see e.g. Chapter 10 of Tallents [110]).

Calculations of radiated electromagnetic power are best carried out in the frame of reference of the radiating particle, atom or ion. Any angular change in this radiated power in other frames of reference (such as the rest frame) can be treated as outlined in Section 3.4 with the particular frequency changes due to the Doppler effect treated as outlined in Section 2.6.

When there are a large number of emitters producing radiation, we can define an emission coefficient ϵ as the sum of the power from individual identical emitters such that in the frame of the emitting particle

$$\epsilon = N_p \frac{dP_e}{dv \, d\Omega}, \qquad (7.1)$$

where N_p is the number density of emitters and $dP_e/(dv \, d\Omega)$ is the power radiated per emitter per unit solid angle per unit frequency. Absorption of light can be regarded as the inverse process to emission whereby impinging electromagnetic radiation accelerates a charged particle.

The general change in intensity dI of light over a small length dz_l along a line of sight through a medium is found by taking into account the emission minus the absorption over the length dz_l. The resulting formula is known as the equation of radiative transfer:

$$dI = \epsilon \, dz_l - KIdz_l \qquad (7.2)$$

The absorption coefficient K gives the rate of absorption of radiation of intensity I per unit length. For example, K can have units of m^{-1}. Another measure of absorption or more generally interaction of radiation with matter is a cross-section. Cross-sections σ_K represent notional areas associated with absorption or radiation interaction with a single particle. If there is a number density of N_p absorbing particles, the absorption coefficient K is related to the absorption cross-section σ_K by

$$K = N_p\,\sigma_K. \tag{7.3}$$

The absorption of radiation is also sometimes characterized by a measure of opacity K/ρ, where ρ is the material density or mass per unit volume of the medium. Cross-sections σ_K per atom or opacities K/ρ have more constant values with changing number density N_p and density ρ and so are often tabulated in the literature rather than absorption coefficients K (see e.g. Colgan et al. [21] for tabulations of plasma opacities and Henke et al. [49] for tabulations of the atomic cross-sections and opacities of room temperature elements).

The inverse of the absorption coefficient $1/K$ gives a measure of the distance traveled by photons in a medium before they are absorbed and is referred to as the mean free path for absorption or the attenuation length for the radiation. Over a distance of the attenuation length $1/K$, the intensity of light in a purely absorbing, uniform medium drops by a factor of e^{-1}. This drop is apparent if Equation 7.2 is solved assuming zero emission ($\epsilon = 0$) and a uniform medium as then

$$I(z_l) = I(0)\,\exp(-Kz_l), \tag{7.4}$$

where $I(z_l)$ is the intensity as a function of distance z_l in propagating through a uniform medium.

The emission coefficient ϵ is the power radiated per unit solid angle per unit volume by a body of gas or plasma. The emission coefficient can be specified for a particular frequency range or a particular spectral line transition so that the full units may be, for example, watts per cubic meter per steradian per unit frequency (W m^{-2} sr^{-1} Hz^{-1}). The intensity I must have comparable units; for example, watts per square meter per steradian per unit frequency. In laser work, the hybrid unit of watts per square centimeter (Wcm^{-2}) is often employed for the laser power per unit area. Other research fields, particularly astrophysics, use centimeter-gram-second (CGS) units so that powers are measured in erg s^{-1} and intensities in erg s^{-1} sr^{-1} Hz^{-1} (1 joule $= 10^7$ erg).

The total emission coefficient at a particular frequency can be evaluated by adding together the emission coefficients due to different process involving free and bound electrons. Some specific processes contributing to the total emission coefficient ϵ for a plasma are considered in this chapter. We consider emission coefficients associated with spectral line emission arising from discrete bound quantum states of atoms or ions in Section 7.3. Continuum emission arising from the transitions of free electrons is considered in Section 7.7. Processes contributing to the absorption coefficient K are considered in Sections 7.4 (spectral line absorption), 7.7 (inverse bremsstrahlung absorption), and 7.11 (light scattering). As for emission, absorption coefficients for different processes can be added to give the total absorption coefficient K.[1]

The optical depth τ can be specified by considering the optical depth change $d\tau = K dz_l$ over an incremental length (dz_l) along a line of sight. We then have

$$\frac{dI}{d\tau} = \frac{\epsilon}{K} - I = S_f - I, \tag{7.5}$$

where $S_f = \epsilon/K$ is known as the source function. Optical depth is Lorentz invariant as it can be measured by counting photon numbers using, for example, Equation 7.4 where for a uniform medium $\tau = K z_l$ and the photon numbers are proportional to the intensities. However, the source function S_f changes with the frame of reference (this change is discussed later in this section).

The radiative transfer equation can be integrated to the form

$$I = \int_0^\tau S_f(\tau_z) \exp(-\tau_z) d\tau_z + I(0) \exp(-\tau), \tag{7.6}$$

where τ is the optical depth measured through the gas or plasma along a line of sight and $I(0)$ is the intensity incident onto the plasma. For many gases or plasmas, there is no external source of radiation so that $I(0) = 0$. If the source function is constant with optical depth then $S_f(\tau_z) = S_f$ everywhere and

$$I = S_f(1 - \exp(-\tau)). \tag{7.7}$$

For a large body of emitting and absorbing gas or plasma with large τ, we see that $I = S_f$.

The density of emitters in a frame S' observed from a stationary frame S increases due to length contraction in the direction of the velocity $\mathbf{v}$ of the frame S' such that

$$N_p = \gamma N'_p \tag{7.8}$$

where N_p is the density in frame S and N'_p is the density of emitters in the moving S' frame. Doppler shifts affect the frequency range of emission in the two frames. Adapting Equation 2.47, the frequency ν detected in the stationary frame is related to the frequency ν' emitted in the moving frame by

$$\nu = \nu' \left(1 + \frac{v}{c} \cos\theta'\right) \gamma$$

---

[1] Emission and absorption coefficients associated with different radiative processes can be added when each emission or absorption process is sufficiently isolated that the light electric fields do not add (or subtract) coherently. Coherent electric field addition can be important for bright light production (see Chapter 5) as light intensity is proportional to the square of the light electric field.

where θ' is the angle between the observer line of sight and the velocity $\mathbf{v}$ in the moving frame (assuming that positive v implies the emitting body is moving towards the observer). For a frequency range dv in the stationary frame

$$dv = dv' \left(1 + \frac{v}{c}\cos\theta'\right)\gamma, \tag{7.9}$$

where dv' is the corresponding frequency range in the moving frame. The emission coefficient ϵ in the stationary frame is consequently related to emission in the moving frame by

$$\epsilon = N\frac{dP_e}{dv\,d\Omega} = \frac{1}{1 + v\cos\theta'/c}N_p'\frac{dP_e}{dv'\,d\Omega} = \gamma^2\left(1 + \frac{v}{c}\cos\theta'\right)^2 N_p'\frac{dP'}{dv'\,d\Omega'} \tag{7.10}$$

upon using Equation 3.55. We use Equation 3.55 which neglects the Doppler effect on dP'/dv' as Doppler changes in the detected frequencies are taken into account by Equation 7.9. Equation 7.10 leads to the relationship between an emission coefficient ϵ' in the moving frame and the emission coefficient ϵ in the stationary frame. Due to the Doppler effect $v/v' = \gamma(1 + v\cos\theta'/c)$, we have

$$\epsilon = \left(\frac{v}{v'}\right)^2\epsilon'. \tag{7.11}$$

Equation 7.11 means that the value of ϵ/v^2 is invariant under a Lorentz transformation.

Other radiative transfer quantities that are invariant under a Lorentz transformation can be deduced. Consider a gas or plasma uniformly streaming with a velocity $\mathbf{v}$ between two planes parallel to the x-axis (see Figure 7.1). If the planes are separated by a distance l and light is directed at an angle θ to the planes in the stationary frame S and at an angle θ' in the frame S' moving with the gas or plasma, then the optical depths of the light through the gas or plasma as measured in the two frames of reference are given by

$$\tau = \frac{Kl}{\sin\theta}$$

$$\tau' = \frac{K'l'}{\sin\theta'}.$$

Optical depth is measured by counting the photons passing along a ray path. The photon number reduces proportionally to $\exp(-\tau)$ and is the same in both frames of reference, so the optical depth τ is Lorentz invariant. Similarly, the Lorentz transformation for distances perpendicular to the velocity $\mathbf{v}$ does not alter the distances,

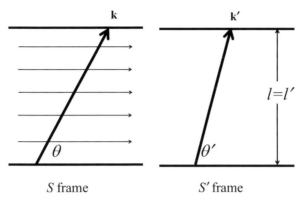

Figure 7.1   A moving gas or plasma between two parallel planes is illustrated. In the stationary frame of reference S, a ray of light propagates at an angle θ to the planes, while in the moving frame of reference S', the ray propagates at an angle θ'. The scenario depicted is used to deduce the Lorentz invariance $K\nu$ of the absorption coefficient K of the gas, where ν is the frequency of the light.

so $l = l'$. The angles θ and θ' of the light in the two frames change according to Equation 2.42 such that

$$\sin \theta = \frac{\sin \theta'}{\gamma(1 + v \cos \theta'/c)}.$$

The absorption coefficients in the two frames are consequently related by

$$K = \frac{K'}{\gamma(1 + v \cos \theta'/c)}. \tag{7.12}$$

Due to the Doppler effect $\nu/\nu' = \gamma(1 + v \cos \theta'/c)$, so that

$$K = \frac{\nu'}{\nu} K'. \tag{7.13}$$

Equation 7.13 indicates that the value $K\nu$ is invariant under a Lorentz transformation.

Due to the ϵ/ν^2 invariance of the emission coefficient and the $K\nu$ invariance of the absorption coefficient under a Lorentz transformation, the source function $S_f = \epsilon/K$ is such that the value of S_f/ν^3 is Lorentz invariant. As the source function occurs in the equation of radiative transfer (Equation 7.5) as $S_f - I$, intensity I/ν^3 is also Lorentz invariant.

It is often convenient to determine the radiative transfer quantities ϵ, K and S_f in the proper frame of the material. Using the results of this section, it is possible to then find the radiative transfer quantities in any other frame. As the transformation of the frequencies ν in the frames varies with the angle to the relative velocity **v**, quantities that are isotropic in one frame, are not necessarily isotropic in another frame.

7.2 The Einstein A and B Coefficients

The emission and absorption of light arises due to the acceleration of charged particles. In Section 1.3 we evaluated the power emitted by an accelerated charged particle in the frame of reference of the particle. Energy is conserved in radiation emission by a loss of energy from the accelerating particle. With free particles, the loss of energy due to radiation emission is associated with a momentum change of the particle and hence the existence of a "radiation reaction force" (see Section 7.12). Absorption of light can be regarded as the inverse process to emission with an incoming electromagnetic wave accelerating a charged particle so that the particle gains energy.

The radiative emission by free particles, such as a free electron in a gas or plasma proceeds as treated in Section 1.3. Electrons in atoms in the potential field of a nuclear charge are, however, in quantized states and can only change energy in quantum leaps between discrete energy gaps. However, the emission and absorption of radiation for free particles and those such as atomic electrons in discrete quantum states with large energy gaps between states both result from charged particle acceleration. With electronic states in atoms, the acceleration processes needs a quantum mechanical treatment (see e.g. Chapter 10 of Tallents [110]).

Free electron and other free particles not in a confining potential field are also in quantum states; the quantum states are just spaced closely together. In this section, we assume two energy states are involved in emission or absorption of radiation of a particular frequency. We consider the possible radiative processes that can occur between the two states following an argument originally proposed by Einstein.

In his treatment of the radiative processes between two energy levels, Einstein introduced the concept of stimulated emission, where a photon interacting with an excited energy state causes de-excitation to a lower energy state and the emission of an identical photon to the stimulating photon [36]. We show in this section that it is not possible to get agreement between the black-body radiation energy density and the form of the equilibrium radiation density arising from a rate equation balance of photo-absorption and spontaneous emission alone. This is the original argument for the existence of stimulated emission.

After Einstein proposed the idea of stimulated emission, for many years stimulated emission was a curiosity necessarily used in models of radiative processes. This changed with the invention of the laser in 1960 and the later development of applications using the extremely bright, collimated and coherent radiation possible

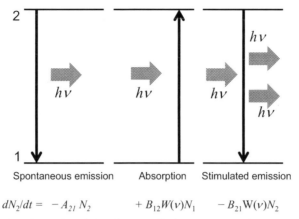

$$dN_2/dt = -A_{21} N_2 \qquad + B_{12}W(v)N_1 \qquad - B_{21}W(v)N_2$$

Figure 7.2   A schematic of the three radiative processes between two generic energy levels. The rate of change of the population of the upper quantum state is shown expressed in terms of the Einstein A and B coefficients.

with lasers. The word laser is an acronym for light amplification by the stimulated emission of radiation.

We undertake an evaluation of the relationships between the rates of the three radiative processes in the frame of reference S of the atoms or particles involved in the emission or absorption processes. The rates of the three radiative optical process-spontaneous emission, absorption and stimulated emission are given by rate equations involving the population densities of the upper (N_2) and lower (N_1) quantum states and the radiation energy density $W(v)$. Energy conservation requires that the relevant radiation energy density is at photon energy $hv = E_2 - E_1$, where E_2 and E_1 are the upper and lower quantum state energies, respectively. We do not need to consider the detail of, for example, atomic or free electron energy levels and just need to consider the upper (labeled 2) and lower (labeled 1) states involved in a set of related radiative processes (see Figure 7.2).

The rate of change of the population densities is linearly dependent on the population density and for absorption and stimulated emission is also linearly dependent on the radiation energy density. The rate equations for spontaneous emission, absorption and stimulated emission can be added to give the total rate of change of, say, the upper quantum state due to radiative processes:

$$\frac{dN_2}{dt} = -A_{21}N_2 + B_{12}W(v)N_1 - B_{21}W(v)N_2 \qquad (7.14)$$

The proportionality constant for the rate of spontaneous emission (A_{21}) is known as the transition probability or Einstein A coefficient, while similar constants (B_{12} and

B_{21}) for absorption and stimulated emission are simply referred to as the Einstein B coefficients.

In the absence of a radiation field $W(\nu)$, Equation 7.14 simplifies to

$$\frac{dN_2}{dt} = -A_{21}N_2$$

which has a solution

$$N_2(t) = N_2(0) \exp(-A_{21}t). \tag{7.15}$$

In the absence of a radiation field, a population initially in the upper quantum state at time $t = 0$ decays exponentially in time with a time constant of $1/A_{21}$.

It is useful to follow Einstein's original treatment and consider what happens for ensembles of atoms which are in an equilibrium state. Equilibrium now means that the populations N_2 and N_1 are in equilibrium with population ratio given by the Boltzmann ratio and that the radiation field is the equilibrium black-body distribution. Equilibrium also means that the populations are in steady state with $dN_2/dt = 0$.

The equilibrium condition that $dN_2/dt = 0$ when the rates of specific related transitions are added together is often referred to as a condition of "detailed balance" for the transitions. Detailed balance arises as an extension of the "microreversibility" of transitions between quantum states [83]. A transition and the inverse transition between states occur at the same rate because the transition processes occur at the same rate in the forward and backward time directions. Later in this book, we also use detailed balance for the reverse processes of bremsstrahlung/inverse bremsstrahlung (Section 7.7) and collisional excitation/de-excitation plus collisional ionization/three-body recombination (Section 8.3). Here we use detailed balance for the three radiative processes: photon absorption, spontaneous emission, and stimulated emission.

Rearranging Equation 7.14 with $dN_2/dt = 0$ for an equilibrium steady-state for the three radiative processes between two quantum states of an ensemble of atoms, we can write an expression for the radiation energy density

$$W(\nu) = \frac{A_{21}}{(N_1/N_2)\,B_{12} - B_{21}}.$$

The Boltzmann population ratio depends on the temperature T of the system such that

$$\frac{N_2}{N_1} = \frac{g_2 \exp(-E_2/k_B T)}{g_1 \exp(-E_1/k_B T)} = \frac{g_2}{g_1} \exp\left(-\frac{h\nu}{k_B T}\right),$$

where g_1 and g_2 are the degeneracies of the energy levels 1 and 2, respectively. It is convenient to group quantum states with similar (or identical energies), so the degeneracies represent the number of quantum states at energy E_1 and E_2, respectively. The above equation for $W(v)$ can be rewritten as

$$W(v) = \frac{A_{21}}{(g_1/g_2) \exp(hv/k_B T) B_{12} - B_{21}}. \tag{7.16}$$

Assuming equilibrium means that the radiation energy density is given by the black-body energy density (see Equation 6.44) with radiation energy density between frequency v and $v + dv$ given by

$$W_p(v) dv = \frac{8\pi h v^3}{c^3} \frac{1}{\exp(hv/k_B T) - 1}.$$

Equating $W_p(v)$ to $W(v)$ (from Equation 7.16) can give us information regarding the relationships between the Einstein coefficients A_{21}, B_{12}, and B_{21}. The radiation energy density and Equation 7.16 can only both be correct if

$$B_{12} \frac{g_1}{g_2} = B_{21} \tag{7.17}$$

and

$$A_{21} = \frac{8\pi h v^3}{c^3} B_{21}. \tag{7.18}$$

If stimulated emission did not exist, Equation 7.16 would have $B_{21} = 0$ in the denominator and it would be impossible to equate Equation 7.16 with the black-body radiation energy density. It was this potential inconsistency that led Einstein to propose the existence of the stimulated emission process.

The relationships between the Einstein coefficients (Equations 7.17 and 7.18) have been obtained by considering a system of radiation and matter in thermal equilibrium. However, the exact rates of A_{21}, B_{12}, and B_{21} are determined by quantum mechanics with parameters determined by the central potential arising from the nucleus of an atom and the electric field of the radiation. Consequently, the relationships (Equations 7.17 and 7.18) are independent of the state in which atoms are embedded. It does not matter if the population distributions of photons and quantum state populations are in equilibrium or not in equilibrium. The relationships between the Einstein coefficients apply for materials not in equilibrium, including the highly nonequilibrium systems used for laser production.

In Section 7.1 we found that light intensity and the value of a source function in radiative transfer vary with a proportionality to frequency of v^3 and are Lorentz invariant. Other atomic physics parameters varying with a proportionality to frequency of v^3 are also Lorentz invariant. Equation 7.18 shows that the ratios A_{21}/B_{21}

and A_{21}/B_{12} are Lorentz invariant as the ratios scale with frequency ν such that, for example, $A_{21}/B_{12} \propto \nu^3$. The ratio of the spontaneous emission transition probability to the Einstein B coefficients for absorption and stimulated emission are the same in all frames of reference.

7.3 Spectral Line Emission Coefficients

The emission coefficients discussed in Section 7.1 can be more explicitly formulated for spectral line transitions between discrete quantum states. In the frame of reference of the emitting atom, ion or molecule, the emission coefficient ϵ for a transition due to spontaneous emission between discrete quantum states can be written as

$$\epsilon = \frac{1}{4\pi} N_2 A_{21} h\nu f(\nu), \tag{7.19}$$

where $f(\nu)$ is known as the lineshape function. Here the emission coefficient ϵ is in units of power per unit volume per unit frequency per steradian. The factor of $1/4\pi$ in Equation 7.19 on the right-hand side converts the total isotropic emission into the emission per unit of steradian. The lineshape function $f(\nu)$ represents the probability of emission per unit frequency assuming that the transition takes place at some frequency. The assumption that radiative emission occurs at some frequency requires the condition that

$$\int_0^\infty f(\nu)d\nu = 1. \tag{7.20}$$

Due to this normalization of the lineshape function, the lineshape function has units of inverse frequency. We consider an example of a lineshape function in Section 7.5 where Doppler broadening is treated.

It is interesting to examine the changes in the terms contributing to a spectral line emission coefficient (Equation 7.19) between the moving frame of reference where emission occurs and a rest frame of reference where spectral line radiation is detected. A scenario here could be an emitting volume of gas or plasma moving in a frame of reference at relative velocity **v**. The number density N_2 of the upper quantum state in the two frames are related by $N_2 = \gamma N_2'$ (see Equation 7.8). As the transition probability A_{21} has units of inverse time, the transition probabilities in the two frames are related by $A_{21} = A_{21}'/\gamma$. The total number of photons $N_2 A_{21}$ emitted into all angles is thus Lorentz invariant.

The frequency ν in Equation 7.19 transforms between frames as given by Equation 7.28. The lineshape function $f(\nu)$ has units of inverse frequency with a value proportional to the inverse of the lineshape function spectral width. As frequency differentials transform between frames in the same way as frequency (see Equation 7.9), we have that $vf(v)$ is Lorentz invariant. The Lorentz transformation of the spectral line emission coefficient ϵ is consequently dependent on the Lorentz transformation of the solid angle of the emission. We found (Equation 3.45) that solid angle $d\Omega$ transforms as

$$d\Omega = \frac{d\Omega'}{\gamma^2(1 + v\cos\theta'/c)^2}.$$ (7.21)

The emission coefficient ϵ is defined here in units of inverse steradians, so that the Lorentz transformation of the emission coefficient is given by

$$\epsilon = \epsilon'\gamma^2(1 + v\cos\theta'/c)^2 = \epsilon'\left(\frac{v}{v'}\right)^2$$ (7.22)

in agreement with the earlier result in Section 7.1 (Equation 7.11).

With a large number of radiative transitions from many randomly oriented atoms or ions, spectral line emission from a gas or plasma is isotropic in the frame of the emitting gas or plasma. However, a body of gas or plasma moving at relativistic velocities $\mathbf{v}$ radiates in the rest frame with enhanced emission in the direction of $\mathbf{v}$. Using Equation 3.52, we have that the emission coefficient ϵ in the rest frame is related to an isotropic emission coefficient ϵ' in the moving frame by

$$\epsilon = \frac{1}{\gamma^4(1 - v\cos\theta/c)^4}\,\epsilon',$$ (7.23)

where θ is the angle of emission in the rest frame to the velocity $\mathbf{v}$. As discussed in Section 3.4, there is a beaming effect, whereby the emission in the rest frame is concentrated in the direction of the velocity $\mathbf{v}$. The concentration of emission in the forward velocity direction is within a half-angle of $1/\gamma$ when v approaches the speed of light c.

7.4 Spectral Line Absorption Coefficients

Absorption and stimulated emission are treated together in radiative transfer calculations. Stimulated emission is equivalent in many ways to negative absorption: photons are added to the radiation field in stimulated emission, rather than being subtracted as occurs in absorption. The population density of an excited quantum states decreases due to stimulated emission, rather than being added to as occurs in absorption. The contribution of stimulated emission to population densities and the radiation field are added to the contribution due to the pure absorption process to give a value for a total or net absorption coefficient K.

The rate of change of the population of energy levels due to absorption and stimulated emission is given by

$$\frac{dN_2}{dt} = (B_{12}N_1 - B_{21}N_2) \int_0^\infty W(\nu)f(\nu)d\nu, \qquad (7.24)$$

where $W(\nu)$ is the radiation energy density per unit volume per unit frequency and $f(\nu)$ is the lineshape function. The radiation energy density $W(\nu)$ can be spectrally broad or narrow. If spectrally broad like the black-body radiation distribution, the radiation energy density $W(\nu)$ varies slowly compared to the lineshape function $f(\nu)$ and can be removed from within the integral. The Equation for dN_2/dt then becomes equal to the absorption and stimulated emission terms in Equation 7.14 where we assumed the incident radiation was the spectrally broad black-body radiation.

Consider light absorbed by the radiative processes between two discrete quantum states in a medium with a population N_2 of the upper quantum state. Each absorbed photon in a volume of unit area and thickness Δz_l increases the upper quantum state density N_2 by one and decreases the light intensity by the photon energy $h\nu$. The change in the number of photons is equal to dN_2/dt multiplied by the volume Δz_l for a unit area of thickness Δz_l. The light intensity integrated in frequency and angle changes by

$$\Delta \left(\int_{\nu=0}^\infty \int_\Omega I \, d\nu d\Omega \right) = -h\nu \frac{dN_2}{dt} \Delta z_l.$$

Taking the limits of differential increments, the frequency and angle integrated intensity rate of change along an increment of beam path is given by

$$\frac{d \left(\int\int I \, d\nu d\Omega \right)}{dz_l} = -h\nu \frac{dN_2}{dt}. \qquad (7.25)$$

Using Equations 7.17 and 7.18, we can equate the Einstein B coefficient values to the transition probability A_{21}. The radiation energy density $W(\nu)$ is related to the angle integrated intensity by $W(\nu) = (1/c) \int I d\Omega$, so that the rate of light absorption is given by

$$\frac{d \left(\int\int I \, d\nu d\Omega \right)}{dz_l} = -\frac{c^3}{8\pi \nu^2} A_{21} \left(\frac{g_2}{g_1}N_1 - N_2 \right) \int_0^\infty \left[\int_\Omega \left(\frac{I}{c} \right) d\Omega \right] f(\nu)d\nu. \qquad (7.26)$$

If we consider a small increment of frequency ν so that the intensity I variation with frequency over the small frequency range is constant compared to the lineshape $f(\nu)$ variation with frequency, the lineshape $f(\nu)$ can be removed from within the integration over frequency. Simplifying terms, we have for all frequencies and angles that

$$\frac{dI}{dz_l} = -KI,$$

where the absorption coefficient K at all frequencies across the line profile and for all angles is given by

$$K = \frac{c^2}{8\pi v^2} A_{21} \left(\frac{g_2}{g_1} N_1 - N_2 \right) f(v). \tag{7.27}$$

The lineshape function $f(v)$ varies rapidly with frequency v with a typical frequency width Δv such that $v/\Delta v \gg 100$. The $1/v^2$ term in Equation 7.27 only varies slowly with frequency and so can be accurately set at the line center frequency v_0. Due to the spectral narrowness of $f(v)$, the $1/v^2$ term can also be accurately assumed to be constant in determining Equation 7.27 from Equation 7.26.

We examined the Lorentz transformation of the absorption coefficient K from the moving frame of reference to the rest frame of reference in Section 7.1. The geometry of the absorption of a moving gas or plasma as observed in the moving and rest frame is illustrated in Figure 7.1. We found that Kv is invariant under a Lorentz transformation.

7.5 Doppler Broadening

A lineshape function represents the probability per unit of frequency of emission in a narrow spectral range assuming emission occurs at some frequency. A useful example of a lineshape function $f(v)$ occurs when emitting particles have different Doppler-shifted frequencies depending on their relative velocity to the observer. The observed frequency v in a rest frame of reference is related to the frequency v' emitted in a moving frame of reference by Equation 2.47. We have

$$v = v' \left(1 + \frac{v}{c} \cos \theta' \right) \gamma, \tag{7.28}$$

where θ' is the angle in the moving frame between the direction of propagation of the light and the velocity $\mathbf{v}$ of the moving frame. The component of the velocity $v_{||}$ of emitting particles along a line of sight is given by $v_{||} = v \cos \theta'$. Rearranging the Doppler shift expression, each value of $v_{||}$ produces a shift in frequency between the frames such that

$$\frac{v}{\gamma v'} - 1 = \frac{v_{||}}{c}. \tag{7.29}$$

Any distribution in the number density $N(v_{||})$ of emitting particles with velocity along the line of sight produces an equivalent distribution in the number of particles with the Doppler frequency shift given by Equation 7.29.

A lineshape function $f_D(v)$ broadened by the Doppler effect from a velocity distribution $N(v_{||})$ has the form

$$f_D(v)dv \propto N(v_{||})dv_{||} = N\big(c[v/(v'\gamma) - 1]\big)\,dv_{||}$$

so that

$$f_D(v) \propto N(v_{||})\frac{dv_{||}}{dv}. \tag{7.30}$$

In a gas or plasma at moderate temperatures, the velocities are much less than the speed of light c, so that γ is close to one and $dv_{||}/dv$ is constant with varying v and $v_{||}$ for emission in the moving frame at a particular frequency $v_0 = v'$. At moderate temperatures, a thermal distribution of atoms or ions along a line of sight has a Maxwellian distribution of speeds along the line of sight with

$$N(v_{||}) \propto \exp\left(-\frac{Mv_{||}^2}{2k_BT}\right), \tag{7.31}$$

where T is the atomic or ionic temperature and M is the mass of the atom or ion. The form of the distribution function of particles has been discussed in Section 6.5. Equation 7.31 represents the form of particle distribution applicable when the velocities $v_{||} << c$ and we need the velocity component in one-direction (the line of sight). Using Equations 7.31 and 7.29, the lineshape function is determined by the relative number of atoms or ions with velocities giving the appropriate Doppler shift of frequency. We have

$$f_D(v) \propto \exp\left(-\frac{Mc^2(v - v_0)^2}{2k_BTv_0^2}\right),$$

where the frequency v_0 corresponds to a velocity $v_{||} = 0$ (and hence $v_0 = v'$). The normalization requirement (Equation 7.20) enables the elimination of the proportionality for the lineshape function. Using the definite integral

$$\int_{-\infty}^{\infty} \exp(-x^2)dx = \pi^{1/2},$$

we have for the normalized line shape

$$f_D(v) = \frac{1}{\pi^{1/2}v_0}\left(\frac{Mc^2}{2k_BT}\right)^{1/2}\exp\left(-\left(\frac{v - v_0}{v_0}\right)^2\left(\frac{Mc^2}{2k_BT}\right)\right). \tag{7.32}$$

This expression is often written in terms of a measure of the spectral width of the line profile:

$$f_D(\nu) = \frac{(4 \ln 2)^{1/2}}{\pi^{1/2} \Delta \nu} \exp\left(-4 \ln 2 \left(\frac{\nu - \nu_0}{\Delta \nu}\right)^2\right), \tag{7.33}$$

where $\Delta \nu$ is a measure of the full width at half maximum of the Gaussian shaped lineshape function $f_D(\nu)$. Comparing Equation 7.33 to Equation 7.34 gives the relationship of the linewidth $\Delta \nu$ to the temperature T of the atoms or ions. We have

$$\Delta \nu = (4 \ln 2)^{1/2} \frac{\nu_0}{c} \sqrt{\frac{2 k_B T}{M}}. \tag{7.34}$$

Measurements of the linewidth $\Delta \nu$ of spectral lines emitted from a plasma can be used as diagnostics of ion temperatures in the plasma. However, other line broadening process can also be important (see Chapter 10 of Tallents [110]). When using spectral linewidths to determine temperatures, it is important to ensure that the spectral line broadening is dominated by the Doppler shift of frequencies associated with a thermal spread of velocities.

As well as using the Doppler effect to measure the temperature of emitting ions in a plasma, Doppler broadening of the spectrum of light produced by Thomson scatter off free electrons in a plasma is used to measure electron temperatures. Thomson scatter from effectively isolated electrons is discussed in Section 4.6. Scatter from an ensemble of electrons with different velocities produces a shift in scattered frequency proportional to the velocity along a direction $\hat{\mathbf{j}} - \hat{\mathbf{i}}$, where $\hat{\mathbf{i}}$ and $\hat{\mathbf{j}}$ are unit vectors in the direction of, respectively, the incident and scattered light propagation directions (see Equation 4.70). If the electrons are thermalized with a Maxwellian distribution, the Thomson scattered light spectral profile follows the form of Equation 7.33 with the spectral width given by Equation 7.34. With Thomson scatter off electrons, the mass M given in Equation 7.34 becomes the electron mass m_0. As $m_0 \ll M$, the scattered spectrum is spectrally broad enabling accurate electron temperature measurements using Thomson scatter.

7.6 Doppler Broadening in Highly Relativistic Plasmas

Doppler broadening of spectral line emission when atoms or ions have thermalized velocities approaching a significant fraction of the speed of light is more complicated than the Doppler broadening considered when all velocities are well below the speed of light (Section 7.3). With relativistic motion, the direction of emission or absorption which is usually isotropic in the atom or ion frame of reference is strongly altered in the laboratory frame of reference. In the laboratory frame, emission from a relativistic particle is beamed in the direction of the velocity (see Section 3.5).

For emitting atoms or ions moving relativistically towards the observer, we are going to assume that the approximation of a one-dimensional distribution of velocities is accurate as only those atoms or ions moving parallel to the observer line of sight produce light that is detected. We find that this approximation leads to an expression for the Doppler broadening from a highly relativistic plasma that is good agreement with more involved evaluations [46], [52], [53], [106].

Doppler broadening when particles have velocities approaching the speed of light needs to be evaluated using the Maxwell-Juttner expression (Equation 6.58) for the distribution of particle energies rather than the Maxwellian distribution of particle velocities (Equation 7.31). For atoms or ions moving towards the observer, the one-dimensional Maxwell-Juttner distribution is appropriate due to the beaming effect of emission from relativistic atoms and ions. The one-dimensional relativistic Maxwell-Juttner particle distribution function is also accurate for atoms or ions with a small velocity along the line of sight.

Equation 6.58 gives the one-dimensional distribution of particle numbers as a function of the Lorentz factor γ. For a distribution of atoms or ions moving at relativistic thermal velocities:

$$N_{MJ}(\gamma) = \frac{1}{Z_{1D}(T)} \gamma \left(\gamma^2 - 1\right)^{-1/2} \exp\left(-\frac{\gamma Mc^2}{k_B T}\right), \qquad (7.35)$$

where we use M for the rest mass of the atoms or ions. We now use Equation 7.29 to find an expression for the frequency v of Doppler-shifted light as a function of the Lorentz parameter γ. We assume a one-dimensional Maxwell-Juttner distribution with the atoms or ions having relativistic velocities in only one direction so that the γ parameter is solely determined by the velocity along the line of sight. Using Equation 7.29 we can write that

$$\gamma = \frac{1}{\left(1 - v^2/c^2\right)^{1/2}} = \frac{\gamma}{\left(\gamma^2 - (v/v_0 - \gamma)^2\right)^{1/2}}$$

on setting the frequency v' in the moving frame of reference to be equal to the unshifted line center frequency v_0 in the laboratory frame of reference. We then have a relationship for the Lorentz parameter to the Doppler-shifted frequencies:

$$\gamma^2 - 1 = \left(\frac{v}{v_0} - \gamma\right)^2.$$

Solving this equation for γ, we have

$$\gamma = \frac{1}{2}\left(\frac{v}{v_0} + \frac{v_0}{v}\right). \qquad (7.36)$$

We can rearrange Equation 7.36 to give the Doppler-shifted frequency ν:

$$\nu = \nu_0(\gamma \pm (\gamma^2 - 1)^{1/2}).\tag{7.37}$$

Differentiation of this expression shows that

$$\frac{d\nu}{d\gamma} = \nu_0\left(1 \pm \frac{\gamma}{(\gamma^2 - 1)^{1/2}}\right).\tag{7.38}$$

The lineshape parameter is related to the particle distribution function by equations similar to Equation 7.30. Using the Maxwell-Juttner expression (Equation 7.35) we have for the Doppler-broadened lineshape function

$$f_{MJ}(\nu) \propto N_{MJ}(\gamma)\frac{d\gamma}{d\nu} \propto \frac{\gamma\left(\gamma^2 - 1\right)^{-1/2}}{\nu_0\left(1 \pm \gamma(\gamma^2 - 1)^{-1/2}\right)}\exp\left(-\frac{\gamma Mc^2}{k_BT}\right)$$

with the relationship of γ to the Doppler-shifted frequency ν given by Equation 7.36.

We want to normalize the expression for the Doppler-broadened lineshape function $f_{MJ}(\nu)$ so that

$$\int_{\nu=0}^{\infty} \frac{\gamma\left(\gamma^2 - 1\right)^{-1/2}}{\nu_0\left(1 \pm \gamma(\gamma^2 - 1)^{-1/2}\right)}\exp\left(-\frac{\gamma Mc^2}{k_BT}\right)d\nu = 1.$$

Using Equation 7.38, we can convert the normalization integral from an integral in frequency ν to an integral in the Lorentz parameter γ:

$$\int_{\nu=0}^{\infty} \frac{\gamma\left(\gamma^2 - 1\right)^{-1/2}}{\nu_0\left(1 \pm \gamma(\gamma^2 - 1)^{-1/2}\right)}\exp\left(-\frac{\gamma Mc^2}{k_BT}\right)d\nu$$

$$= 2\int_{\gamma=1}^{\infty} \frac{\gamma\left(\gamma^2 - 1\right)^{-1/2}}{\nu_0\left(1 \pm \gamma(\gamma^2 - 1)^{-1/2}\right)}\exp\left(-\frac{\gamma Mc^2}{k_BT}\right)\frac{d\nu}{d\gamma}d\gamma$$

$$= 2\int_{\gamma=1}^{\infty} \gamma\left(\gamma^2 - 1\right)^{-1/2}\exp\left(-\frac{\gamma Mc^2}{k_BT}\right)d\gamma$$

The factor two multiplying the integrations is introduced when changing integration from frequency ν to Lorentz parameter γ because there are two frequencies associated with each γ (an up-shifted frequency and a down-shifted frequency). The integral with respect to the Lorentz parameter γ was obtained before in Section 6.6 (Equation 6.57):

$$\int_{\gamma=1}^{\infty} \gamma\left(\gamma^2 - 1\right)^{-1/2}\exp\left(-\frac{\gamma Mc^2}{k_BT}\right)d\gamma = K_1(Mc^2/k_BT),\tag{7.39}$$

where $K_1(Mc^2/k_BT)$ is a modified Bessel function of the second kind and order one evaluated at an inverse temperature value of Mc^2/k_BT. The normalized lineshape

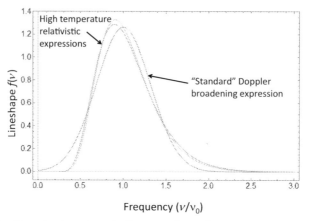

Figure 7.3 Doppler-broadened line profiles for an extreme temperature $k_B T/Mc^2 = 0.1$. The "standard" Doppler broadening expression is plotted using Equation 7.34, while the two almost superimposed curves are plotted with a more rigorous Doppler evaluation using Equations 7.40 and 7.43. All the profiles are normalized so that their integrals over frequency are equal to one.

function for Doppler broadening with particles at significant relativistic energies directed in a single direction becomes

$$f_{MJ}(\nu) = \left(\frac{1}{2K_1(Mc^2/k_BT)} \right) \left(\frac{\gamma}{\nu_0 \left((\gamma^2 - 1)^{1/2} \pm \gamma \right)} \right) \exp\left(-\frac{\gamma Mc^2}{k_BT} \right)$$

$$= \left(\frac{1}{2K_1(Mc^2/k_BT)} \right) \left(\frac{\gamma}{\nu} \right) \exp\left(-\frac{\gamma Mc^2}{k_BT} \right) \qquad (7.40)$$

with the Lorentz parameter γ given by Equation 7.36. A sample profile plotted using Equation 7.40 is given in Figure 7.3. Figure 7.3 exemplifies that at extreme temperatures, the observed Doppler-broadened profile has a maximum at a lower frequency to the unshifted frequency ν_0 (the frequency of interaction in the atomic or ionic frame of reference). The shift in the peak of Doppler-broadened profiles is examined in Exercise 7.3.

As well as being applicable to relativistic Doppler broadening of the emission from thermalized distributions of atoms or ions in a plasma, Equation 7.40 can be applied as a spectral line profile for the absorption of radiation. Absorption is enhanced at a higher shifted frequency for atom or ions moving towards the light source with the angular variation in the atom or ion frame moving to higher angles in an analogous fashion to the beaming of emission along the velocity direction.

Equation 7.40 is strictly applicable for observations of line profiles in the line center and high frequency line wing associated with thermalized atom or ions with a relativistic velocity distribution. Doppler treatments assuming a plasma with atoms or ions moving with relativistic velocities in three-dimensions allowing for beaming effects in emission or absorption produce a similar proportionality $f_{MJ}(v) \propto \exp\left(-\gamma Mc^2/k_B T\right)$ as shown in Equation 7.40. However, different published treatments present expressions with differing small amounts from the γ/v term multiplying $\exp\left(-\gamma Mc^2/k_B T\right)$ [46], [52], [53], [106].

Considering our result for the highly relativistic modification of Doppler-broadened line profiles (e.g. Figure 7.3), it seems that the treatment of highly relativistic atomic or ionic emission in this section (and in the literature) is largely for intellectual thoroughness as the temperatures required for significant departures from the treatment of Section 7.3 are so extreme that no bound electrons exist. Hydrogen has the smallest rest mass energy of any element at 938.27 MeV, so temperatures $k_B T/Mc^2 = 0.1$ (see Figure 7.3) imply temperatures for hydrogen ions of 93.827 MeV equivalent to 1.09×10^{12} degrees Kelvin. As the ionization energy of hydrogen is 13.6 eV, spectral line emission at such extreme electron temperatures does not occur as there are no bound electrons, only hydrogen nuclei (protons) and free electrons.

Equation 7.40 reduces to the low velocity Doppler expression (Equation 7.34) when $k_B T << Mc^2$. The normalizing modified Bessel function of the second kind for large $Mc^2/k_B T$ becomes

$$K_1(Mc^2/k_B T) \approx \frac{\sqrt{\pi}}{2} \left(\frac{2k_B T}{Mc^2}\right)^{1/2} \exp\left(-\frac{Mc^2}{k_B T}\right). \tag{7.41}$$

This approximation is accurate for all temperatures $k_B T < Mc^2$ (see Figure 7.4). The expression for the Doppler-broadened profile when $k_B T << Mc^2$ then simplifies to

$$f_{MJ}(v) \approx \frac{1}{\sqrt{\pi}} \left(\frac{Mc^2}{2k_B T}\right)^{1/2} \left(\frac{1}{v}\right) \exp\left(-\frac{(\gamma - 1)Mc^2}{k_B T}\right). \tag{7.42}$$

The $\gamma - 1$ term in the exponential can be expanded in terms of frequency using Equation 7.36. We have

$$\gamma - 1 = \frac{1}{2}\left(\frac{v}{v_0} + \frac{v_0}{v}\right) - 1$$

$$= \frac{v^2 + v_0^2 - 2vv_0}{2vv_0}$$

$$= \frac{(v - v_0)^2}{2vv_0}.$$

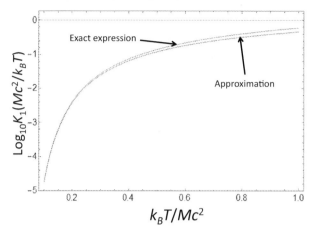

Figure 7.4   The normalization parameter $K_1(Mc^2/k_BT)$ for highly relativistic Doppler line profiles as a function of the atom or ion temperature in units of the atom or ion rest mass energy. The exact expression is shown alongside an approximation given by Equation 7.41 (as labeled).

Substituting into Equation 7.42 gives an expression for Doppler broadening when the atoms or ions have velocities much less than the speed of light:

$$f_{MJ}(v) \approx \frac{1}{\sqrt{\pi} v} \left(\frac{Mc^2}{2k_BT} \right)^{1/2} \exp \left(-\frac{(v - v_0)^2 Mc^2}{2vv_0k_BT} \right) \qquad (7.43)$$

Due to the near-identical profiles of Equations 7.40 and 7.43 for $k_BT < 0.1Mc^2$, Equation 7.43 can be used to accurately represent high temperature Doppler profiles when $k_BT < 0.1Mc^2$ (see Figure 7.3). At lower temperatures, the frequencies v are approximately equal to the line center frequency v_0 as the Doppler broadening effect is small. For an accurate representation of the line profile at temperatures $k_BT << 0.1Mc^2$, we only need to consider the frequency difference $v - v_0$ from the line center frequency v_0 within the exponential and can set $v = v_0$ elsewhere. Equation 7.43 for low velocity/low temperature Doppler broadening then becomes equivalent to Equation 7.34.

7.7 Continuum Radiation

Free electrons in a plasma can undergo acceleration in the potential field of the ions and emit radiation. The observed spectrum of emission arising from an initially free electron only varies slowly with frequency and so is known as "continuum radiation." The free electrons are also referred to as "continuum" electrons as the free electron quantum states are close together in energy. The deacceleration or "braking" of free electrons as they approach the positive potential near an ion leads to continuum radiation emission. From the German for "braking radiation," this continuum emission process is also known as "bremsstrahlung" for free-free transitions when the electron remains free after interaction with an ion. Continuum emission is referred to as recombination radiation when the electron finishes in a bound quantum state of the ionic potential.

We found in Section 1.3 that any charged accelerating body emits electromagnetic radiation. The emission arising when a free electron interacts with the potential field of an ion can be obtained using the expression for the electromagnetic radiation produced by a charge that is accelerated (see Equation 1.15). Integrating the emission of many electron-accelerating collisions over a Maxwellian distribution of electron velocities leads to an expression for the radiation at different frequencies for a plasma with a particular electron temperature T. The expression for the emission when the free electron ends in another free electron state can be extended to the situation where free electrons recombine into bound quantum states of the ions. Considering the power radiated by an accelerating charge to calculate the emission coefficient due to free electron collisions with the potential field of an ion is explained in Section 5.2 of Tallents [110].

Rather than follow the treatment of my earlier book [110] in the evaluation of bremsstrahlung emission, it is interesting to investigate another technique to calculate the rate of bremsstrahlung. We consider a method for evaluating electron collisional cross-sections and continuum radiation emission known as the "method of virtual photons" or the Weizsaker-Williams method [11]. This approach uses ideas regarding the equivalence of frames of reference so that light emission in an electron-ion collision occurs due to the creation of a pulse of electromagnetic radiation in the electron frame as the electric field of the ion passes the electron. This radiation scatters via Compton or Thomson scatter (depending on the frequency of the radiation) off the electron leading to bremsstrahlung radiation emission.

The collision of an electron with an ion can be considered in the frame of reference of the electron. Consider a collision between an electron and an ion of charge $Z_i + 1$ in the frame of reference of the electron with the electron at rest. We can assume that the ion moves with velocity **v** parallel to the x-axis on a trajectory with a constant y-coordinate such that $y = b$ with the electron at the origin of the coordinate system. The parameter b defined in this way is known as the "impact parameter" in collision theory. It is the closest distance between the colliding ion and electron if the particles did not deviate their paths during the collision. For high velocity collisions, the impact parameter b represents approximately the distance of closest approach of the two particles (see Figure 7.5).

Following the discussion regarding the Lorentz transformation of electric and magnetic fields in Section 3.2, we can show that the electrostatic field of the moving ion is transformed into a transverse pulse of electromagnetic radiation in the

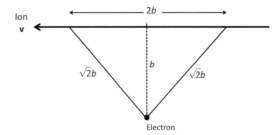

Figure 7.5   A schematic of the collision of an ion with an electron in the frame of the electron. For an impact parameter b, the electric field from the ion seen by the electron varies from half the peak value to the peak and then back to half the peak as the ion is at distances $\sqrt{2}b$, b, and then $\sqrt{2}b$ again. The schematic is for $\gamma \approx 1$ as the Lorentz contraction to $2b/\gamma$ of the distance of closest approach for the ion is not shown. As occurs for most electron-ion collisions, the path deviation of the particles due to the collision is shown as negligible.

frame of the electron. Electric and magnetic fields perpendicular to the velocity **v** transform between frames of reference according to Equation 3.27. Adapting Equation 3.27 to a transformation from the moving frame of the ion to the rest frame of the electron, the electric and magnetic fields in the direction perpendicular to **v** seen by the electron are given by

$$\mathbf{E}_{\perp} = \gamma \mathbf{E}'_{\perp}, \qquad \mathbf{B}_{\perp} = \gamma (1/c^2)(\mathbf{v} \times \mathbf{E}'_{\perp}).$$

The magnetic field $\mathbf{B}_{\perp}$ is directed perpendicular to the electric field $\mathbf{E}_{\perp}$ and the velocity **v**. Using $\beta = v/c$, the relative amplitudes of the electric and magnetic fields in the rest frame of the electron are given by

$$\frac{E_{\perp}}{B_{\perp}} = \frac{c}{\beta}.$$

As the directions of the electric and magnetic fields are perpendicular and the relative magnitudes $E_{\perp}/B_{\perp}$ are close to c for a high velocity collision ($E/B = c$ for electromagnetic radiation in vacuum; see Equation 1.11), the pulse of electric and magnetic field created in the electron frame due to the ion-electron collision appears as a pulse of light to the electron (see Figure 7.5). The β parameter modifies the phase velocity of the light to c/β in a similar way to the refractive index modification for light propagation in a medium other than vacuum (see Exercise 4.5).

The maximum electric field experienced by the electron in a close-encounter or "collision" with an ion of charge $Z_i + 1$ is given by

$$E_{\perp} = \gamma \frac{1}{4\pi \epsilon_0} \frac{(Z_i + 1)e}{b^2}. \tag{7.44}$$

In the electron frame, the time duration of an electron-ion collision can be estimated as $2b/(\gamma v)$. As illustrated in Figure 7.5 for $\gamma \approx 1$, this is approximately the time

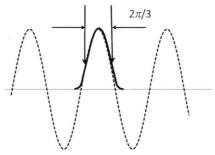

Figure 7.6   A schematic of the electric field seen by an electron due to the ion in an electron-ion collision. The electric field oscillation occupies approximately a $2\pi/3$ phase oscillation of a sinusoidal wave (superimposed).

duration $2b/v$ for the electric field of the ion seen by the electron to vary from half the peak value up to the peak value and then back to half the peak. Allowing for a Lorentz length contraction in $2b$ (to $2b/\gamma$) along the direction of the velocity $\mathbf{v}$ changes the electron-ion collision time to $2b/(\gamma v)$. In terms of the "phase" of the electric field variation, an electric field varying from half the peak value up to the peak value and then back to half the peak represents a $2\pi/3$ phase change in a sinusoidal wave (see Figure 7.6). The maximum angular frequency of the pulse of light is consequently approximately $(2\pi/3)\gamma v/2b \approx \gamma v/b$ giving a maximum frequency in units of hertz of $\gamma v/(2\pi b)$. The spectrum of the light produced in the frame of the electron is examined in Exercise 7.4.

We can estimate the energy contained in the pulse of light arising from the electron-ion collision by multiplying the energy density of electromagnetic radiation ($(1/2)\epsilon_0 E_\perp^2$; see Equation 1.12) by a measure $2\pi b^3/\gamma$ for the volume of the collision interaction. We assume that the volume of interaction is a Lorentz contracted length $2b/\gamma$ along the direction of the velocity $\mathbf{v}$ over an area of πb^2 perpendicular to the velocity (see Figure 7.7). The intensity of this energy of light is found by dividing by the frequency range of the light and multiplying by c (see the discussion leading to Equation 1.12 regarding the relationship between radiation energy density and intensity).

The spectrum of light seen by the electron varies from zero up to a maximum at the frequency $\gamma v/(2\pi b)$. Consequently, the average intensity I_e of the light pulse at the electron in the electron frame of reference in units of power per unit area per unit of frequency is

$$I_e = \frac{(1/2)\epsilon_0 c E_\perp^2\, 2\pi b^3/\gamma}{\gamma v/(2\pi b)} = 2\pi^2\epsilon_0 c \left(\frac{1}{4\pi\epsilon_0}\right)^2 \frac{(Z_i+1)^2 e^2}{v}. \tag{7.45}$$

This average intensity I_e is independent of the impact parameter b and the relativistic γ factor.

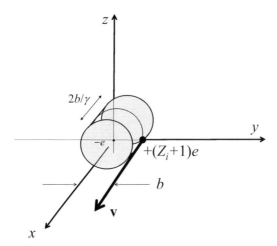

Figure 7.7   A schematic outline of an ion of charge $Z_i + 1$ moving at velocity **v** in the direction of the x-axis colliding with an electron. The illustration is in the frame of reference of the electron for an impact parameter b. The volume of interaction is a Lorentz contracted $2b/\gamma$ in the x-direction and an area of πb^2 perpendicular to the yz–plane.

To evaluate bremsstrahlung radiation, the method of virtual photons assumes that the intensity I_e of light in the electron rest frame is scattered by the electron. For photon energies much less than the electron rest mass energy $m_0 c^2 = 511$ keV, the scattering cross-section for an isolated electron is the Thomson scatter cross-section. At higher photon energies, the correct Compton scattering cross-section is needed (see Section 7.8).

The Thomson scattering cross-section σ_T is determined by evaluating the radiation emitted by an electron oscillating (and hence accelerating) in the electric field of an electromagnetic wave (see Section 4.6). We have

$$\sigma_T = \frac{8\pi}{3} r_e^2, \tag{7.46}$$

where r_e is the classical electron radius. The value of r_e is determined by the distance where the electrostatic potential energy of a sphere $e^2/(4\pi\epsilon_0 r_e)$ around the electron charge e is equal to the electron rest mass energy $m_0 c^2$. We have

$$r_e = \frac{e^2}{4\pi\epsilon_0} \frac{1}{m_0 c^2}. \tag{7.47}$$

The intensity of the Thomson scattered light pulse from one electron is $I_e \sigma_T$ which is evaluated by combining Equations 7.45, 7.46, and 7.47. We have

$$I_e \sigma_T = \frac{4\pi^2}{3} \left(\frac{e^2}{4\pi\epsilon_0}\right)^3 \frac{1}{m_0^2 c^3} \frac{(Z_i + 1)^2}{v}. \tag{7.48}$$

Equation 7.48 represents the power radiated into 4π steradian per unit frequency in a single electron-ion collision for an electron of velocity $\mathbf{v}$. If electrons with a single velocity $\mathbf{v}$ and electron density n_e collide with a plasma containing ions of density n_{Z_i+1}, the power radiated per unit volume per unit steradian per unit frequency in the electron frame can be written in the form

$$\frac{dP'}{d\nu' dV' d\Omega'} = \frac{4\pi^2}{3} \left(\frac{e^2}{4\pi\epsilon_0}\right)^3 \frac{1}{m_0^2 c^3} \frac{(Z_i+1)^2}{\nu} n_e (\gamma n_{Z_i+1}) \frac{2G_f}{\sqrt{3}} \left[\frac{3}{8\pi} \sin^2 \theta_A'\right],$$

(7.49)

where $2G_f/\sqrt{3}$ is a quantum mechanics correction of magnitude ≈ 1 known as a Gaunt factor and we have allowed for a density increase in the rest frame ion density n_{Z_i+1} due to length contraction. The acceleration of the electrons due to the virtual quanta of radiation is perpendicular to the velocity $\mathbf{v}$, so the factor $(3/8\pi) \sin^2 \theta_A'$ within the square brackets of Equation 7.49 shows the variation with the angle θ_A' of emission to the electric field of the virtual quanta in the frame of reference of the electron (according to Equation 1.15). The angle to the velocity $\mathbf{v}$ is $\pi/2 - \theta_A'$ so $\sin^2 \theta_A' = \cos^2 \theta'$, where θ' is the angle of the radiation propagation relative to the velocity $\mathbf{v}$ in the electron frame.

For electron velocities v a significant fraction of the speed of light, the power radiated in the laboratory frame is found using the transformation given by Equation 3.59. The power radiated in the laboratory frame for electrons with a single velocity $\mathbf{v}$ varies with angle θ to the velocity as

$$\frac{dP}{d\nu dV d\Omega} = \frac{8\pi^2}{3} \left(\frac{e^2}{4\pi\epsilon_0}\right)^3 \frac{1}{m_0^2 c^3} \frac{(Z_i+1)^2}{\nu} n_e n_{Z_i+1} \frac{G_f}{\sqrt{3}} \left[\frac{3}{8\pi}\right]$$
$$\left(\frac{1}{1 - v\cos\theta/c}\right)^4 \left(1 - \left(\frac{\sin\theta \, \cos\phi}{\gamma(1 - v\cos\theta/c)}\right)^2\right),$$

(7.50)

where ϕ is an azimuthal angle (see Figure 3.1). As velocities v approach the speed of light c, the emission becomes more localized within angles $< 1/\gamma$ around the direction of the velocity $\mathbf{v}$ (see the discussion in Section 3.4).

Equation 7.50 is applicable if the photon energy $h\nu$ is below the electron rest mass energy $m_0 c^2$. The Thomson scatter cross-section can then be assumed without consideration of the Compton cross-section change with frequency (see Section 4.8). Scattering where the Compton scatter cross-section is required is treated in Section 7.8.

A plasma usually comprises a distribution of electron velocities v with magnitude much less than the speed of light. The electron frame of reference signified by primed symbols in Equation 7.49 then becomes equivalent to the laboratory frame of reference and we can set $\gamma \approx 1$. With a large number of electron bremsstrahlung events, we only need to consider the angle integrated radiated power. The integration over all angles $\int (3/8\pi) \sin^2 \theta_A' d\Omega = 1$ in Equation 7.49. An angle integrated radiated power for a density of n_e electrons and n_{Z_i+1} ions in the laboratory frame for electron velocity $v << c$ becomes

$$\frac{dP}{dvdV} = \frac{8\pi^2}{3} \left(\frac{e^2}{4\pi\epsilon_0} \right)^3 \frac{1}{m_0^2 c^3} \frac{(Z_i+1)^2}{v} n_e n_{Z_i+1} \frac{G_f}{\sqrt{3}}. \tag{7.51}$$

In order to determine the bremsstrahlung emission coefficient ϵ for a distribution of electron speeds we need to integrate Equation 7.51 over the electron speed distribution function. In thermal equilibrium at nonrelativistic velocities, the electron velocity distribution is given by a Maxwellian distribution. The distribution of electrons per unit volume with speed between v and $v + dv$ for a Maxwellian distribution is given by

$$f_v(v)dv = \left(\frac{m_0}{2\pi k_B T} \right)^{3/2} 4\pi v^2 \exp\left(-\frac{m_0 v^2}{2k_B T} \right) dv. \tag{7.52}$$

To obtain the emission coefficient $\epsilon_{ff}(v)$, we need to undertake the integration

$$\epsilon_{ff}(v) = \int_{(2h\nu/m_0)^{1/2}}^{\infty} \frac{dP}{dvdV} f_v(v)dv.$$

The lower limit on the electron speed $(2h\nu/m_0)^{1/2}$ is set so that the free electrons considered have sufficient kinetic energy $(1/2)m_0 v^2$ to produce emission of a photon of energy $h\nu$. Substituting Equations 7.52 and 7.48, the emission coefficient in units of radiated power per unit volume per unit frequency in hertz assuming nonrelativistic electron velocities becomes

$$\epsilon_{ff}(v) = \frac{32\pi^3}{3} n_e n_{Z_i+1} (Z_i+1)^2 \left(\frac{e^2}{4\pi\epsilon_0} \right)^3 \frac{1}{m_0^2 c^3} \left(\frac{m_0}{2\pi k_B T} \right)^{3/2}$$

$$\int_{\sqrt{\frac{2h\nu}{m_0}}}^{\infty} \frac{G_f}{\sqrt{3}} v \exp\left(-\frac{m_0 v^2}{2k_B T} \right) dv$$

$$= \frac{16\pi^2}{3} n_e n_{Z_i+1} (Z_i+1)^2 \left(\frac{e^2}{4\pi\epsilon_0} \right)^3 \frac{1}{m_0^2 c^3} \left(\frac{m_0}{2\pi k_B T} \right)^{1/2} \frac{\tilde{G}_f}{\sqrt{3}} \int_{\frac{h\nu}{k_B T}}^{\infty} \exp(-x) \, dx$$

$$= \frac{16\pi^2}{3} n_e n_{Z_i+1} (Z_i+1)^2 \left(\frac{e^2}{4\pi\epsilon_0} \right)^3 \frac{1}{m_0^2 c^3} \left(\frac{m_0}{2\pi k_B T} \right)^{1/2} \frac{\tilde{G}_f}{\sqrt{3}} \exp\left(-\frac{h\nu}{k_B T} \right), \tag{7.53}$$

where $\tilde{G}_f$ signifies a Gaunt factor averaged over the electron velocity distribution. Equation 7.53 is identical to equations where bremsstrahlung emission is evaluated from an expression (Equation 1.15) for the radiation produced by an accelerating charge.

Expressions for free-free bremsstrahlung emission and free-bound recombination emission coefficients have been determined in Chapter 5 of Tallents [110]. The total continuum emission associated with an arbitrary ion of number density n_{Z_i+1} and charge $Z_i + 1$ associated with bremsstrahlung and recombination to ion charge Z_i is given by

$$\epsilon(v) = \epsilon_{ff}(v) + \epsilon_{fb}(v)$$

so that

$$\epsilon(v) = \frac{16\pi^2}{3} n_e n_{Z_i+1}(Z_i + 1)^2 \left(\frac{e^2}{4\pi\epsilon_0}\right)^3 \frac{1}{m_0^2 c^3} \left(\frac{m_0}{2\pi k_B T}\right)^{1/2} \exp\left(-\frac{hv}{k_B T}\right) C_0,$$

(7.54)

where

$$C_0 = \frac{1}{\sqrt{3}} \left[\tilde{G}_{ff} + \sum_n \tilde{G}_{fb} \frac{2E_n}{n} \left(\frac{u_{Z_i}}{g_{Z_i}}\right) \frac{e^{\frac{E_n}{k_B T}}}{k_B T} \right].$$

The energy E_n represents the ionization energy for an energy level of principal quantum number n in the C_O expression. The degeneracy g_{Z_i} of the lower bound level and the number of holes u_{Z_i} in this level before the recombination are included in Equation (7.54). The rate of recombination for ions which are not hydrogen-like is reduced by u_{Z_i}/g_{Z_i} as some of the quantum states are already occupied for nonhydrogen-like ions. (For hydrogen-like ions $u_{Z_i}/g_{Z_i} = 1$.) The Gaunt factors $\tilde{G}_{ff}$ and $\tilde{G}_{fb}$ represent the appropriate quantum mechanical corrections for free-free and free-bound processes respectively averaged over a Maxwellian velocity distribution. For photon energies $hv < E_n$, the Gaunt factor $\tilde{G}_{fb} = 0$ as it is not energetically possible to have a free-bound photon energy less than the ionization energy E_n. Otherwise the Gaunt factors $\tilde{G}_{ff}$ and $\tilde{G}_{fb}$ are dependent on detailed quantum mechanical calculations with values not too far from unity. The summation in the evaluation of C_0 is over all levels n with nonzero holes u_{Z_i} per level.

Important features of continuum emission due to free-free bremsstrahlung and free-bound recombination radiation include the variation of the emission coefficients $\propto \exp(-hv/k_B T)$ with photon energy hv and the proportionality of emission with $(Z_i + 1)^2 n_e n_{Z_i+1}$ which means that highly ionized, high density plasmas radiate strongly. The variation $\propto \exp(-hv/k_B T)$ enables continuum emission measurements as a function of photon energy to be used as diagnostics of electron temperature T in plasmas.

The process of inverse bremsstrahlung occurs when an electron in the field of an ion absorbs energy from an electromagnetic wave. When the radiation field is in equilibrium, the rates of bremsstrahlung and inverse bremsstrahlung are in detailed balance with bremsstrahlung emission balancing the grouping together of photon absorption and stimulated emission. An equilibrium radiation field is the black-body radiation field discussed in Section 6.4. We can obtain an expression for the absorption coefficient K_{ib} for inverse bremsstrahlung from the power emitted by bremsstrahlung using

$$K_{ib}I_{bb}(\nu) = \epsilon_{ff}(\nu),$$

where $I_p(\nu)d\nu$ is the black-body radiation intensity between frequency ν and $\nu + d\nu$ (see Equation 6.45). We have that

$$I_{bb}(\nu) = \frac{8\pi h\nu^3}{c^2}\frac{1}{\exp{(h\nu/k_BT)} - 1}.$$

The absorption coefficient for inverse bremsstrahlung can now be readily calculated. We have

$$K_{ib} = 8\sqrt{\pi}\left(\frac{e^2}{4\pi\epsilon_0}\right)^3\frac{4}{3c^3}\frac{Z_i^2}{m_0^2}n_en_{Z_i}\left(\frac{m_0}{2k_BT}\right)^{1/2}\frac{c^2}{8\pi h\nu^3}\left(1 - \exp\left(-\frac{h\nu}{k_BT}\right)\right).$$

(7.55)

The absorption of radiation due to free-free inverse bremsstrahlung varies as $n_en_{Z_i}\nu^{-3}T^{-1/2}$. The exponential term in the bracket represents the negative effect on absorption (i.e. the production of photons) associated with stimulated emission.

7.8 Continuum Emission with Relativistic Electrons

In this section we evaluate bremsstrahlung for relativistic electron-ion collisions by considering the virtual photons created in the electron frame of reference when the electric field of the ion moves rapidly past the electron. For relativistic velocities, we now extend the treatment of Section 7.7 to allow the virtual photons to have sufficient photon energy that the Compton scatter cross-section becomes relevant, rather than the frequency independent Thomson cross-section. A complete evaluation of moderate to highly relativistic bremsstrahlung including the Compton scatter cross-section, Compton frequency shifts, the Maxwell-Juttner electron energy dependence and Doppler effects necessarily involves numerical evaluations [96], [120]. We present an analytic approach and find an approximate relativistic correction to bremsstrahlung emission rates valid for electron temperatures k_BT less than the electron rest mass energy $m_0c^2 = 511$ keV.

In the electron frame where the electron has a relativistic velocity relative to the ions in a plasma, the scattered frequency v'_s is Compton shifted from the incident frequency v' (see Equation 4.92). We have for the frequency of light v'_s produced by bremsstrahlung in the electron frame that

$$v'_s = \frac{v'}{1 + (1 - \cos \theta')hv'/(m_0c^2)}, \tag{7.56}$$

where θ' is the angle between the incident virtual quanta propagation direction (parallel to the velocity $\mathbf{v}$ of the ions relative to the electron) and the scattered photon propagation direction. We want an expression relating the incident frequency v' in terms of the scattered frequency v'_s, so rearrange Equation 7.56 to

$$v' = \frac{v'_s}{1 - (1 - \cos \theta')hv'_s/(m_0c^2)}. \tag{7.57}$$

The differential Compton cross-section variation is given by the Klein-Nishina equation (Equation 4.95). We have

$$\frac{d\sigma_C}{d\Omega'} = \frac{r_e^2}{2} \left(\frac{v'}{v'_s} + \frac{v'_s}{v'} - \sin^2 \theta' \right), \tag{7.58}$$

where r_e is the Thomson scatter radius. Combining Equations 7.57 and 7.58 gives a differential Compton scatter cross-section given by

$$\frac{d\sigma_C}{d\Omega'} = \frac{r_e^2}{2} \left(1 - (1 - \cos \theta')hv'_s/(m_0c^2) + \frac{1}{1 - (1 - \cos \theta')hv'_s/(m_0c^2)} - \sin^2 \theta' \right). \tag{7.59}$$

The power radiated per unit volume per unit steradian per unit frequency in the electron frame considering the Compton scatter cross-section is found by modifying Equation 7.49. Expressing the relativistic power as a ratio R'_{rel} of the power radiated when relativistic effects are negligible, we can write that

$$\left(\frac{dP'}{dv'dV'd\Omega'} \right)_{Rel} = \left(\frac{dP'}{dv'dV'd\Omega'} \right)_0 R'_{Rel},$$

where the subscript $_{Rel}$ denotes the relativistic radiated power and the subscript $_0$ denotes the power radiated ignoring relativistic effects. The ratio of the relativistic to nonrelativistic bremsstrahlung power is given by

$$R'_{Rel} = \gamma \frac{3}{16\pi} \left(\mu'^2 - (1 - \mu')hv'_s/(m_0c^2) + \frac{1}{1 - (1 - \mu')hv'_s/(m_0c^2)} \right) \tag{7.60}$$

after substituting $\mu' = \cos \theta'$, including the Lorentz factor γ due to the relativistic increase of ion density seen by the electrons associated with a $1/\gamma$ reduction of distance in the ion velocity direction (see Equation 7.49) and dividing by the angle integrated Thomson scatter cross-section $\sigma_S = (8\pi/3)r_e^2$ (see Equation 4.68).

The angle average $<R'_{Rel}>$ of the ratio of the relativistic to nonrelativistic bremsstrahlung power in the electron frame is found by integrating over all values of μ'. The ratio R'_{Rel} is azimuthally symmetric (dependent only on the angle θ' to the velocity $\mathbf{v}$), so we have for the angle average

$$<R'_{Rel}> = 2\pi \int_{\theta'=0}^{\pi} R_{Rel} \sin\theta' d\theta' = 2\pi \int_{\mu'=-1}^{1} R'_{Rel} d\mu'$$

$$= \gamma \left(\frac{1}{4} - \frac{3h\nu'_s}{4m_0c^2} - \frac{3}{8}\frac{m_0c^2}{h\nu'_s} \ln\left(1 - 2\frac{h\nu'_s}{m_0c^2}\right) \right). \qquad (7.61)$$

The angle integrated radiated power for a density n_e of electrons of velocity v and density n_{Z_i+1} of ions in the laboratory frame is found by modifying Equation 7.51. In the electron frame of reference, the relativistic angle averaged bremsstrahlung becomes

$$\frac{dP'}{dv'_s dV'} = \frac{8\pi^2}{3} \left(\frac{e^2}{4\pi\epsilon_0} \right)^3 \frac{1}{m_0^2 c^3} \frac{(Z_i+1)^2}{c\gamma((\gamma^2-1)^{1/2}} n_e n_{Z_i+1} \frac{G_f}{\sqrt{3}} <R'_{Rel}> \qquad (7.62)$$

upon substituting $v = c\gamma(\gamma^2 - 1)^{1/2}$ for the electron velocity.

Due to the relativistic Doppler effect, the radiated power in the laboratory frame is increased by the Lorentz factor γ in moving from the electron frame (see the discussion in Section 4.6). In addition, the electron density n_e is increased by a relativistic factor of γ in the laboratory frame associated with relativistic length shortening in the electron velocity direction. Allowing for these two effects, we have that the power radiated in the laboratory frame (unprimed variables) is related to the power radiated in the electron frame (primed variables) by

$$\frac{dP}{dv_s dV} = \gamma^2 \frac{dP'}{dv'_s dV'}. \qquad (7.63)$$

We need to integrate the radiated bremsstrahlung power over the Maxwell-Juttner expression for the different electron energies (Equation 6.55). We can undertake the integration in energy by integrating as a function of the Lorentz γ parameter. The Maxwell-Juttner particle distribution function for free relativistic particles with an isotropic (three-dimensional) statistical distribution is given by

$$f_{MJ}(\gamma) = \frac{1}{Z(T)}\gamma\left(\gamma^2 - 1\right)^{1/2} \exp\left(-\frac{\gamma m_0 c^2}{k_B T}\right), \qquad (7.64)$$

where the Maxwell-Juttner normalization constant $Z(T)$ is defined by Equation 6.54. The minimum Lorentz parameter γ_{min} in the integration over all γ needed to have sufficient energy in the Bremsstrahlung collision to produce photons of energy $h\nu_s'$ is given by

$$\gamma_{min} = 1 + \frac{h\nu_s'}{m_0c^2} \approx 1 + \frac{h\nu_s}{m_0c^2} \tag{7.65}$$

for $h\nu' << m_0c^2$. We can approximate the frequency ν_s' of bremsstrahlung in the electron frame to the frequency ν_s in the laboratory frame when $h\nu_s' << m_0c^2$. The Doppler shift in frequency from the electron to laboratory frame is given by Equation 4.99:

$$\nu_s' = \nu_s\gamma_{min}\left(1 - \frac{v}{c}\cos\theta\right) \approx \nu_s\left(1 - \frac{v}{c}\cos\theta\right) \tag{7.66}$$

Taking the limits of the angle θ of the bremsstrahlung emission (scatter off the electron) to the electron velocity $\mathbf{v}$ from $\theta = 0$ to $\theta = \pi$, fractional Doppler shifts in the frequency transform from the electron to laboratory frame range over $1 \pm v/c$ for $v/c << 1$, so approximately average to an unshifted frequency.

Using Equations 7.62 and 7.63, the integration of the bremsstrahlung for different Lorentz γ parameters in the electron frame gives a relativistic expression for the free-free emission coefficient in the laboratory frame:

$$\epsilon_{Rel}(\nu_s) = \int_{\gamma_{min}}^{\infty} \gamma^2 \frac{dP'}{d\nu_s'dV'} f_{MJ}(\gamma)d\gamma$$

$$= \frac{8\pi^2}{3}\left(\frac{e^2}{4\pi\epsilon_0}\right)^3 \frac{1}{m_0^2c^3}\frac{(Z_i+1)^2}{c}n_e n_{Z_i+1}\frac{\tilde{G}_f}{\sqrt{3}}\frac{1}{Z(T)}$$

$$\left(\frac{1}{4} - \frac{3h\nu_s'}{4m_0c^2} - \frac{3}{8}\frac{m_0c^2}{h\nu_s}\ln\left(1 - 2\frac{h\nu_s}{m_0c^2}\right)\right)\int_{\gamma_{min}}^{\infty} \gamma^3 \exp\left(-\frac{\gamma m_0c^2}{k_BT}\right)d\gamma$$

Setting $x = \gamma m_0c^2/k_BT$, the integral over the Lorentz factor γ becomes

$$\int_{\gamma_{min}}^{\infty} \gamma^3 \exp\left(-\frac{\gamma m_0c^2}{k_BT}\right)d\gamma = \left(\frac{k_BT}{m_0c^2}\right)^4 \int_{x_{min}}^{\infty} x^3 \exp(-x)dx,$$

where $x_{min} = \gamma_{min}m_0c^2/k_BT = m_0c^2/k_BT + h\nu_s/k_BT$. The definite integral has the following form:

$$\int_{x_{min}}^{\infty} x^2 \exp(-x)dx = (x_{min}^3 + 3x_{min}^2 + 6x_{min} + 6)\exp(-x_{min})$$

We approximate a solution for the integral over the Lorentz factor γ:

$$\int_{\gamma_{min}}^{\infty} \gamma^3 \exp\left(-\frac{\gamma m_0c^2}{k_BT}\right)d\gamma \approx \left(\frac{k_BT}{m_0c^2}\right)\left(1 + 3\frac{k_BT}{m_0c^2} + 3\frac{h\nu_s}{m_0c^2}\right)\exp\left(-\frac{m_0c^2 + h\nu_s}{k_BT}\right)$$

valid when $k_B T < m_0 c^2$ and $h\nu_s < m_0 c^2$. The relativistic bremsstrahlung in the laboratory frame then becomes

$$\epsilon_{Rel}(\nu_s) \approx \frac{8\pi^2}{3} \left(\frac{e^2}{4\pi\epsilon_0}\right)^3 \frac{1}{m_0^2 c^3} \frac{(Z_i+1)^2}{c} n_e n_{Z_i+1} \frac{\tilde{G}_f}{\sqrt{3}} \exp\left(-\frac{m_0 c^2}{k_B T}\right) \frac{1}{Z(T)}$$

$$\left(\frac{k_B T}{m_0 c^2}\right) \exp\left(-\frac{h\nu_s}{k_B T}\right) \left[1 + 3\frac{k_B T}{m_0 c^2} + 3\frac{h\nu_s}{m_0 c^2}\right]$$

$$\left[\frac{1}{4} - \frac{3h\nu_s}{4m_0 c^2} - \frac{3}{8}\frac{m_0 c^2}{h\nu_s} \ln\left(1 - 2\frac{h\nu_s}{m_0 c^2}\right)\right]. \tag{7.67}$$

The change in the emission coefficient due to relativistic effects can be judged by comparing Equations 7.53 and 7.67. We have for the ratio of the relativistic bremsstrahlung and nonrelativistic bremsstrahlung emission coefficients

$$\frac{\epsilon_{Rel}(\nu_s)}{\epsilon_{ff}(\nu_s)} = \left(\frac{\pi}{2}\right)^{1/2} \exp\left(-\frac{m_0 c^2}{k_B T}\right) \frac{1}{2Z(T)} \left(\frac{k_B T}{m_0 c^2}\right)^{3/2}$$

$$\left[1 + 3\frac{k_B T}{m_0 c^2} + 3\frac{h\nu_s}{m_0 c^2}\right] \left[\frac{1}{4} - \frac{3h\nu_s}{4m_0 c^2} - \frac{3}{8}\frac{m_0 c^2}{h\nu_s} \ln\left(1 - 2\frac{h\nu_s}{m_0 c^2}\right)\right]. \tag{7.68}$$

The Maxwell-Juttner normalization constant $Z(T)$ is defined by Equation 6.54:

$$Z(T) = \int_{\gamma=1}^{\infty} \gamma \left(\gamma^2 - 1\right)^{1/2} \exp\left(-\frac{\gamma m_0 c^2}{k_B T}\right) d\gamma = \frac{k_B T}{m_0 c^2} K_2(m_0 c^2/k_B T), \tag{7.69}$$

where $K_2(m_0 c^2/k_B T)$ is a modified Bessel function of the second kind and order two evaluated at an inverse temperature value of $m_0 c^2/k_B T$. The large argument approximation for the Bessel function $K_2(m_0 c^2/k_B T)$ has a form (see Appendix B)

$$K_2(m_0 c^2/k_B T) \approx \left(\frac{\pi}{2}\right)^{1/2} \left(\frac{k_B T}{m_0 c^2}\right)^{1/2} \exp\left(-\frac{m_0 c^2}{k_B T}\right) \left[1 + \frac{15}{8}\frac{k_B T}{m_0 c^2}\right].$$

Using this approximation for the modified Bessel function, the ratio of the relativistic bremsstrahlung and nonrelativistic bremsstrahlung emission given by Equation 7.68 is given by

$$\frac{\epsilon_{Rel}(\nu_s)}{\epsilon_{ff}(\nu_s)} \approx \left[\frac{1 + 3k_B T/m_0 c^2 + 3h\nu_s/m_0 c^2}{1 + (15/8)k_B T/m_0 c^2}\right] \left[\frac{1}{4} - \frac{3h\nu_s}{4m_0 c^2} - \frac{3}{8}\frac{m_0 c^2}{h\nu_s} \ln\left(1 - 2\frac{h\nu_s}{m_0 c^2}\right)\right]. \tag{7.70}$$

Assuming that $k_B T \ll m_0 c^2$, we approximate

$$-\frac{3}{8}\frac{m_0 c^2}{h\nu_s} \ln\left(1 - 2\frac{h\nu}{m_0 c^2}\right) \approx \frac{3}{4}$$

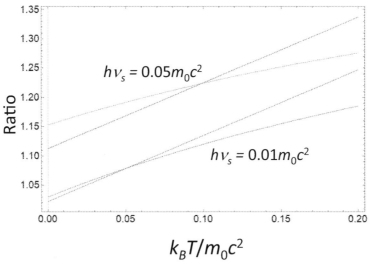

Figure 7.8 The relativistic correction $\epsilon_{Rel}(v_s)/\epsilon_{ff}(v_s)$ for bremsstrahlung and inverse bremsstrahlung as a function of electron temperature in units of the electron rest mass energy $m_0c^2 = 511$ keV. Plots at photon energies of $hv_s = 0.05m_0c^2$ and $0.01m_0c^2$ (as labeled) are shown. The straight line variations are the estimated small relativity corrections using Equation 7.71, while the curved variations are plotted using Equation 7.70.

and drop all terms involving the squares of hv/m_0c^2 and k_BT/m_0c^2 or products $(hv/m_0c^2)(k_BT/m_0c^2)$. The ratio of the relativistic bremsstrahlung and nonrelativistic bremsstrahlung emission then simplifies to

$$\frac{\epsilon_{Rel}(v_s)}{\epsilon_{ff}(v_s)} \approx 1 + \frac{9}{8}\frac{k_BT}{m_0c^2} + \frac{9}{4}\frac{hv_s}{m_0c^2}. \tag{7.71}$$

We find that there is a multiplying $1 + (9/8)k_BT/m_0c^2 + (9/4)hv/m_0c^2$ correction to the bremsstrahlung rate of emission when the electron temperature is sufficiently large for relativistic effects to start to become important. Example of the relative change of bremsstrahlung due to relativity effects are shown in Figure 7.8.

Detailed treatments of relativistic electron-ion bremsstrahlung show an increase in relativistic bremsstrahlung compared to nonrelativistic bremsstrahlung approximately proportional to the electron temperature. A numerical treatment found that relativistic bremsstrahlung relative to a nonrelativistic treatment increases approximately linearly with electron temperature and photon energy $\epsilon_{Rel}(v)/\epsilon_{ff}(v) \propto 1 + k_BT/m_0c^2 + 0.7hv/m_0c^2$ [96]. Based on a fit to relativistic bremsstrahlung emission coefficients, Rybicki and Lightman [101] give a photon energy independent temperature dependent correction for the emission coefficient $\epsilon_{Rel}(v)$ for bremsstrahlung where relativity effects are important such that $\epsilon_{Rel}(v)/\epsilon_{ff}(v) \propto 1 + 2.6k_BT/m_0c^2$.

When electrons have extreme temperatures such that most of the electrons have relativistic velocities, electron-electron bremsstrahlung can become significant. A multiplicative correction for bremsstrahlung emission at moderate electron temperatures ($k_BT < m_0c^2$) including the effect of electron-electron collisions has been obtained [57], [75]. For a hydrogen plasma, the bremsstrahlung emission coefficient including emission from electron-electron collisions is given by the following fit to a detailed evaluation of the bremsstrahlung emission:

$$\frac{\epsilon_{ee/ei}(v)}{\epsilon_{Rel}(v)} = 1 + \left(0.85 + 1.35 \left(\frac{hv}{k_BT}\right)^{1/2} + 0.38 \left(\frac{hv}{k_BT}\right)\right) \frac{k_BT}{m_0c^2}, \qquad (7.72)$$

where $\epsilon_{Rel}(v)$ is the electron-ion emission coefficient calculated including relativistic effects. Equation 7.72 is valid for electron temperatures $k_BT < m_0c^2$. For electrons with temperature k_BT much greater than the electron rest mass energy m_0c^2, electron-electron collisions radiate twice as much energy as electron-proton collisions [2]. In electron-electron collisions at extreme temperatures both electrons radiate, whereas with an electron-proton collision, only the less massive electron radiates.

Inverse bremsstrahlung absorption can be deduced assuming detailed balance of photon absorption and stimulated emission with spontaneous emission as considered in Section 7.7. We obtain an expression for the absorption coefficient K_{Rib} for relativistic inverse bremsstrahlung from the power $\epsilon_{Rel}(v)$ emitted by relativistic bremsstrahlung using

$$K_{Rib} \, I_{bb}(v) = \epsilon_{Rel}(v),$$

where $I_p(v)dv$ is the black-body radiation intensity between frequency v and $v + dv$ (see Equation 6.45). The intensity of black-body radiation is invariant under a Lorentz transformation as $I_{bb}(v) \propto v^3$ (see Section 7.1). The variation of the relativistic inverse bremsstrahlung absorption coefficient from the nonrelativistic expression consequently follows the variation of the relativistic bremsstrahlung emission compared to the nonrelativistic emission (Equations 7.70 and 7.71 and Figure 7.8). A relativistic treatment of inverse bremsstrahlung has been discussed by Iglesias and Rose [56], [57].

Black-body radiation peaks at photon energies $hv = 2.8k_BT$. Using Equation 7.71, the bremsstrahlung emission and inverse bremsstrahlung absorption for a relativistic plasma at the peak of the black-body spectrum can be expected to increase approximately $\propto 1 + 7k_BT/m_0c^2$.

7.9 Inverse Compton Scattering

When the photon energy increases during light scattering (rather than decreases), the process is called "inverse Compton scattering." If photons are scattered by electrons with a high velocity, the photon energies can increase on average, with energy transferred from the electrons to the photons. Compton scattering by free electrons of light propagating isotropically at all angles is a feature of many astrophysical plasmas and is considered in this section. An example where this is important is the effect on the spectrum of black-body cosmic microwave background scattering in "clouds" of low density plasma [14]. As a starting point, we consider single electron scattering in this section where a photon only interacts once with an electron. We follow the treatment of Blumenthal and Gould [11] (see also Chapter 7 of Rybicki and Lightman [101]).

Single electron Thomson and Compton scatter was considered in Sections 4.6 and 4.8. For the treatment in this section, we use the total scattering cross-section σ_S as obtained in Equation 4.68 which is independent of the photon energy for incident photon energies much less than the electron rest mass energy (see Figure 4.7). Even allowing for Doppler shifts of frequency into the electron rest frame, such a condition is well satisfied, for example, when ultra-violet light to microwaves scatter off relativistic electrons.

In unit time, photons from a distance c (the distance light travels in unit time) can scatter off a single electron, so $c\,N(E_p)dE_P$ photons of energy E_p to $E_p + dE_p$ are incident, where $N(E_p)$ is the number density of photons with photon energy E_p to $E_p + dE_p$. Multiplying by the energy E_p of the photons and integrating over all photon energies gives the total incident power of all the photons. Considering all quantities in the frame of the electron and signifying this using a prime on the variables, the total power P' scattered in the electron frame of reference by a single electron is

$$P' = \sigma_S\, c \int E'_p\, N(E'_p)\, dE'_p. \tag{7.73}$$

The number density $N(E_p)$ of photons per unit of energy per unit of volume is Lorentz invariant. In transforming from the moving to rest frame, length contraction ensures that volumes scale as $1/\gamma$, while energy scales as γ so that a measure per unit volume per unit energy is independent of γ. Similarly, power P (energy per unit time) is Lorentz invariant as energy and time both scale as γ when transforming from the moving to rest frame. A similar argument shows that the quantity dE_p/E_p is Lorentz invariant. Utilizing the Lorentz invariance of P, $N(E_p)$ and dE_p/E_p, we write that

$$P = P' = \sigma_S\, c \int E'^2_p \frac{N(E'_p)\, dE'_p}{E'_p} = \sigma_S\, c \int E'^2_p \frac{N(E_p)\, dE_p}{E_p}. \tag{7.74}$$

We assume that photons have the same photon energy before and after scattering in the electron frame of reference. This means that in the electron frame, the photons are assumed to be in the Thomson scatter regime where the total cross-section σ_S is constant with photon energy (see Figure 4.7). We need to take account of the Doppler shifts in photon energy between the laboratory and electron frames of reference. Using Equation 2.46, scattered photon energies E'_p in the moving electron frame are related to scattered photon energies E_p in the rest frame by

$$E'_p = E_p \, \gamma \left(1 - \frac{v}{c} \cos \theta \right), \tag{7.75}$$

where θ is the angle between the rest frame observer line of sight and the velocity $\mathbf{v}$ of the electron. Using Equation 7.74, the scattered power P per electron in the laboratory or rest frame can be written as

$$P = \sigma_S \, c \, \gamma^2 \int \left(1 - \frac{v}{c} \cos \theta \right)^2 N(E_p) E_p \, dE_p. \tag{7.76}$$

All the quantities expressed in Equation 7.76 are now evaluated in the laboratory or rest frame of reference.

To consider an isotropic distribution of photons interacting with an isotropic distribution of electron velocities, we need to average the angular variation of the photon propagation directions in Equation 7.76. Quantities that vary only with the angle θ to an axis have a cylindrical symmetry. With cylindrical symmetry, increments of solid angle $d\Omega$ are found by integrating solid angles $d\Omega = 2\pi \sin \theta d\theta$. We have

$$<\cos^2 \theta> = \frac{\int \cos^2 \theta d\Omega}{\int d\Omega} = \frac{2\pi \int_{\theta=0}^{\pi} \cos^2 \theta \sin \theta d\theta}{2\pi \int_{\theta=0}^{\pi} \sin \theta d\theta} = \frac{1}{2} \int_{u=-1}^{1} u^2 du = \frac{1}{3},$$

$$<\cos \theta> = \frac{\int \cos \theta d\Omega}{\int d\Omega} = \frac{2\pi \int_{\theta=0}^{\pi} \cos \theta \sin \theta d\theta}{2\pi \int_{\theta=0}^{\pi} \sin \theta d\theta} = \frac{1}{2} \int_{u=-1}^{1} u du = 0$$

upon substituting $u = \cos \theta$. The average over all solid angles of the angular variation of photon propagation directions in Equation 7.76 is given by

$$<(1 - (v/c) \cos \theta)^2> = 1 + \left(\frac{v}{c} \right)^2 <\cos^2 \theta> -2 \left(\frac{v}{c} \right) <\cos \theta> = 1 + \frac{1}{3} \left(\frac{v}{c} \right)^2. \tag{7.77}$$

Substituting Equation 7.77 into Equation 7.76, the scattered power P per electron becomes

$$P = \sigma_S \, c \, \gamma^2 \left(1 + \frac{1}{3} \left(\frac{v}{c} \right)^2 \right) U_p, \tag{7.78}$$

where U_p is the photon energy density as given by

$$U_p = \int N(E_p) E_p \, dE_p. \tag{7.79}$$

The inverse Compton scattered power P per electron given by Equation 7.78 is independent of the spectrum of photon energies and varies only with the spectrally integrated photon energy density U_p and the electron velocity v. Equation 7.78 is also valid for arbitrary values of the electron velocity v and holds for nonrelativistic and relativistic electron velocities. However, it does assume that the Thomson scattering cross-section σ_S is valid in the frame of reference of the scattering electron. For Equation 7.78 to be valid, the relativistic boost to photon energy γE_p in the electron frame for photons of energy E_p in the laboratory frame needs to be much less than the electron rest mass energy $m_0 c^2$ for σ_S to remain constant with photon energy and for the photon energies in the electron frame to remain unchanged before and after scattering.

The inverse Compton scattering process can be regarded as consisting of electrons initially absorbing photon energy and then reradiating energy as scattered photons. In the laboratory or rest frame, an electron initially gains energy due to the photons at a rate

$$P_{abs} = \sigma_S c \int N(E_p) E_p \, dE_p = \sigma_S c \, U_p. \tag{7.80}$$

The net power P_{loss} lost by an electron to the photons is given by the difference between the radiated power P per electron and the rate of electron energy "absorption":

$$P_{loss} = P - P_{abs} = \sigma_S c \, U_p \left[\gamma^2 \left(1 + \frac{1}{3} \left(\frac{v}{c} \right)^2 \right) - 1 \right]$$

The square of the Lorentz parameter γ^2 can be expanded:

$$\gamma^2 - 1 = \frac{1}{1 - (v/c)^2} - 1 = \frac{(v/c)^2}{1 - (v/c)^2} = \gamma^2 \left(\frac{v}{c} \right)^2$$

The net power lost by an electron to photons in inverse Compton scatter is then given by

$$P_{loss} = \frac{4}{3} \sigma_S c \, \gamma^2 \, U_p \left(\frac{v}{c} \right)^2. \tag{7.81}$$

The total scattering cross-section σ_S is 6.524×10^{-29} m^2, so the rate of electron power loss P_{loss} due to inverse Compton scatter is small even for highly relativistic electrons with large γ. Interestingly, it is shown in Section 5.3 that the power of cyclotron emission associated with electrons orbiting a magnetic field (see Equation 5.31) has the same dependency as Equation 7.81 if the radiation energy density U_p is replaced by the energy density U_m of the magnetic field causing the electron orbit.

The total rate of loss of electron energy P_{tot} per unit volume due to inverse Compton scatter is found by considering the electron energy distribution $f(\gamma)$. Here $f(\gamma)d\gamma$ is the number of electrons per unit volume with Lorentz factors between γ and $\gamma + d\gamma$. Treatments of equilibrium nonrelativistic and relativistic particle distribution functions are given in Sections 6.5 and 6.6, respectively. We are using the Lorentz factor to represent the electron energy as the kinetic energy in units of the electron rest mass energy m_0c^2 is given by $\gamma - 1$. The power of electron energy loss per unit volume due to inverse Compton scatter is given by

$$P_{tot} = \int_{\gamma=1}^{\infty} P_{loss} f(\gamma)d\gamma, \tag{7.82}$$

where P_{loss} is given by Equation 7.81.

For nonrelativistic electrons where $\gamma \approx 1$, the total rate of loss of electron energy P_{tot} is better evaluated in terms of velocity v:

$$P_{tot} = \int_{v=0}^{\infty} P_{loss} f_v(v)dv = \frac{4}{3}\sigma_S c U_p \int_{v=0}^{\infty} \left(\frac{v}{c}\right)^2 f_v(v)dv \tag{7.83}$$

The equilibrium distribution of electron speeds $f_v(v)$ is a Maxwellian as expressed in Equation 6.51. Substituting Equation 6.51 for the Maxwellian distribution of electrons of temperature T in Equation 7.83:

$$P_{tot} = \frac{4}{3}\sigma_S c U_p \left(\frac{m_0}{2\pi k_B T}\right)^{3/2} \frac{4\pi}{c^2} \int_{v=0}^{\infty} v^4 \exp\left(-\frac{m_0 v^2}{2k_B T}\right) dv \tag{7.84}$$

Letting $x = m_0 v^2/(2k_B T)$, we have

$$P_{tot} = \frac{4}{3}\sigma_S c U_p \frac{4k_B T}{\pi^{1/2} m_0 c^2} \int_{x=0}^{\infty} x^{3/2} \exp(-x)dx.$$

The definite integral has a value of $3\pi^{1/2}/4$. The total power per electron exchanged between the electrons and photons during inverse Compton scatter for nonrelativistic electrons with an equilibrium Maxwellian distribution becomes

$$P_{tot} = \left(\frac{4k_B T}{m_0 c^2}\right)\sigma_S c U_p. \tag{7.85}$$

Assuming single photon-electron scattering, the power per unit volume exchanged between electrons and photons for an ensemble of electrons with number density n_e is the power given in Equation 7.85 multiplied by n_e.

Equation 7.85 is valid for any photon distribution and only requires knowledge of the total photon energy density U_p (see Equation 7.79). As expressed in Equation 7.85, the power P_{tot} added to the photons by a single scattering event can be evaluated using the electron temperature $k_B T$ expressed in units of the electron rest mass energy $m_0 c^2 = 511$ keV. The total scattering cross-section σ_S is 6.524×10^{-29} m^2, so assuming the electron temperature $k_B T << m_0 c^2$, we see that single photon-electron scattering usually adds only a small amount of energy to the photons even for large electron densities n_e.

7.10 Inverse Compton Scattering in a Plasma

Isotropic radiation impinging into a plasma with free electrons can undergo multiple scattering events before the radiation emerges from the plasma. In this section, we extend the treatment of inverse Compton scatter of an isotropic radiation field as given in Section 7.9 to consider the effect on photon energy densities where a photon scatters more than once in a plasma of free electrons and ions. The ions in a plasma maintain charge neutrality, but due to their much greater mass, have much smaller cross-sections for direct interaction with electromagnetic radiation compared to the electrons, so we only need to consider scatter off electrons.[2]

The change in the energy of the photons in a photon-electron scatter event is found by dividing the power P_{tot} per electron exchanged in inverse Compton scatter between electrons and photons (Equation 7.85) by the rate at which photon-electron scatter occurs. The photon density of all photon energies per unit volume is given by $\int N(E_p)dE_p$, so the rate of photon collisions with a single electron is $\sigma_S c \int N(E_p)dE_p$. The change in photon energy density ΔU_p when a photon scatters off a density of n_e electrons is then

$$\Delta U_p = \frac{P_{tot}}{\sigma_S c \int N(E_p)dE_p} n_e. \tag{7.86}$$

The total photon energy density U_p is given by Equation 7.79 which we can rearrange in terms of an average photon energy $<E_p>$:

$$U_p = \int N(E_p)E_p \, dE_p = <E_p> \int N(E_p)dE_p \tag{7.87}$$

---

[2] Bound electrons in atoms and ions can interact with electromagnetic radiation with large cross-sections over narrow spectral ranges due to "resonances" associated with transitions between discrete quantum states (see Section 7.4). Such spectral line absorption can remove energy density from a photon field, but can often be neglected if ions are fully stripped of electrons as happens in low density hydrogen gas plasmas of astrophysical interest or the photon energies are too high or too low to be significantly absorbed by atomic or ionic bound states. Our treatment of inverse Compton scatter assumes that other energy transfer processes, such as the transfer of energy due to collisions between electrons and ions is negligible.

Combining Equation 7.85 for the rate P_{tot} of electron-photon energy exchange per electron and Equation 7.87 for the radiation energy density U_p into Equation 7.86, we can write that the increase in photon energy density ΔU_p relative to the average photon energy $<E_p>$ is given by

$$\frac{\Delta U_p}{<E_p>} = \frac{4k_BT}{m_0c^2}n_e, \tag{7.88}$$

where k_BT is the electron temperature, m_0c^2 is the electron rest mass energy and n_e is the electron density. Using Equation 7.87, we can equate a change in photon energy density ΔU_p to the change in average photon energy $\Delta<E_p>$:

$$\Delta U_p = \Delta<E_p> \int N(E_p)dE_p$$

Equation 7.88 can then be expressed as

$$\frac{\Delta<E_p>}{<E_p>} = \frac{4k_BT}{m_0c^2}n_e \left[\int N(E_p)dE_p\right]^{-1}. \tag{7.89}$$

As the number density n_e of electrons and the number of photons remains constant during scattering, Equation 7.89 indicates that after one scattering collision between photons and electrons, the average photon energy $<E_p>$ increases by a factor $(1 + 4k_BT/(m_0c^2))$. If there are N scattering collisions between photons and electrons in a plasma before the photons emerge from the plasma, then the average photon energy increases as

$$<E_p(N)> = <E_p(0)> \left(1 + \frac{4k_BT}{m_0c^2}\right)^N, \tag{7.90}$$

where $<E_p(0)>$ represents the initial average photon energy before the radiation enters the plasma and $<E_p(N)>$ is the photon energy density after a mean number N of scatterings. For a large number N of scatterings, the growth of the average photon energy becomes exponential with N. We approximate

$$\left(1 + \frac{4k_BT}{m_0c^2}\right)^N \approx \exp\left(\frac{4k_BTN}{m_0c^2}\right).$$

The Compton y parameter is a measure of the average fractional photon energy change per scattering multiplied by the number N of scatterings. We write for the Compton y parameter:

$$y = \frac{k_BTN}{m_0c^2} \tag{7.91}$$

The average photon energy variation using the Compton y parameter becomes

$$<E_p(y)>=<E_p(0)> e^{4y}. \tag{7.92}$$

The average photon energy $<E_p(y)>$ grows exponentially with the Compton y parameter from the initial average photon density $<E_p(0)>$ present before the photons entered the scattering medium.

The increase of photon energy given by Equation 7.92 is independent of the spectrum of radiation, but is particularly relevant when the spectrum is that of a black-body. Detailed theoretical examinations of the black-body spectrum of photon energies produced by inverse Compton scatter confirm that the increase in photon energy density occurs due to the number of lower energy photons decreasing and the number of higher photon energies increasing so that the average photon energy $<E_p(y)>$ increases as shown in Equation 7.92 [97].

In the astrophysical literature, the spectral distortion of the black-body cosmological background radiation caused by inverse Compton scatter off plasma clouds containing free electrons is known as the Sunyaev-Zel'dovich effect. The Sunyaev-Zel'dovich effect on the cosmological background radiation is used to map large plasma clouds existing in the early universe immediately after the decoupling of the cosmological background radiation black-body temperature from the temperature of matter [14], [99], [108].

We want to make an estimate of the number N of scattering events associated with the optical depth τ for photons scattering in a plasma. If the Compton scatter is optically thin (small τ and a small number N of scatterings), then $\tau = N$. If the Compton scatter is optically thick (large τ and a large number N of scatterings), then photons escape from the plasma via a random walk with $N = \tau^2$. To approximate both optically thin and thick scenarios, we can write

$$N \approx \tau(1 + \tau). \tag{7.93}$$

The Compton y parameter then becomes

$$y \approx \frac{k_B T}{m_0 c^2} \tau(1 + \tau). \tag{7.94}$$

7.11 Compton Scatter of a Light Beam

In Section 7.9 the single electron scattering of isotropic radiation such as a black-body radiation field with light propagating in all directions was considered. The effect of multiple scattering events in a plasma was examined in Section 7.10. In this section, we consider light scatter from a beam of light as occurs with a laser beam incident into a plasma.

For nonrelativistic electron velocities, the scattered light spectrum of incoherent Thomson scatter has a spectral variation proportional to the electron velocity distribution. With the scatter of light, there is a Doppler shift in both the absorption and re-emission processes such that the frequency v_s observed after scattering is related to the incoming frequency v by an expression found in Section 2.6 (Equation 2.49):

$$v_s = v \, \gamma^2 \left(1 + (\hat{\mathbf{j}} - \hat{\mathbf{i}}) \cdot \mathbf{v}/c \right), \tag{7.95}$$

where $\hat{\mathbf{i}}$ and $\hat{\mathbf{j}}$ are unit vectors in the direction of the incident light beam and the scattered light direction, respectively. If $v << c$, there is a simple proportionality of the shift in frequency $\Delta v_s = v_s - v$ on scattering such that

$$\frac{\Delta v_s}{v} \approx \frac{(\hat{\mathbf{j}} - \hat{\mathbf{i}}) \cdot \mathbf{v}}{c}. \tag{7.96}$$

When $v << c$, the scattered light spectrum $\Delta v_s/v$ directly reflects the number of of electrons with velocity $\mathbf{v}$ along a direction determined by the difference $\hat{\mathbf{j}} - \hat{\mathbf{i}}$ in the unit vectors. The scatter frequencies are up- and down-shifted depending on the velocity of the scattering electrons. A Maxwellian distribution of velocities results in a Gaussian spectral variation of the scattered spectrum with the spectral width proportional to the square root of the electron temperature (see the discussion in Section 7.5). Thomson scatter of incident laser light in nonrelativistic plasmas is a useful technique for measuring electron temperatures.

The scattered power $P(\gamma)$ per electron in the laboratory or rest frame due to Thompson scatter was obtained as a function of the electron Lorentz parameter γ in Section 7.9. Adapting Equation 7.76, we can write that

$$P(\gamma) = \sigma_S \, \gamma^2 \left(1 - \frac{v}{c} \cos\theta \right)^2 P_0, \tag{7.97}$$

where σ_S is the Thomson scatter cross-section, P_0 is the incident light power and θ is the angle of the electron velocity $\mathbf{v}$ to the light beam direction. The scattered power per electron when an electron of velocity v scatters light was investigated further in Section 7.9 with an angle θ averaging procedure to allow for the random orientation of electron velocities $\mathbf{v}$ to the light beam directions. Adapting Equation 7.78 using the relationship between velocity and the Lorentz parameter given by $(v/c)^2 = 1 - 1/\gamma^2$, the scattered power $P(\gamma)$ in the laboratory frame per electron of speed v and Lorentz parameter γ can be written as

$$P(\gamma) = \sigma_S \, \gamma^2 \left(1 + \frac{1}{3} \left(\frac{v}{c} \right)^2 \right) P_0 = \sigma_S \frac{1}{3} \left(4\gamma^2 - 1 \right) P_0. \tag{7.98}$$

The scattered spectrum of light from a beam passing through relativistic plasma depends on the number of electrons as a function of the Lorentz parameter γ which for an equilibrium relativistic plasma is given by the Maxwell-Juttner distribution

(Equation 6.55). The Maxwell-Juttner particle distribution function for free relativistic particles with a statistical distribution at a high temperature is given by

$$f_{MJ}(\gamma) = \frac{1}{Z(T)}\gamma\left(\gamma^2 - 1\right)^{1/2}\exp\left(-\frac{\gamma m_0 c^2}{k_B T}\right), \tag{7.99}$$

where $Z(T) = k_B T/(m_0 c^2)K_2(m_0 c^2/k_B T)$ is a normalization term used so that $\int_{\gamma=1}^{\infty} f_{MJ}(\gamma) = 1$ (see Equation 6.54). Here $K_2(m_0 c^2/k_B T)$ is a modified Bessel function of the second kind and order two (see Appendix B). The scattered power per unit length $dP_S(\gamma)/dz$ as a function of the Lorentz parameter γ for the electrons with Lorentz parameter γ is then given by

$$\frac{dP_S(\gamma)}{dz} = -\frac{1}{3}n_e\sigma_S\left(4\gamma^2 - 1\right)f_{MJ}(\gamma)P_0$$

$$= -\frac{1}{3}n_e\sigma_S\left(4\gamma^2 - 1\right)\frac{1}{Z(T)}\gamma\left(\gamma^2 - 1\right)^{1/2}\exp\left(-\frac{\gamma m_0 c^2}{k_B T}\right)P_0, \tag{7.100}$$

where n_e is the electron density. The rate of incident laser power change dP_l/dz with distance z due to scatter off all the electrons is given by an integration over all Lorentz parameters γ:

$$\frac{dP_l}{dz} = -\frac{1}{3}n_e\sigma_S P_l\int_{\gamma=1}^{\infty}\left(4\gamma^2 - 1\right)f_{MJ}(\gamma)d\gamma$$

$$= -\frac{1}{3}n_e\sigma_S P_l\int_{\gamma=1}^{\infty}\left(4\gamma^2 - 1\right)\frac{1}{Z(T)}\gamma\left(\gamma^2 - 1\right)^{1/2}\exp\left(-\frac{\gamma m_0 c^2}{k_B T}\right)d\gamma \tag{7.101}$$

With a light beam passing through a plasma, we can neglect multiple scattering events if the solid angle of detection of the light is sufficiently small and the opacity of the plasma is not too great. Photons are unlikely to scatter off two or more electrons back into the direction of the original beam.

If we express Equation 7.101 in terms of an absorption coefficient K_l and write

$$\frac{P_l}{dz} = -K_l P_l,$$

we have an expression for the "absorption coefficient" due to scattering:

$$K_l = \frac{1}{3}n_e\sigma_S\int_{\gamma=1}^{\infty}\left(4\gamma^2 - 1\right)\frac{1}{Z(T)}\gamma\left(\gamma^2 - 1\right)^{1/2}\exp\left(-\frac{\gamma m_0 c^2}{k_B T}\right)d\gamma$$

$$= \frac{n_e\sigma_S}{3K_2(m_0 c^2/k_B T)}\frac{m_0 c^2}{k_B T}\int_{\gamma=1}^{\infty}\left(3\gamma(\gamma^2-1)^{1/2}+4\gamma(\gamma^2-1)^{3/2}\right)\exp\left(-\frac{\gamma m_0 c^2}{k_B T}\right)d\gamma$$

$$= n_e\sigma_S\left(1 + 4\left(\frac{k_B T}{m_0 c^2}\right)\frac{K_3(m_0 c^2/k_B T)}{K_2(m_0 c^2/k_B T)}\right), \tag{7.102}$$

where $K_3(m_0c^2/k_bT)$ is the Bessel function of the second kind of order 3. Using standard expansions for small k_BT/m_0c^2 (see Appendix B), the two Bessel functions are approximately equal (to first order in k_BT/m_0c^2) giving a simplified expression for the absorption coefficient due to scattering:

$$K_l \approx n_e \sigma_S \left(1 + 4 \left(\frac{k_BT}{m_0c^2} \right) \right)$$
(7.103)

Equation 7.103 shows that there is a relativistic increase in scatter by a factor of $1 + 4k_BT/m_0c^2$ for a directed light beam incident into electrons of temperature T. This relative increase is consistent with the calculation of the relative increase in energy of isotropic radiation due to a single Compton scatter event in a plasma of relativistic electrons (considered in Section 7.10, see Equation 7.90). Photons scattered out of a directed beam decrease the intensity of the beam and become part of the isotropic distribution of photons in the scattering medium. The similarity of Equations 7.90 and 7.103 demonstrates that the photons scattered when a beam of light is incident into a relativistic plasma add their energy to the isotropic radiation field in the plasma at the same rate as the radiation field increase in energy with an incident isotropic radiation field.

The treatment of scatter in this section and in Sections 7.9 and 7.10 assumes that photon energies in the scattering electron frame are below the rest mass energy $m_0c^2 = 511$ keV of the electron so that the Thomson scatter cross-section σ_S is valid. This is not too hard a constraint. The photon energies are upshifted by γ in moving from the laboratory to electron frame, but the photon energies are upshifted again by γ after scatter in moving back from the electron to laboratory frame. Scattered photon energies up to $\approx 0.2\gamma m_0c^2$ have Thomson scatter cross-sections σ_S.

7.12 The Radiation Reaction Force

The energy radiated by a free charged particle must come from the kinetic energy of the particle or from a force maintaining the kinetic energy of the particle. A reduction in kinetic energy of a particle after radiation emission requires that there is a force associated with radiation emission. This force is known as the radiation reaction force or alternatively the Abraham-Lorentz force.

We can develop a formula for the radiation reaction force using energy balance during the radiation emission of charged particles. A particle with a velocity $\mathbf{u}$ and charge q radiates with a power P after integration over all angles (see Equation 1.16):

$$P = \frac{q^2}{4\pi\epsilon_0} \frac{2}{3c} \left(\frac{|\dot{\mathbf{u}}|}{c} \right)^2,$$
(7.104)

where $\dot{\mathbf{u}}$ is the acceleration of the particle. The rate of change of the kinetic energy of the particle is determined by Equation 7.104. We set this rate of energy change to be equivalent to $-\mathbf{F}_{rad} \cdot \mathbf{u}$ which is the rate of change of the work performed by a radiation reaction force $\mathbf{F}_{rad}$.

The equivalence between Equation 7.104 and $-\mathbf{F}_{rad} \cdot \mathbf{u}$ may not generally hold at any instant of time as, for example, there may be energy distributed to virtual radiation (as discussed in Section 7.7) or in nonradiative electric or magnetic fields. However, averaging in time where the particle returns to the same energy state is valid as the energy of virtual radiation or in nonradiative fields remains constant. Integrating between time t_1 and t_2, we have a balance between the rate of work done by the radiation force and the power radiated due to the charged particle acceleration:

$$-\int_{t_1}^{t_2} \mathbf{F}_{rad} \cdot \mathbf{u}\, dt = \frac{q^2}{4\pi\epsilon_0} \frac{2}{3c^3} \int_{t_1}^{t_2} \dot{\mathbf{u}} \cdot \dot{\mathbf{u}}\, dt \qquad (7.105)$$

as $|\dot{\mathbf{u}}|^2 = \dot{\mathbf{u}} \cdot \dot{\mathbf{u}}$. The right-hand side of this equation can be rearranged using integration by parts such that

$$-\int_{t_1}^{t_2} \mathbf{F}_{rad} \cdot \mathbf{u}\, dt = \frac{q^2}{4\pi\epsilon_0} \frac{2}{3c^3} \left([\dot{\mathbf{u}} \cdot \mathbf{u}]_{t_1}^{t_2} - \int_{t_1}^{t_2} \ddot{\mathbf{u}} \cdot \mathbf{u}\, dt \right),$$

where $\ddot{\mathbf{u}}$ is the second derivative of the velocity $\mathbf{u}$ with respect to time (which is equal to the derivative of the acceleration). Assuming that the particle initial state at time t_1 and final state at time t_2 are the same simplifies the analysis. In periodic motion where the particle returns to the same state at regular intervals, we can set $t_2 - t_1$ to be equal to the time period. An electron accelerated in the electric field of incident light undergoes periodic motion (see Chapter 4). With periodic motion $\dot{\mathbf{u}} \cdot \mathbf{u}$ can be set to be the same at times t_2 and t_1 so that

$$[\dot{\mathbf{u}} \cdot \mathbf{u}]_{t_1}^{t_2} = 0.$$

This analysis implies that the radiation reaction force is such that

$$\int_{t_1}^{t_2} \left(\mathbf{F}_{rad} - \frac{q^2}{4\pi\epsilon_0} \frac{2}{3c^3} \ddot{\mathbf{u}} \right) \cdot \mathbf{u}\, dt = 0$$

so that the appropriate expression for the temporally averaged radiation reaction force is given by

$$\mathbf{F}_{rad} = \frac{q^2}{4\pi\epsilon_0} \frac{2}{3c^3} \ddot{\mathbf{u}}. \qquad (7.106)$$

This formula for the radiation reaction force is proportional to $\ddot{\mathbf{u}}$ which is the derivative of acceleration.[3] Such a dependence increases the degree of the equation of

---

[3] The derivative of acceleration is known as "jerk," so $\ddot{\mathbf{u}}$ is the "rate of jerk."

motion of a particle to the third derivative of position if radiation emission is a significant energy loss process.

We considered the transformation of acceleration between frames of reference in Section 3.3. An acceleration $\dot{u}_{\parallel}$ parallel to the relative velocity between frames transforms from a frame S to a frame S' with $\dot{u}'_{\parallel} = \gamma^3 \dot{u}_{\parallel}$ (see Equation 3.33), while an acceleration $\dot{u}_{\perp}$ perpendicular to the relative velocity transforms from frame S to frame S' with $\dot{u}'_{\perp} = \gamma^2 \dot{u}_{\perp}$ (see Equation 3.34), where $\gamma = 1/(1 - u^2/c^2)^{1/2}$ is the Lorentz parameter. In order to determine the radiation reaction force in the frame S', we need to evaluate the derivative $\ddot{\mathbf{u}}'$ of the acceleration in the S' frame using the relationships for the transformation of the acceleration. We write

$$\ddot{\mathbf{u}}' = \frac{d}{dt}\left(\gamma^3 \dot{u}_{\parallel}\hat{\mathbf{i}}_{\parallel} + \gamma^2 \dot{u}_{\perp}\hat{\mathbf{i}}_{\perp}\right), \tag{7.107}$$

where $\hat{\mathbf{i}}_{\parallel}$ and $\hat{\mathbf{i}}_{\perp}$ are unit vectors in the directions parallel and perpendicular to the velocity $\mathbf{u}$, respectively. Evaluating Equation 7.107, we find

$$\ddot{\mathbf{u}}' = \frac{1}{1 - u^2/c^2}\left(\ddot{u}_{\parallel}\frac{1}{(1 - u^2/c^2)^{1/2}}\hat{\mathbf{i}}_{\parallel} + \ddot{u}_{\perp}\hat{\mathbf{i}}_{\perp} + \frac{u}{c^2}\left(\frac{3\dot{u}_{\parallel}}{(1 - u^2/c^2)^{3/2}}\hat{\mathbf{i}}_{\parallel} + 2\dot{u}_{\perp}\hat{\mathbf{i}}_{\perp}\right)\right). \tag{7.108}$$

The radiation reaction force $\mathbf{F}'_{rad}$ in a frame S' for a charged particle moving at velocity $\mathbf{u}$ is found using Equation 7.108 and the Equation 7.106 relationship

$$\mathbf{F}'_{rad} = \frac{q^2}{4\pi\epsilon_0}\frac{2}{3c^3}\ddot{\mathbf{u}}'.$$

A simpler relationship for the radiation reaction force in the frame S' is found if we define a vector $\mathbf{K}'$ equal to the derivative of the expression for a transformed acceleration assumed to be parallel to the particle velocity:

$$\mathbf{K}' = \frac{d}{dt}\left(\frac{\dot{\mathbf{u}}}{(1 - u^2/c^2)^{3/2}}\right) \tag{7.109}$$

Following Clemmow and Dougherty [17] (p. 133), the derivative of the acceleration $\ddot{\mathbf{u}}'$ observed in frame S' after some manipulation can be shown to be given by

$$\ddot{\mathbf{u}}' = \mathbf{K}' + \left(\frac{1}{(1 - u^2/c^2)^{1/2}} - 1\right)\frac{\mathbf{K}' \cdot \mathbf{u}}{u^2}\mathbf{u}.$$

The radiation reaction force in a frame S' for a charged particle moving at velocity $\mathbf{u}$ is then

$$\mathbf{F}'_{rad} = \frac{q^2}{4\pi\,\epsilon_0}\frac{2}{3c^3}\left(\mathbf{K}' + \left(\frac{1}{(1-u^2/c^2)^{1/2}} - 1\right)\frac{\mathbf{K}'\cdot\mathbf{u}}{u^2}\mathbf{u}\right). \qquad (7.110)$$

This is the relativistic form of the radiation reaction force in the laboratory frame for an accelerating charge q when the speed u of the charged particle approaches the speed of light c. When $u << c$, the radiation reaction force $\mathbf{F}'_{rad}$ can be quickly shown to be equal to the force given by the nonrelativistic expression (Equation 7.106).

Equation 7.110 illustrates that with our treatment allowing for relativistic effects and assuming periodic charged particle acceleration, the radiation reaction force is zero when $\mathbf{K}' = 0$. Many charged particle acceleration processes are periodic. For example, cyclotron emission (Section 5.3) and electron motion in light scattering (Section 4.6) are periodic.

It is feasible that an acceleration occurs at a constant rate and is not periodic. The radiation reaction force on a charge is then predicted by Equations 7.106 and 7.110 to be zero, even though energy is radiated by the accelerating charge. The inconsistency of the absence of a radiation reaction force in our treatment for constant charged particle acceleration is partly explained by the following argument. A constant acceleration does need to start at some time. Assuming a "jerk" on a charge from zero acceleration to a constant acceleration in a short time leads to identical relationships for the radiation reaction force during the jerk as given by Equations 7.106 and 7.110 (see Exercise 7.10). A charged particle undergoing constant acceleration after an initial jerk then feels a radiation reaction force which decreases rapidly in time (see Exercise 7.10). Equations 7.106 and 7.110 are quantitatively accurate representations of the radiation reaction force during periodic acceleration and short periods of jerk, though some more conceptual issues associated with radiation reaction are unresolved (see Chapter 7 of Jackson [58] for an earlier discussion of issues related to radiation reaction).

At the high electromagnetic wave intensities possible with high power focused lasers, the laser electric field is sufficiently strong that the quantum nature of radiation from plasma material in the laser focus needs to be considered. The radiation emitted by an accelerated charged particle in a strong electric field becomes stochastic rather than continuous with an intermittent reaction force associated with the creation of each photon [10]. In addition, in a strong electric field, virtual photons are not all reabsorbed by a radiating electron, but are emitted as "real" photons. The radiation reaction effect is then enhanced with a larger emission of scattered photons [10], [20]. Modeling, simulation and some experiments on the radiation reaction of electrons in focused lasers where the reduced vector potential a_o of the laser electric field significantly exceeds unity are being undertaken. The reduced vector potential a_o is defined in Section 4.1 (see Equation 4.13).

Exercises

7.1 A tokamak has hydrogen ions with an ion temperature of 1 keV at a particular position in the machine. The Balmer alpha transition of hydrogen at 656.28 nm is observed due to a charge exchange process with a neutral beam injected into the tokamak. Determine the full width at half maximum in wavelength of the Balmer alpha line profile. [1.58 nm]

7.2 For the same conditions as Exercise 7.1, determine the full width at half maximum for the Balmer alpha line profile for deuterium ions in the tokamak. [1.12 nm]

7.3 Equation 7.43 is an expression for a Doppler broadened line profile valid for low and high atomic and ionic temperatures:

$$f_{MJ}(v) \approx \frac{1}{\sqrt{\pi} v} \left(\frac{Mc^2}{2k_B T} \right)^{1/2} \exp\left(-\frac{(v - v_0)^2 Mc^2}{2vv_0 k_B T} \right)$$

Show that the maximum of this Doppler line profile is at a frequency v_{max} different from the frequency v_0 of interaction in the frame of the atom or ions by $v_{max} - v_0$, where

$$v_{max} - v_0 \approx -\frac{v_0 k_B T}{Mc^2}$$

provided $k_B T << Mc^2$.

7.4 The method of virtual photons discussed in Section 7.7 shows that the electric field variation from an ion of charge $Z_i + 1$ as it passes an electron is equivalent to a pulse of light in the frame of the electron. Extending Equation 7.44, the electric field seen by the electron varies approximately in time as

$$E(t) = \gamma \frac{1}{4\pi \epsilon_0} \frac{(Z_i + 1)e}{b^2 + (\gamma v t)^2}.$$

To obtain the frequency variation of the light produced, consider the Fourier transform of this temporal variation given by

$$E(\omega) = \int_{-\infty}^{\infty} E(t) \exp(i\omega t) dt.$$

Show that

$$E(\omega) = \frac{1}{4\epsilon_0} \frac{(Z_i + 1)e}{b v} \exp\left(-\frac{b\omega}{\gamma v} \right).$$

[You may need the definite integral result:

$$\int_0^{\infty} \frac{\cos x}{1 + (x/B)^2} dx = \frac{\pi}{2} B \exp(-B).]$$

7.5 Using the result from Exercise 7.4, show that the spectral intensity of light seen by an electron in its frame of reference during an electron-ion collision is given by

$$I(v) = \frac{\pi^2 c}{2\epsilon_0} \frac{(Z_i + 1)^2 e^2}{\gamma^2 v^4} v_0^2 \exp\left(-\frac{v}{v_0}\right),$$

where $v_0 = \gamma v/(4\pi b)$.

7.6 The solar opacity is largely determined by inverse bremsstrahlung absorption of a black-body radiation distribution. An important frequency averaged opacity is the Rosseland mean opacity K_{Ros} which varies with temperature T proportionally to $1/T^{3.5}$. Given the temperature at the center of the Sun is 1.3 keV, estimate the relativistic correction to the Rosseland mean opacity. How much would the temperature at the center of the Sun need to change to compensate for the relativistic effect if the opacity is kept constant? [Black-body radiation peaks at $hv/k_BT \approx 2.8$. Using Equation 7.71 relativistic effects produce a 1.8% increase of opacity at the black-body peak. The temperature needs to reduce by $\approx 0.5\%$.]

7.7 If electron kinetic energies are significantly below the electron rest mass energy, show that the kinetic energy of the electrons varies due to inverse Compton scatter such that

$$T_k(t) = T_k(0) \exp\left(-\frac{t}{t_0}\right)$$

with

$$t_0 = \frac{3}{8} \frac{1}{\sigma_S c U_p},$$

where the radiation energy density U_p is measured in units of the electron rest mass.

7.8 A large body of gas associated with a galaxy in the early universe has a free electron density of 3×10^6 m^{-3} and a thickness of 10^4 parsec (3×10^{20} m). Estimate the optical depth of the body due to Thomson scatter for the transmitted cosmic microwave background. [0.06]

7.9 The electrons in the body of gas for Exercise 7.8 have a temperature of 1 keV. Estimate the change in average photon energy and hence the change from the average cosmic microwave background temperature of 2.725 K due to inverse Compton scatter in the body of gas. [+1.2 mK. An apparent temperature change in the cosmic microwave background due to inverse Compton scatter is known as the Sunyaev-Zel'dovich effect.]

7.10 Consider a charge q initially at rest which is jerked to an acceleration $\Delta \dot{u}$ over a short time $t = 0$ to time $t = \Delta t$ with the acceleration then kept constant at times $t > \Delta t$. The radiation reaction force F_{rad} can be assumed to be given by a balance of the radiated power P with the work done by the radiation reaction force $F_{rad}\, u$, where u is the velocity of the charge. Show that during the short acceleration jerk:

$$F_{rad} = \frac{q^2}{4\pi \epsilon_0} \frac{2}{3c^3} \frac{\Delta \dot{u}}{\Delta t}$$

After the initial acceleration when $t > \Delta t$, show that the radiation reaction force decreases with time t such that

$$F_{rad} = \frac{q^2}{4\pi \epsilon_0} \frac{2}{3c^3} \frac{\Delta \dot{u}}{t}.$$

[Taking the limits of $\Delta \dot{u} \to d\dot{u}$ and $\Delta t \to dt$, the first equation for F_{rad} has $F_{rad} \propto d\dot{u}/dt = \ddot{u}$ and is identical to Equation 7.106.]

8

Collisional Processes

Collisions between particles enable the transfer of energy between particles and the establishment of equilibrium population distributions. Atoms, ions and electrons undergo energy transferring ion-ion, electron-electron and electron-ion collisions, while photons distribute or receive energy in interactions with atoms, ions or electrons. Though photons and particularly high energy photons exhibit particle behavior, in this chapter we concentrate on particle collisions, where particles are usually the electrons and ions present in a plasma. The interaction of photons with matter has been discussed in Chapters 4 and 7.

Some general comments regarding collisions, including their importance in particle physics are first considered. Nonrelativistic and relativistic collisional processes causing excitation and ionization of electrons in the bound quantum states of an atom or ion are then treated. Plasma collisional processes can determine the rate of light emission. Collisional excitation and ionization can determine the amount of radiation emitted and the degree of ionization when collisional de-excitation processes are less than radiative decay processes and departures from statistical equilibrium occur. In the "coronal" limit at low density where collisional de-excitation and collisional three-body recombination are small, the population of excited quantum states are determined by a balance of collisional excitation and radiative decay, while the degree of ionization is determined by a balance between collisional ionization and radiative recombination.

In high density plasmas, collisional de-excitation and collisional three-body recombination can dominate over radiative de-excitation and radiative recombination. If the radiative decay processes are smaller than the collisional decay processes, a rate equation balance occurs between the collisional excitation/collisional ionization and collisional de-excitation/three-body recombination processes. Detailed balance (see Section 7.2) then ensures that with no other competing population changing processes, the ionization populations will be determined by the Saha-Boltzmann equation and the excited quantum state populations in the ions by the Boltzmann ratio (see Section 6.2). The scenario where the populations of a plasma are given by equilibrium relationships, though the radiation field may not be in equilibrium (i.e. not a black-body distribution) is known as local thermodynamic equilibrium. Local thermodynamic equilibrium is universally abbreviated to LTE in the astrophysics and plasma literature [83].

8.1 Relativistic Collision Kinematics

High energy particle collisions can produce new particles with different rest masses. Highly energetic particle collisions in the early universe, some existing astrophysical objects and in particle accelerator laboratories can create elementary particles of matter with rest mass energies much higher than the low rest mass, more stable, particles such as protons and electrons found now in most of the universe. Studies of high energy particle collisions have led to the standard model of particle physics which explains the elementary particles observed along with their dynamics [81].

In the standard model of particle physics, protons and neutrons are made of three quarks, while electrons are an example of particles collectively known as leptons. All particles comprising an odd number of quarks are classed as baryons and have a half-integer (or multiple of one-half) spin. With a half-integer spin, baryons such as the neutron and proton are fermions. As discussed in Section 6.1, fermions have an occupancy of either one or zero in a quantum state, a criterion known as the Pauli exclusion principle. Leptons such as the electron also have a spin of one half and are fermions. The rest mass energy of protons, neutrons and electrons are the lowest (or next lowest) of the two classes (baryons and leptons) and hence they are the most stable.[1]

Other standard model particles are gauge bosons which have spin of one and in the standard model are the "force carriers" between particles, causing, for example, quarks to bind together (the W^+, W^-, and Z gauge bosons) and electric charges to attract or repel each other (photons here are the gauge bosons). We saw in Section 7.7 that the "method of virtual photons" enables the interaction between an electron and ion to be interpreted as the exchange of a photon with scattering of the photon producing bremsstrahlung. All bosons, including gauge bosons have no limit on the occupancy of a quantum state (as discussed for photons in Section 6.4).

For particle physics experiments, charged particles are accelerated along a line (a linear accelerator) or in a large diameter circle so as to minimize radiation losses

---

[1] The proton is made up of two "up" quarks and one "down" quark, while the neutron comprises two "down" and one "up" quark. An isolated neutron (mass $1.674927471 \times 10^{-27}$ kg) decays with a half-life of just over 10 minutes to a proton (mass $1.6726219 \times 10^{-27}$ kg) and a W^- boson. The W^- boson then rapidly decays into an electron and an antineutrino. The isolated proton has been observed to be stable over the lifetime of the universe. The neutron exists stably in some atomic and ionic nuclei, provided the number of protons and neutrons in the nucleus are approximately equal and the total number of protons ≤ 92.

(see Chapter 5). The energies and angular distribution of the particles resulting from collisions between particles, along with the manner of the resultant particle interaction in a detector enable the dynamics of collisional reactions and the nature of the particles created by collisions to be determined.

Particle collisions typically involve two particles of, say, rest mass m_1 and m_2 producing particles of rest mass m_a and m_b. The total energy (the rest mass energy and the kinetic energy) is conserved during the collision. Conservation of total energy gives the relationship:

$$\gamma_1 m_1 c^2 + \gamma_2 m_2 c^2 = \gamma_a m_a c^2 + \gamma_b m_b c^2, \tag{8.1}$$

where γ_1, γ_2, γ_a, and γ_b are the Lorentz parameters associated with the particles in a frame of reference, usually either the laboratory or the center of mass frame of reference. Kinetic energy can be converted into rest mass energy of a new particle (and vice versa). The conservation of total energy as illustrated in Equation 8.1 goes some way to explaining why particles with extremely high kinetic energy as produced in particle accelerators are needed to create exotic particles with high rest mass energy. The kinetic energy of an impinging low rest mass particle can only be converted to a high rest mass energy particle if the kinetic energy is sufficient. High rest mass energy particles typically decay to two lower rest mass energy particles with an energy balance as for Equation 8.1, but with only one energy term on the left-hand side.

Particle collisions must also conserve momentum. We showed in Section 2.7 that the momentum **p** of particles with velocity **v** relative to an observer is given by Equation 2.62:

$$\mathbf{p} = \gamma m_0 \mathbf{v}$$

Conservation of momentum requires that

$$\gamma_1 m_1 \mathbf{v}_1 + \gamma_2 m_2 \mathbf{v}_2 = \gamma_a m_a \mathbf{v}_a + \gamma_b m_b \mathbf{v}_b. \tag{8.2}$$

As well as conserving energy and momentum, collisional processes also conserve charge. High energy collisions where kinetic energy is converted into rest mass energy also conserve quantities involved in the standard model of particle physics, such as the baryon number, the lepton number and spin.

8.2 Collisions in Plasmas

Collisions in plasmas may be "elastic" and simply distribute kinetic energy and momentum between colliding electrons and ions, thus enabling electrons and ions to have thermal energy distributions characterized by a temperature (see Section 6.1).

Electron-ion collisions in plasmas can also be "inelastic" (as discussed in Section 8.4 of this chapter) where the collisions induce changes in the quantum state of atoms or ions. Bound electrons can move to higher energy bound excited states or to the continuum of free electrons (i.e. ionization) due to a free electron-ion collision. The inverse processes of collisional de-excitation and three-body recombination can also occur. Collisional or three-body recombination involves two free electrons interacting with an ion so that one electron recombines to a bound quantum state of a lower charged ion and the other electron remains as a free electron with a higher kinetic energy.

When the electrons, ions and atoms in a plasma undergo elastic collisions they can share energy and momentum so that a thermal distribution characterized by a temperature is produced. We have considered the nonrelativistic Maxwellian thermal distribution (Section 6.5) and the relativisitic Maxwell-Juttner distribution (Section 6.6). A thermal distribution of particle energy characterized by a temperature is produced if the rate of collisions between particles is sufficiently fast and the plasma conditions remain constant for sufficient time. In elastic collisions between similar mass particles such as electron-electron and ion-ion collisions, the charged particle momentum and energy can be readily exchanged between the colliding particles. Collisions between electrons and ions also transfer energy, but at a slower rate causing electrons and ions to often have different temperatures. Electrons equilibrate energy with ions on a timescale $\approx (1/2) \times 1,836 M_A$ longer than they equilibrate energy to other electrons. Here M_A is the atomic mass of the ion species and 1,836 is the mass ratio between protons and electrons (see Exercise 8.2 and Spitzer [107]).

The idea of a collision of an electron with an ion was considered in Section 7.7. In the frame of the ion, the acceleration of the electron produces bremsstrahlung emission. The approach in Section 7.7 evaluated the amount of bremsstrahlung emission using the idea of virtual photons in the frame of the electron. In the virtual photon model, bremsstrahlung is produced by the rapid change of electric and magnetic field from an ion near the electron being equivalent to light that then scatters from the electron.

An important measure with all collisions is the value of the cross-section for the collision. The cross-section has the dimensions of area and can be thought of as an area around the ion or atom associated with the necessary proximity of approach of the other colliding particle sufficient to cause the particular change of the ion quantum state (for inelastic collisions) or a particular deviation of the colliding particles (for elastic collisions). An impact factor b is the closest distance of approach of the two colliding particles if they continue undeviated by Coulomb (or other) forces as the collision ensues. The cross-section $d\sigma$ for a particular impact parameter range from b to $b + db$ is given by

$$d\sigma = 2\pi b \, db. \tag{8.3}$$

The maximum impact parameter b_{max} still resulting in the appropriate collisional change is related to the total cross-section σ by $\sigma = \pi b_{max}^2$. Impact parameters b and cross-sections σ are measured perpendicular to the relative velocity of colliding particles so are not affected by relativistic length contraction. Collisional impact parameters and cross-sections are both Lorentz invariant.

Inelastic collisions between electrons and ions or atoms result in changes in the quantum state of the ions or atoms. Bound electrons in an ion can transition between quantum states as a result of an electron-ion collision. The number of electron-ion collisions per unit time is proportional to the number of ions in the initial quantum state and the number of electrons approaching each ion in unit time with a sufficiently small impact parameter. Assuming that a population density N_1 of a particular ionic or atomic quantum state exists, the rate of change dN_1/dt of this population due to a beam of electrons of velocity v and density n_e can be written as

$$\frac{dN_1}{dt} = -N_1 \gamma n_e \, v \, \sigma, \tag{8.4}$$

where σ is the appropriate cross-section for the transition and γ is the Lorentz parameter for the electrons. In the frame of reference of the ions or atoms, length contraction by $1/\gamma$ in the direction of the electron velocity $\mathbf{v}$ causes the electron density n_e in the frame of the ions or atoms to increase to γn_e. For electrons with velocities $v \ll c$, the Lorentz parameter γ approaches unity and Equation 8.4 becomes the nonrelativistic expression with a rate $N_1 n_e v \sigma$. Usually there is a Maxwellian or Maxwell-Juttner distribution of electron velocities v which requires an appropriate integration over all velocities (see Sections 8.3 and 8.4).

The angular momentum of two interacting charges remains constant in an elastic collision. Consider an electron of initial momentum p in a collision with an ion. In the frame of reference of the ion, the angular momentum $J = p \, b$ of the electron is conserved in an elastic collision, while in an inelastic collision, any change in angular momentum of the electron must be compensated by a change in the angular momentum of the ion. A small range of angular momentum dJ is related to a small range of impact parameter by $dJ = p \, db$, so we can write for the cross-section $d\sigma$ with angular momentum from J to $J + dJ$ that

$$d\sigma = 2\pi b \, db = 2\pi \frac{J}{p} \frac{dJ}{p} = \frac{2\pi J \, dJ}{p^2}.$$

In an inelastic collision, the maximum change of angular momentum of the colliding electron will be from a value J_{max} before the collision to zero after the collision. This implies a maximum value of the cross-section σ_{max} for an inelastic collision such that

$$\sigma_{max} = \frac{2\pi}{p^2} \int_0^{J_{max}} J dJ = \frac{\pi}{p^2} J_{max}^2. \tag{8.5}$$

By considering the change in angular momentum of the quantum states of the ion before and after the collision, we can evaluate J_{max} and hence obtain a maximum value for the cross-section σ_{max}.

Interestingly, we also see immediately from Equation 8.5 that the maximum cross-section for an inelastic collision where the free electron kinetic energy is not conserved has a dependence inversely proportional to p^2 or proportional to $1/E$ in the nonrelativistic limit and $1/E^2$ in the highly relativistic limit, where E is the initial kinetic energy of the colliding electron. The value of cross-sections given by Equations 8.3 and 8.5 is a maximum possible value that can be used for a cross-section upper bound estimate.

8.3 Nonrelativistic Collisional Rates

As nonrelativistic cross-sections for excitation by inelastic electron collisions tend to have a variation with the kinetic energy E of the incident electron approximately proportionally to $1/E$, the cross-section for collisional excitation between two quantum levels of an ion involved in an inelastic collision is often written in terms of a collision strength $\Omega(E)$ such that

$$\sigma(E) = \frac{\Omega(E)\,\pi a_0^2}{g_1\,E}, \tag{8.6}$$

where πa_0^2 is an approximate cross-section for the ground state of the hydrogen atom (taken as the area associated with the Bohr radius a_0) and g_1 is the degeneracy of the initial quantum level (the number of quantum states and hence electron occupancy in the initial quantum level). We label the lower energy quantum state as 1 and the upper quantum state as 2 for convenience. The Bohr radius $a_0 = 0.53 \times 10^{-10}$ m is the most probable distance of the electron from the nucleus in the hydrogen atom. It is more accurate, for example, to tabulate the small variation of the collision strength $\Omega(E)$ variation with electron energy E rather than the more rapidly varying cross-section $\sigma(E)$.

The number of inelastic collisions per unit time is proportional to the cross-section $\sigma(v)$, the number $f_v(v)\,n_e$ of electrons with a particular speed and the electron speed v (as faster electrons have more collisions per unit time). The dependence $\propto 1/E$ of cross-sections with the energy E of incident electrons enables the averaging of cross-sections when there is a distribution of electron energies to proceed more readily. For an electron collision-induced transition between bound states 1 and 2 in an atom or ion with cross-section $\sigma_{12}(E)$, we define a rate coefficient K_{12} given by:

$$K_{12} = \int_{\sqrt{2\,\Delta E/m_0}}^{\infty} \sigma_{12}(v)\, v f_v(v)\, dv$$

$$= \sqrt{\frac{2}{m_0}} \int_{\Delta E}^{\infty} \sigma_{12}(E)\, E^{1/2} f_E(E) dE, \tag{8.7}$$

where ΔE is the energy difference between the lower and upper atom or ion energy states and $f_v(v)$ and $f_E(E)$ are the distribution of speeds and energy respectively normalized so that $\int_0^{\infty} f_v(v) = \int_0^{\infty} f_E(E) = 1$ (see Section 6.5). We assume throughout that ΔE is a positive quantity.

Employing the idea of a rate coefficient K_{12} enables the electron density n_e to be explicitly included in rate equations for the change of population densities. The rate dN_1/dt of electron collision-induced transitions from the lower 1 state to the upper state 2 is given by

$$\frac{dN_1}{dt} = n_e N_1 K_{12}, \tag{8.8}$$

where N_1 is the population density of the initial quantum state and n_e is the number density of electrons.

To evaluate Equation 8.7 we need expressions for the distribution of speeds $f_v(v)$ or energy $f_E(E)$. For equilibrium distributions characterized by a temperature, the Maxwellian distribution $f_E(E)$ of electron energies normalized so that $\int_0^{\infty} f_E(E) = 1$ was determined in Chapter 6 (Equation (6.52):

$$f_E(E) = \frac{2}{\sqrt{\pi}} \left(\frac{1}{k_B T} \right)^{3/2} E^{1/2} \exp\left(-\frac{E}{k_B T} \right) \tag{8.9}$$

We can expand Equation 8.7 by substituting for the Maxwellian distribution of electron energies and Equation 8.6 for the collision strength $\Omega(E)$. We obtain

$$K_{12} = \sqrt{\frac{2}{m_0}} \int_{\Delta E}^{\infty} \sigma_{12}(E)\, E^{1/2} f_E(E) dE$$

$$= 2\sqrt{\frac{2}{\pi m_0}} \left(\frac{1}{k_B T} \right)^{\frac{3}{2}} \int_{\Delta E}^{\infty} \sigma_{12}(E)\, E \exp\left(-\frac{E}{k_B T} \right) dE$$

$$= 2\sqrt{\frac{2}{\pi m_0}} \left(\frac{1}{k_B T} \right)^{\frac{3}{2}} \frac{\pi a_0^2}{g_1} \int_{\Delta E}^{\infty} \Omega_{12}(E)\, \exp\left(-\frac{E}{k_B T} \right) dE. \tag{8.10}$$

Following the convenient approximation that the collision strength $\Omega_{12}(E) = \Omega_{12}$ is approximately constant with electron kinetic energy E (see, for example, Van Regemorter [113]), we obtain

$$K_{12} = 2\sqrt{\frac{2}{\pi m_0}} \left(\frac{1}{k_B T}\right)^{\frac{1}{2}} \frac{\pi a_0^2}{g_1} \Omega_{12} \, \exp\left(-\frac{\Delta E}{k_B T}\right). \qquad (8.11)$$

The value of the rate coefficient K_{12} for collisional excitation between two atomic or ionic quantum states varies with the temperature $k_B T$ of the particles colliding with the ion, the energy gap ΔE between the atomic or ionic energy states, the degeneracy g_1 of the lower energy state and the atomic physics of the transition encapsulated in the approximately constant value of the collision strength Ω_{12} (see Chapter 12 of Tallents [110]).

We can use the principle of detailed balance to find an expression for the rate of collisional de-excitation where an atom or ion undergoes a transition from our generic energy level 2 to a lower level 1 as a result of an inelastic collision. The principle of detailed balance requires that inverse collisional and radiative processes affecting the quantum state of an atom or ion occur at equal rates for a system in thermal equilibrium. We considered an example of such a detailed balance for radiative processes in Section 7.2 where a relationship between the Einstein A and B coefficients for spontaneous emission, stimulated emission and absorption was found. We now apply detailed balance to the processes of collisional excitation and collisional de-excitation between two quantum states of an atom or ion.

The ratio of the equilibrium population between two bound quantum states depends on temperature. Knowing the rate coefficient for one process, it is possible to find the rate coefficient for the inverse process. As we discussed in Section 7.2 for the radiative Einstein coefficients, the collisional rate coefficients are simply atomic parameters and so the rate coefficients found in this way are, therefore, independent of the state of the plasma and apply to plasmas not in equilibrium (provided the electron energy distribution is the equilibrium thermal distribution).

In thermal equilibrium electron collisional excitation from quantum state 1 to 2 is balanced by electron collisional de-excitation from quantum state 2 to 1 so that the rate of change of the quantum state populations are such that $dN_2/dt = dN_1/dt$. We have:

$$n_e N_1 K_{12} = n_e N_2 K_{21}, \qquad (8.12)$$

where the populations of the quantum states N_1 and N_2 are the equilibrium populations. For nonrelativistic temperatures, the electrons of total density n_e often have a Maxwellian distribution. Using the Boltzmann relation for the quantum state population ratio N_1/N_2 (see Equation 6.27), we can relate the rate coefficient for collisional de-excitation K_{21} to that for collisional excitation K_{12} by rearranging Equation 8.12:

$$K_{21} = K_{12} \frac{g_1}{g_2} \exp\left(\frac{\Delta E}{k_B T}\right), \qquad (8.13)$$

where g_1, g_2 are the statistical weights and ΔE is the energy difference between the energy levels 1 and 2.

Electron collisional de-excitation rates vary as $n_e N_2 K_{21}$ (Equation 8.12). The competing radiative decay rate between two quantum states is given by $N_2 A_{21}$ (see Equation 7.2). There is always some low density limit where the radiative decay rate varying $\propto N_2$ (proportional to density) is greater than the electron collisional decay rate varying $\propto n_e N_2$ (proportional to the square of density). Such low density regimes where radiative decay dominates are said to be in "coronal" equilibrium due to the relevance to the solar corona. In coronal equilibrium, the emission coefficient ϵ of a spectral line is found by adapting Equation 7.19. For the total emission coefficient integrated over a spectral line for a plasma in coronal equilibrium, we have

$$<\epsilon> = \frac{1}{4\pi} N_2 A_{21} h\nu = \frac{1}{4\pi} n_e N_1 K_{12} h\nu. \tag{8.14}$$

The emission from the spectral line is proportional to the rate coefficient K_{12} for collisional excitation.

The cross-section for collisional ionization by free electrons varies in a similar way to collisional excitation exhibiting a $1/E$ dependence on the energy of the incident electron. Following Lotz [69] we can write for the cross-section for collisional ionization from a quantum state with ionization energy ΔE_{ion} that

$$\sigma_{1C}(E) = C_{ion} \, g_1 \frac{\ln(E/\Delta E_{ion})}{\Delta E_{ion} E}, \tag{8.15}$$

where C_{ion} is a proportionality constant and g_1 is the degeneracy of the initial quantum state. For electron energy $E \leq \Delta E_{ion}$, $\sigma_{1C}(E) = 0$. Experimentally, C_{ion} is a constant between 2.6 and 4.5×10^{-18} m^2 (eV)^2 [69]. The rate coefficient for collisional ionization can be written as

$$
\begin{aligned}
K_{1C} &= \sqrt{\frac{2}{m_0}} \int_{\Delta E_{ion}}^{\infty} \sigma_{1C}(E) \, E^{1/2} f_E(E) \, dE \\
&= 2\sqrt{\frac{2}{\pi m_0}} \left(\frac{1}{k_B T}\right)^{\frac{3}{2}} \int_{\Delta E_{ion}}^{\infty} \sigma_{pC}(E) \, E \, \exp\left(-\frac{E}{k_B T}\right) dE \\
&= 2\sqrt{\frac{2}{\pi m_0}} \left(\frac{1}{k_B T}\right)^{\frac{3}{2}} \frac{C_{ion} \, g_1}{\Delta E_{ion}} \int_{\Delta E_{ion}}^{\infty} \ln(E/\Delta E_{ion}) \, \exp\left(-\frac{E}{k_B T}\right) dE \\
&= 2\sqrt{\frac{2}{\pi m_0}} \left(\frac{1}{k_B T}\right)^{\frac{3}{2}} C_{ion} \, g_1 \int_{1}^{\infty} \ln(x) \, \exp(-xy) \, dx \\
&= 2\sqrt{\frac{2}{\pi m_0}} \left(\frac{1}{k_B T}\right)^{\frac{1}{2}} \frac{C_{ion} g_1}{\Delta E_{ion}} E_i(y), \tag{8.16}
\end{aligned}
$$

where $y = \Delta E_{ion}/k_B T$ and the exponential integral $E_i(y)$ is given by

$$E_i(y) = \int_y^\infty \frac{e^{-x}}{x}\, dx.$$

An approximate evaluation of the exponential integral for small $y \ll 1$ is given by

$$E_i(y) \approx -\ln y.$$

The rate coefficient for collisional ionization when $k_B T \gg \Delta E_{ion}$ is then given by

$$K_{1C} \approx 2\sqrt{\frac{2}{\pi m_0}} \frac{C_{ion} g_1}{\Delta E_{ion} (k_B T)^{1/2}} \ln\left(\frac{k_B T}{\Delta E_{ion}}\right). \tag{8.17}$$

The inverse of collisional ionization is known as three-body recombination. In collisional ionization, an impinging electron colliding with an ion (or atom) causes an electron to be released from the ion (or atom) so that after the collision, there are two free electrons and an ion. The inverse process involves two free electrons colliding with an ion so that one electron recombines to form a lower charged ion (or neutral atom) and one electron remains free. The inverse of collisional ionization involves two electrons and an ion – leading to the name of three-body recombination.

Detailed balance can be used to determine the inverse rate coefficient K_{Cp} for three-body recombination. In equilibrium the rate of collisional ionization will be equal to the rate of three-body recombination. We have

$$n_e N_1 K_{1C} = n_e n_{Z_i+1} K_{C1},$$

where n_{Z_i+1} is the population density of the $Z_i + 1$ charged ion, N_1 is the population density of the state in the Z_i-charged ion, and n_e is the electron density which is assumed to have a Maxwellian distribution. Rearranging and using the Saha-Boltzmann equation (Equation 6.26) for the equilibrium ratio of N_1/n_{Z_i+1} gives

$$K_{C1} = \frac{g_1}{g_{Z_i+1}} \frac{n_e}{2} \left(\frac{h}{m_0}\right)^3 \left(\frac{m_0}{2\pi k_B T}\right)^{3/2} \exp\left(\frac{\Delta E_{ion}}{k_B T}\right) K_{1C}. \tag{8.18}$$

The rate of three-body recombination scales $\propto n_e^2 n_{Z_i+1}$ as the rate coefficient for three-body recombination $K_{C1} \propto n_e$ (Equation 8.18). The rate of radiative recombination proceeds at a rate $n_e n_{Z_i+1} \beta_{rec}$, where β_{rec} is a radiative recombination rate which is independent of density. At low densities in coronal equilibrium the rate of radiative recombination dominates three-body recombination and the ratio of the ionization between a quantum state of population density N_1 in a Z_i charged ion and the population in a $Z_i + 1$ charged ion is given by the balance of collisional ionization and radiative recombination:

$$n_e N_1 K_{1C} = n_e n_{Z_i+1} \beta_{rec} \tag{8.19}$$

We have for the ionization balance in a coronal plasma that

$$\frac{n_{Z_i+1}}{N_1} = \frac{K_{1C}}{\beta_{rec}}.$$ (8.20)

8.4 Relativistic Collisional Rates

In this section, we extend the rate coefficient treatment for inelastic electron-ion collisions to include the effects of relativistic changes of collision rates. We found back in Section 2.8 that the total energy W of a particle is related to the particle momentum p by

$$W = \sqrt{(pc)^2 + (m_0 c^2)^2},$$ (8.21)

where $m_0 c^2$ is the rest mass energy. The kinetic energy E of a particle is related to the total energy W by $E = W - m_0 c^2$. Rearranging Equation 8.21, we obtain the relativistically correct relationship between momentum p and kinetic energy E:

$$\begin{aligned}(pc)^2 &= W^2 - (m_0 c^2)^2 \\ &= (E + (m_0 c^2))^2 - (m_0 c^2)^2 \\ &= E^2 + 2E\, m_0 c^2\end{aligned}$$ (8.22)

The cross-section for collisional excitation between two quantum levels of an ion involved in an inelastic collision with nonrelativistic electrons was written in Section 8.3 in terms of a collision strength $\Omega(E)$ as given by Equation 8.6:

$$\sigma(E) = \frac{\Omega(E)\,\pi a_0^2}{g_1\, E}$$

In an electron-ion collision, cross-sections vary $\propto 1/p^2$ (Equation 8.5). Rather than convert the kinetic energy E to momentum using the nonrelativsitic expression $E = p^2/2m_0$, we use Equation 8.22. The appropriate relativistic expression for the inelastic electron-ion collision cross-section $\sigma(E)$ becomes

$$\sigma(E) = \frac{\Omega(E)\,\pi a_0^2}{g_1}\frac{2m_0 c^2}{E^2 + 2E\, m_0 c^2}.$$ (8.23)

When $E \ll m_0 c^2$, this expression reduces to Equation 8.6 as used for the nonrelativistic evaluation of rate coefficients. We can convert the variation of kinetic energy E to a variation of the Lorentz γ factor using $E = (\gamma - 1)m_0 c^2$. Equation 8.23 then gives the cross-section $\sigma(\gamma)$ as a function of γ:

$$\sigma(\gamma) = \frac{\Omega(\gamma)\,\pi a_0^2}{g_1\, m_0 c^2}\frac{2}{(\gamma - 1)^2 + 2(\gamma - 1)} = \frac{\Omega(\gamma)\,\pi a_0^2}{g_1\, m_0 c^2}\frac{2}{\gamma^2 - 1}$$ (8.24)

We define a relativistic rate coefficient K_{12}^* so that the relativistic rate equations retain the same form as the nonrelativistic rate equations (Equation 8.8), viz

$$\frac{dN_1}{dt} = n_e N_1 K_{12}^*.$$

The relativistic rate coefficient K_{12}^* becomes

$$
\begin{aligned}
K_{12}^* &= \int_{\gamma=\gamma_0}^{\infty} \gamma \, v \, \sigma(\gamma) f_{MJ}(\gamma) d\gamma \\
&= \int_{\gamma=\gamma_0}^{\infty} \gamma \, c \left(1 - \frac{1}{\gamma^2}\right)^{1/2} \sigma(\gamma) f_{MJ}(\gamma) d\gamma \\
&= \int_{\gamma=\gamma_0}^{\infty} c \left(\gamma^2 - 1\right)^{1/2} \sigma(\gamma) f_{MJ}(\gamma) d\gamma,
\end{aligned}
\tag{8.25}
$$

where $\gamma_0 = 1 + \Delta E / m_0 c^2$ is the minimum Lorentz parameter equivalent to a sufficient kinetic energy ΔE to enable excitation from a lower to higher energy state separated by energy ΔE. The Lorentz parameter γ included explicitly in the first expression of the integration allows for the relativistic increase of electron density moving from the electron frame to the ion frame. The velocity v in Equation 8.25 is related to the Lorentz parameter γ by

$$v = c \left(1 - \frac{1}{\gamma^2}\right)^{1/2}.$$

The probability distribution of electrons is represented by $f_{MJ}(\gamma)$ in Equation 8.25. We assume a Maxwell-Juttner thermal variation as considered in Section 6.6. We have for the Maxwell-Juttner distribution (Equation 6.55):

$$f_{MJ}(\gamma) = \frac{1}{Z(T)} \gamma \left(\gamma^2 - 1\right)^{1/2} \exp\left(-\frac{\gamma m_0 c^2}{k_B T}\right),
\tag{8.26}$$

where

$$Z(T) = \int_{\gamma=1}^{\infty} \gamma \left(\gamma^2 - 1\right)^{1/2} \exp\left(-\frac{\gamma m_0 c^2}{k_B T}\right) d\gamma = \frac{k_B T}{m_0 c^2} K_2(m_0 c^2 / k_B T),
\tag{8.27}$$

where $K_2(m_0 c^2 / k_B T)$ is a modified Bessel function of the second kind and order two evaluated at an inverse temperature value of $m_0 c^2 / k_B T$. Substituting Equation 8.24 and the Maxwell-Juttner expression into Equation 8.25, we obtain for the relativistic rate coefficient for electron-ion collisional excitation within an ion or atom that

$$K_{12}^* = \frac{\Omega_{12} \, \pi a_0^2 \, c}{g_1 \, m_0 c^2 Z(T)} \int_{\gamma=\gamma_0}^{\infty} 2\gamma \exp\left(-\frac{\gamma m_0 c^2}{k_B T}\right) d\gamma \tag{8.28}$$

assuming that the collision strength Ω_{12} is constant with electron energy. This integral for the relativistic rate coefficient is analytically solvable:

$$K_{12}^* = \frac{2 \, \Omega_{12} \, \pi a_0^2 \, c}{g_1 \, m_0 c^2 K_2(m_0 c^2/k_B T)} \left(1 + \frac{k_B T}{m_0 c^2} + \frac{\Delta E}{m_0 c^2}\right) \exp\left(-\frac{m_0 c^2 + \Delta E}{k_B T}\right) \tag{8.29}$$

We can evaluate the electron temperatures $k_B T$ needed for relativistic effects to be important for the collisional rate coefficient K_{12}^* by considering a large amplitude asymptotic expression for the modified Bessel function $K_2(m_0 c^2/k_B T)$ (see Appendix B). When $k_B T < m_0 c^2$, we have that

$$K_2(m_0 c^2/k_B T) \approx \frac{\sqrt{\pi}}{2} \left(\frac{2k_B T}{m_0 c^2}\right)^{1/2} \exp\left(-\frac{m_0 c^2}{k_B T}\right) \left[1 + \frac{15}{8} \frac{k_B T}{m_0 c^2}\right]. \tag{8.30}$$

Substituting Equation 8.30 into the expression for the collision excitation rate coefficient K_{12}^* yields the nonrelativistic expression for the rate coefficient with a relativistic correction multiplier:

$$\begin{aligned}
K_{12}^* &\approx 2\sqrt{\frac{2}{\pi m_0}} \left(\frac{1}{k_B T}\right)^{\frac{1}{2}} \frac{\pi a_0^2}{g_1} \Omega_{12} \exp\left(-\frac{\Delta E}{k_B T}\right) \left[\frac{1 + k_B T/m_0 c^2 + \Delta E/m_0 c^2}{1 + (15/8)k_B T/m_0 c^2}\right] \\
&= K_{12} \left[\frac{1 + k_B T/m_0 c^2 + \Delta E/m_0 c^2}{1 + (15/8)k_B T/m_0 c^2}\right] \\
&\approx K_{12} \left[1 - \frac{7}{8} \frac{k_B T}{m_0 c^2} + \frac{\Delta E}{m_0 c^2}\right]
\end{aligned} \tag{8.31}$$

where K_{12} is the nonrelativistic rate coefficient as given by Equation 8.11. A numerical treatment [5] of relativistic effects on the rate of collisional excitation shows an increases in the rate $\propto 1 + \Delta E/m_0 c^2$ and a decrease $\propto 1 - k_B T/m_0 c^2$ as shown (approximately) in Equation 8.31.

The rate of collisional de-excitation is given by Equation 8.13. The rate coefficients K_{21}^* for collisional de-excitation are similarly increased $\propto 1 + \Delta E/m_0 c^2$ or decreased $\propto 1 - k_B T/m_0 c^2$ from the nonrelativistic values as given by Equation 8.31.

Following Lotz [69], we presented Equation 8.15 for the cross-section for collisional ionization from a quantum state with ionization energy ΔE_{ion}:

$$\sigma_{1C}(E) = C_{ion} \, g_1 \frac{\ln(E/\Delta E_{ion})}{\Delta E_{ion} E},$$

where C_{ion} is a proportionality constant and g_1 is the degeneracy of the initial quantum state. For the energy E in the denominator of the cross-section expression, the nonrelativistic relationship between kinetic energy E and momentum p given by $E = p^2/2m_0$ can be used to convert the ionization cross-section to a variation in momentum p and then the relativistic momentum relationship given by Equation 8.22 used to obtain the cross-section for relativistic electron-ion collisional ionization. This is appropriate as electron-ion collisional cross-sections vary $\propto 1/p^2$ (see Section 8.2). We obtain

$$\sigma_{1C}(E) = C_{ion}\, g_1 \frac{\ln(E/\Delta E_{ion})}{\Delta E_{ion}(E^2 + 2Em_0c^2)/(2m_0c^2)}. \tag{8.32}$$

The Lotz cross-section can be expressed in terms of the Lorentz γ factor using the relationship $E = (\gamma - 1)m_0c^2$ for the electron kinetic energy E. The cross-section for collisional ionization becomes

$$\sigma_{1C}(\gamma) = C_{ion}\, g_1 \frac{2}{\Delta E_{ion}m_0c^2(\gamma^2 - 1)} \ln\left(\frac{m_0c^2}{\Delta E_{ion}}(\gamma - 1)\right). \tag{8.33}$$

The rate coefficient K_{1C}^* for collisional ionization for thermalized relativistic electrons is obtained by integrating the cross-section $\sigma_{1C}(\gamma)$ over the Maxwell-Juttner distribution. The relativistic rate coefficient K_{1C}^* becomes

$$K_{1C}^* = \int_{\gamma=\gamma_0}^{\infty} \gamma\, v\, \sigma_{1C}(\gamma) f_{MJ}(\gamma) d\gamma$$

$$= \int_{\gamma=\gamma_0}^{\infty} c\left(\gamma^2 - 1\right)^{1/2} \sigma_{1C}(\gamma) f_{MJ}(\gamma) d\gamma, \tag{8.34}$$

where $\gamma_0 = 1 + \Delta E_{ion}/m_0c^2$ is the minimum Lorentz parameter corresponding to the ionization energy ΔE_{ion}. The Lorentz parameter γ included explicitly in the first expression of the integration allows for the relativistic increase of electron density moving from the electron frame to the ion frame. The velocity v in Equation 8.25 is related to the Lorentz parameter γ by

$$v = c\left(1 - \frac{1}{\gamma^2}\right)^{1/2}.$$

Substituting the Maxwell-Juttner expression (Equation 8.26) gives an expression for the relativistic electron collisional ionization rate coefficient such that:

$$K_{1C}^* = \frac{2C_{ion}\, g_1 c}{\Delta E_{ion}k_BT\, K_2(m_0c^2/k_BT)} \int_{\gamma=\gamma_0}^{\infty} \gamma\, \ln\left(\frac{m_0c^2}{\Delta E_{ion}}(\gamma - 1)\right) \exp\left(-\frac{\gamma m_0c^2}{k_BT}\right) d\gamma, \tag{8.35}$$

where $K_2(m_0c^2/k_BT)$ is a modified Bessel function of the second kind and order two evaluated at an inverse temperature value of m_0c^2/k_BT.

The logarithm in the integration in Equation 8.35 varies slowly compared to the other parameters. For an approximate evaluation of the relativistic electron collisional ionization rate coefficient K_{1C}^*, we can remove the logarithmic term from within the integration and assume an average value for the Lorentz parameter $\gamma = 1 + k_B T/m_0 c^2$ within the logarithmic evaluation. Equation 8.35 then simplifies to

$$K_{1C}^* \approx \frac{2C_{ion}\, g_{1C}}{\Delta E_{ion} k_B T\, K_2(m_0 c^2/k_B T)} \ln\left(\frac{k_B T}{\Delta E_{ion}}\right) \int_{\gamma=\gamma_0}^{\infty} \gamma \exp\left(-\frac{\gamma m_0 c^2}{k_B T}\right) d\gamma. \quad (8.36)$$

The integration of Equation 8.36 proceeds as for the integration in Equation 8.25. The modified Bessel function $K_2(m_0 c^2/k_B T)$ can be expanded using Equation 8.30 provided $k_B T < m_0 c^2$. The rate coefficient for collisional ionization when $k_B T \gg \Delta E_{ion}$ and relativity effects are important is then given by the approximate relationship

$$K_{1C}^* \approx 2\sqrt{\frac{2}{\pi m_0}} \frac{C_{ion}\, g_1}{\Delta E_{ion}\, (k_B T)^{1/2}} \ln\left(\frac{k_B T}{\Delta E_{ion}}\right) \left[\frac{1 + k_B T/m_0 c^2 + \Delta E_{ion}/m_0 c^2}{1 + (15/8)k_B T/m_0 c^2}\right].$$
$$(8.37)$$

The relativistic rate coefficient for collisional ionization K_{1C}^* is related to the nonrelativistic rate coefficient K_{1C} by the same multiplier as found for the collisional excitation rate coefficients (Equation 8.31). We have

$$K_{1C}^* = K_{1C}\left[\frac{1 + k_B T/m_0 c^2 + \Delta E_{ion}/k_B T}{1 + (15/8)k_B T/m_0 c^2}\right]$$
$$\approx K_{1C}\left[1 - \frac{7}{8}\frac{k_B T}{m_0 c^2} + \frac{\Delta E_{ion}}{k_B T}\right] \quad (8.38)$$

for $k_B T \ll m_0 c^2$.

The rate coefficient K_{C1}^* for three-body recombination is found assuming detailed balance of the rate of electron-ion ionization with the rate of three-body recombination as undertaken for the nonrelativistic relationship (Equation 8.18). Using the relativistic form of the Saha-Boltzmann for the ratio N_{Z_i}/N_{Z_i+1} of equilibrium populations (Equation 6.38), we have for the rate coefficient K_{C1}^* for three-body recombination:

$$K_{C1}^* = K_{1C}^* \frac{N_{Z_i}}{N_{Z_i+1}}$$
$$= K_{C1}\left[\frac{1 + k_B T/m_0 c^2 + \Delta E_{ion}/k_B T}{(1 + (15/8)k_B T/m_0 c^2)^2}\right]$$
$$\approx K_{C1}\left[1 - \frac{11}{4}\frac{k_B T}{m_0 c^2} + \frac{\Delta E_{ion}}{k_B T}\right] \quad (8.39)$$

for $k_B T \ll m_0 c^2$. A numerical treatment [5] of relativistic effects on the rate of collisional ionization and three-body recombination shows an increases in the rate

$\propto 1 + \Delta E_{ion}/k_B T$ and a decrease $\propto 1 - k_B T/m_0 c^2$ for collisional ionization and $\propto 1 - 2k_B T/m_0 c^2$ for three-body recombination as found (approximately) above (see Equation 8.38 for ionization and Equation 8.39 for three-body recombination).

The treatments presented in this section give a relativistic correction to collisional rate coefficients. General expression for the rate coefficients applicable for any electron energy are followed by approximate expressions showing relativistic rate coefficient deviations from nonrelativistic values. The approximate expressions are applicable when the electron temperature in energy units is less than the rest mass energy (511 keV) of electrons. We employed rate coefficients to evaluate the relativistic effects on collisional rate coefficients following the approximations of Van Regemorter [113] for collisional excitation and Lotz [69] for collisional ionization. The Lotz and Van Regemorter general expressions for rate coefficients are not as accurate as specific detailed atomic physics evaluations (see e.g. the comments by Sampson and Zhang [102]), but are sufficiently accurate to model the relative rate coefficient changes due to relativistic free electron effects. The relativistic corrections can be applied to rate coefficients determined explicitly and accurately for particular ionic excitation or ionization.

Exercises

8.1   Consider a collision between a particle of mass m_0 and a stationary particle of mass M, where the particle of mass m_0 changes the direction of its velocity by an angle of $\pi/2$ as a result of the collision. Show that the ratio R of the energy of the particle of mass m_0 before the collision to the energy of the particle of mass M after the collision is given by

$$R = \frac{M}{2m_0} + \frac{1}{2}.$$

[When $M \gg m_0$, there is little energy exchange between the mass m_0 particle and the particle of mass M.]

8.2   In Coulomb interactions between ions and electrons in plasmas, a "collision" is usually defined as occurring when the electron velocity changes direction by an angle of $\pi/2$ as considered in Exercise 8.1. If the frequency of such collisions is ν_{ei}, use the result of Exercise 8.1 to show that high energy electrons of mass m_0 with energy E_e add energy to low energy ions of mass M such that the ion energy increases at rate

$$\frac{dE_I}{dt} \approx \frac{E_e}{\tau_{ei}},$$

where the timescale

$$\tau_{ei} = \frac{1}{2\nu_{ei}} \left(\frac{M}{m_0} + 1 \right).$$

[This result illustrates how electron-ion collisions transfer energy a factor slower by $(1/2)(M/m_0) = (1/2) \times 1836 M_A$ compared to electron-electron and ion-ion collisions where the two masses are equal $(m_0 = M)$. Here M_A is the atomic mass of the ions.]

8.3 Consider emission on two spectral lines from an ion in a "coronal" plasma. In coronal plasmas, emission from excited quantum states is directly proportional to the rate of electron collisional excitation from the ground state of the ion. If the excited states of the ion have energies ΔE_{12} and ΔE_{13} above the ground state energy, show that emission coefficients ϵ_{12} and ϵ_{13} for the two lines are related by

$$\frac{\epsilon_{13}}{\epsilon_{12}} = \frac{\Omega_{13}}{\Omega_{12}} \left(\frac{1 + k_B T/m_0 c^2 + \Delta E_{13}/m_0 c^2}{1 + k_B T/m_0 c^2 + \Delta E_{12}/m_0 c^2} \right) \exp\left(-\frac{\Delta E_{13} - \Delta E_{12}}{k_B T} \right),$$

where T is the electron temperature and Ω_{13} and Ω_{12} are the respective collision strengths for electron collisional excitation from the ground to the states 3 and 2, respectively.

8.4 The solar corona is a low density plasma atmosphere surrounding the Sun with a temperature ≈ 200–1000 eV. A solar flare exhibits as a rapid increase in brightness of a localized part of the corona with temperatures up to 10 keV. Hydrogen-like iron $(Z = 26)$ can emit strongly in a solar flare. The ionization energies E_n of hydrogen-like ion quantum states vary as

$$|E_n| = 13.6\frac{Z^2}{n^2} \quad \text{eV},$$

where Z is the atomic number of the ion and n is the principal quantum number. Give an estimate of the change of the ratio of hydrogen-like ion emission for transitions $n = 3$ to 1 compared to $n = 2$ to 1 between a solar flare of temperature 5 keV and 10 keV. [The emission ratio increases by 14%.]

8.5 Tokamak plasmas are in coronal equilibrium and achieve electron temperatures of 20 keV. Iron is a common impurity to an otherwise hydrogen plasma in a tokamak. Estimate the change in the emission of the hydrogen-like iron $n = 2$ to 1 transition considering relativistic effects on the rate coefficient compared to a calculation with the same atomic physics neglecting relativistic effects. $[-2\%]$

9

Relativistic Optics

Optics is the branch of physics that studies the behavior and properties of light, including in particular, the effects on light of interactions with matter. Other chapters have considered the effect of light on matter affected by relativistic motion. This chapter deals with relativistic optics, namely the effect of moving matter on light.

The science of optics is often habitually restricted to describe the behavior of visible, ultraviolet, and infrared light, or sometimes just visible light (wavelength 400–700 nm). However, as light is an electromagnetic wave, other forms of electromagnetic radiation such as X-rays, microwaves, and radio waves exhibit identical properties in vacuum and many similar properties when interacting with a medium. It is a central tenet of relativity that the velocity of light c in vacuum is the same for all frames of reference and is the same for all of the spectral range. The velocity of light, however, in a medium is not constant in all frames of reference. In a medium, light has a phase velocity given by c/η, where η is the refractive index of the medium for the propagating light at the particular frequency of the light (see Section 4.4). Any velocity varying from the speed c of light in vacuum is not Lorentz invariant.

The changing velocity of light in moving media was demonstrated by Fizeau in 1851, well before Einstein's special theory of relativity (published in 1905). Fizeau's experiment is discussed in Section 9.1 and a more general treatment of the phase velocity of light in a medium is then given. Reflections from a rapidly moving mirror are considered. Such mirrors can be created in the laboratory using high power lasers to produce a plasma surface with electron density above the critical density for reflection. The important issue of the Lorentz invariance of the polarization of light is then addressed.

9.1 The Fizeau Effect

The phase velocity of light in a moving transparent medium with refractive index $\eta \neq 1$ is observed to be different from the velocity in the same medium when it is stationary. This can be shown to be a relativistic effect where the stationary observer measures a different velocity to the velocity in the frame of the medium. The effect was demonstrated by Fizeau [42] in 1851 using flowing water and a white light interferometer before the development of the theory of relativity. When light propagates parallel to the velocity of the medium, the change in phase velocity due to the movement of the medium is known as the Fizeau effect.

Using the expressions we found for the relativistic addition of velocities in Section 2.3, it is possible to consider the transformation of the phase velocity $\mathbf{u}'$ of light in the frame of a medium moving at velocity $\mathbf{v}$ to the phase velocity $\mathbf{u}$ in the laboratory or stationary frame. Using relativistic equations for particles to transform the phase velocity of light can be justified by recourse to the correspondence principle that particles and light exhibit wave or particle properties depending on the measurement. Using relativistic equations for particles to transform the phase velocity of light in a medium is demonstrated to be valid in Section 9.2.

Equation 2.27 relates the velocity $u'_{||}$ parallel to the frame velocity $\mathbf{v}$ in a moving frame to the velocity $u_{||}$ in a stationary frame. Rewriting Equation 2.27, the velocity parallel to $\mathbf{v}$ in the laboratory frame of reference is given by

$$u_{||} = \frac{u'_{||} + v}{1 + vu'_{||}/c^2},$$ (9.1)

where $u'_{||}$ is the parallel velocity in the moving frame of reference. In the frame of a medium of refractive index η, the phase velocity of light is c/η. Substituting $u'_{||} = c/\eta$, a transparent medium with light propagating parallel to the velocity $\mathbf{v}$ of the medium will have a phase velocity of light u as observed in the laboratory frame given by

$$u = \frac{c/\eta + v}{1 + v/(\eta c)}.$$ (9.2)

For velocities of the medium $v \ll c$, we can expand this expression for the laboratory phase velocity of the light as follows:

$$u = c\frac{1/\eta + v/c}{1 + v/(c\eta)} \approx c\left(\frac{1}{\eta} + \frac{v}{c}\right)\left(1 - \frac{v}{c\eta}\right) = \frac{c}{\eta} + v\left(1 - \frac{1}{\eta^2}\right) - \frac{v^2}{c\eta}$$

If we assume $v \ll c$, the last term is negligible and we obtain

$$u \approx \frac{c}{\eta} + v\left(1 - \frac{1}{\eta^2}\right).$$ (9.3)

Equation 9.3 is the formula found experimentally by Fizeau [42]. Fizeau and others at the time (1851) of his experiments determined that changes in the velocity of light with changes of the medium velocity were consistent with the vacuum propagation of light in the "aether." His 1851 paper [42] is entitled "The hypotheses relating to the luminous [sic] aether and an experiment which appear to demonstrate that the motion of bodies alters the velocity with which light propagates itself in their interior." The $v(1 - 1/\eta^2)$ term in Equation 9.3 is known as the "Fresnel drag coefficient."

The phase velocity of light in a moving medium can be affected by the Doppler shift of the frequency of light ω from the laboratory frame to the frame of the moving medium. The shift in frequency in the moving medium frame can cause the value of the refractive index to change from what it would be at the frequency in the laboratory frame.[1]

Doppler shifts in frequency are treated in Section 2.6. In the limit of velocity $v \ll c$, Equation 2.48 gives a frequency shift of $\Delta\omega$ when the medium velocity v is parallel and in the direction of light propagation such that

$$\frac{\Delta\omega}{\omega} \approx -\frac{\eta v}{c} \tag{9.4}$$

upon adjusting the speed of light from the vacuum value c to the velocity c/η in the medium. The effect of a Doppler shift $\Delta\omega$ in frequency on the light velocity c/η in a medium of a refractive index η which varies in frequency is found by noting that

$$\frac{d(c/n)}{d\omega} = \frac{d(c/n)}{d\eta}\frac{d\eta}{d\omega} = -\frac{c}{\eta^2}\frac{d\eta}{d\omega}.$$

Increments of the light velocity due to the Doppler shift of the light frequency in the medium are then given by

$$\Delta\left(\frac{c}{\eta}\right) = -\frac{c}{\eta^2}\frac{d\eta}{d\omega}\Delta\omega = \frac{\omega}{\eta}\frac{d\eta}{d\omega}v$$

upon using Equation 9.4. Including the Doppler shift effect, the velocity of light u in a moving medium becomes

$$u \approx \frac{c}{\eta} + v\left(1 - \frac{1}{\eta^2} + \frac{\omega}{\eta}\frac{d\eta}{d\omega}\right) \tag{9.5}$$

after adding the extra Doppler shift term to Equation 9.3. The phase velocity u is affected by the Doppler shift of frequency if the refractive index in the frame of the medium changes with frequency.

---

[1] A medium has a refractive index $\eta = c/u$, where u is the phase velocity of light in the medium. Refractive indices are almost always quoted in the frame of the medium and evaluated using, for example, plasma physics (see Section 4.4). In this chapter, we evaluate phase velocities u and u' in different frames of reference. The appropriate refractive indices if required are given by c/u and c/u'.

Following Equations 9.2 and 9.3, light propagating in a moving plasma (where $\eta < 1$) with a plasma velocity in the direction of light propagation has a slower phase velocity in the laboratory frame than light propagating through a stationary plasma. In the Fizeau experiment, the phase velocity of light propagating parallel to the water velocity in flowing water is found to be greater as the refractive index of water $\eta = 1.33$.

9.2 The Phase Velocity of Light in a Moving Medium

In this section, we examine changes in the phase velocity and angle of propagation of light as measured by an observer in a nominally stationary (laboratory) frame of reference compared to observations in the frame of reference of a moving medium of refractive index η, where the light propagation is not parallel to the velocity between the observers. Section 9.1 determined the phase velocity of a medium moving parallel to the direction of propagation of light. A more general phase velocity can be determined for situations where the direction of propagation (the direction of the light wavevector $\mathbf{k}$) is at an angle of θ to the velocity $\mathbf{v}$ of the medium. We finish by showing that light propagating through a rapidly moving sheet of material of any refractive index propagates at the same angle in the laboratory frame before and after passing through the sheet.

To start our consideration of light propagation at different angles to the velocity of a relativistic medium, we use the expressions for the Lorentz transformation of the frequency and wavevector of an electromagnetic wave as given in Section 3.1. Equation 3.1 gives the Doppler frequency shift from ω in the laboratory frame to ω' in the frame of the medium:

$$\omega' = \gamma(\omega - k_{||}v), \tag{9.6}$$

where $k_{||}$ is the component of wavevector in the direction of $\mathbf{v}$ in the laboratory frame. Equation 3.2 gives the change in the components of the wavevector parallel to $\mathbf{v}$ and perpendicular to $\mathbf{v}$ from the laboratory frame (unprimed $k_{||}$ and $k_{\perp}$) to the frame of the medium (primed $k'_{||}$ and $k'_{\perp}$):

$$k'_{||} = \gamma(k_{||} - v\omega/c^2), \tag{9.7}$$

$$k'_{\perp} = k_{\perp} \tag{9.8}$$

The phase velocity $\mathbf{u}'$ in the medium is found using Equations 9.6, 9.7, and 9.8. Converting the phase velocity $u' = \omega'/k'$ to a vector form, we write

$$\mathbf{u}' = \frac{\omega'}{k'^2}\mathbf{k}'. \tag{9.9}$$

Separating the wavevector $\mathbf{k}'$ into components parallel $k'_{||}$ and perpendicular $k'_{\perp}$ to $\mathbf{v}$ means that

$$k'^2 = k'^2_{||} + k'^2_{\perp}.$$

Substituting Equation 9.6 for ω', Equation 9.7 for $k'_{||}$, and Equation 9.8 for $k'_{\perp}$ into Equation 9.9, the component of the phase velocity in the medium parallel to the velocity **v** is given by

$$u'_{||} = \frac{\omega' k'_{||}}{k'^2_{||} + k'^2_{\perp}} = \frac{\gamma^2(\omega - k_{||}v)(k_{||} - v\omega/c^2)}{k^2_{\perp} + \gamma^2(k_{||} - v\omega/c^2)^2}. \tag{9.10}$$

Similarly, the component of the phase velocity in the medium perpendicular to the velocity **v** is given by

$$u'_{\perp} = \frac{\omega' k'_{\perp}}{k'^2_{||} + k'^2_{\perp}} = \frac{\gamma(\omega - k_{||}v)k_{\perp}}{k^2_{\perp} + \gamma^2(k_{||} - v\omega/c^2)^2}. \tag{9.11}$$

We want to relate the phase velocity of light propagating in the frame of the moving medium to that in the laboratory frame (and vice versa). There are identical phase relationships for light propagation between quantities measured in the laboratory frame (using unprimed symbols) and quantities measured in the frame of reference of the medium (using primed symbols). Rearranging Equation 9.9 and removing the primes on the symbols for the wavevector **k** and phase velocity **u** gives

$$\mathbf{k} = \frac{k^2}{\omega}\mathbf{u}.$$

Substituting $k^2 = \omega^2/u^2$ means that

$$\mathbf{k} = \frac{\omega}{u^2}\mathbf{u}.$$

The components of wavevector and phase velocity parallel and perpendicular to the medium velocity **v** are then as follow:

$$k_{||} = \frac{\omega}{u^2}u_{||}, \tag{9.12}$$

$$k_{\perp} = \frac{\omega}{u^2}u_{\perp} \tag{9.13}$$

Substituting Equations 9.12 and 9.13 into Equations 9.10 and 9.11 gives the parallel and perpendicular components of the phase velocity measured in the frame of the moving medium in terms of the velocity components measured in the frame of the laboratory:

$$u'_{||} = \frac{\gamma^2 c^2(u^2 - u_{||}v)(c^2 u_{||} - u^2 v)}{c^4 u^2_{\perp} + \gamma^2(c^2 u_{||} - u^2 v)^2} \tag{9.14}$$

$$u'_\perp = \frac{\gamma c^4 (u^2 - u_{||} v) u_\perp}{c^4 u_\perp^2 + \gamma^2 (c^2 u_{||} - u^2 v)^2} \tag{9.15}$$

The magnitude u' of the phase velocity can be determined from Equations 9.14 and 9.15. We have

$$u' = (u_{||}'^2 + u_\perp'^2)^{1/2} = \frac{\gamma c^2 (u^2 - u_{||} v)}{\left(c^4 u_\perp^2 + \gamma^2 (c^2 u_{||} - u^2 v)^2\right)^{1/2}}. \tag{9.16}$$

Dividing the numerator and denominator by the phase velocity u and using $\sin \theta = u_\perp/u$ and $\cos \theta = u_{||}/u$ where θ represents the angle of the phase velocity of the light to the velocity $\mathbf{v}$ direction in the laboratory frame, we have

$$u' = \frac{\gamma (u - v \cos \theta)}{\left(\sin^2 \theta + \gamma^2 (\cos \theta - uv/c^2)^2\right)^{1/2}}. \tag{9.17}$$

It is useful to consider the special case where the wave propagation is parallel to the relative velocity $\mathbf{v}$ of the medium. Setting $u_\perp = 0$ and $u_{||} = 0$, we can quickly obtain from Equation 9.15 that $u'_\perp = 0$ and from Equation 9.14 that

$$u' = u'_{||} = \frac{u - v}{1 - uv/c^2}. \tag{9.18}$$

Setting the angle θ of the phase velocity of light to zero in Equation 9.17 also gives this result. For propagation parallel to the velocity of the medium, the phase velocity given by Equation 9.18 transforms in an identical way to the Lorentz transformation of the relativistic velocity of a particle (see Equation 2.27). We used Equation 2.27 as a starting point to examine the Fizeau effect for light propagating parallel to the velocity of the medium in Section 9.1 (see Equation 9.1).

In the frame of the medium, the phase velocity of light is c/η, where η is the refractive index of the medium as measured in the (proper) frame of the medium. Exchanging frames of reference in Equation 9.17 so that the velocity v changes sign and primed and unprimed variables interchange, we can obtain an expression for the phase speed u of light measured in the laboratory frame in terms of the refractive index η of the moving medium. The laboratory frame phase speed u of light propagating at an angle θ' to $\mathbf{v}$ in the moving frame becomes

$$u = \frac{\gamma (c/\eta + v \cos \theta')}{\left(\sin^2 \theta' + \gamma^2 (\cos \theta' + v/(\eta c))^2\right)^{1/2}} \tag{9.19}$$

upon setting the light phase velocity $u' = c/\eta$.

The angle θ' of light propagation to the velocity in the frame of the medium can be deduced from Equations 9.14 and 9.15. We have

$$\tan \theta' = \frac{u'_\perp}{u'_{||}} = \frac{u_\perp}{\gamma (u_{||} - u^2 v/c^2)}. \tag{9.20}$$

The angle θ of propagation in the laboratory frame is determined by using $\tan \theta = u_\perp/u_{||}$ and $\cos \theta = u_{||}/u$. Equation 9.20 can then be rearranged to a form

$$\tan \theta' = \frac{\sin \theta}{\gamma (\cos \theta - uv/c^2)}. \tag{9.21}$$

When the phase velocity of light u is equal to the vacuum speed c in the laboratory frame, Equation 9.21 for the angle θ' of propagation in the medium becomes

$$\tan \theta' = \frac{\sin \theta}{\gamma (\cos \theta - v/c)}. \tag{9.22}$$

The "beaming" of radiation emitted in the moving frame and observed in the laboratory frame was discussed in Section 3.5. Our treatment here also shows that light from a stationary frame propagates in a relativistically moving frame with propagation paths at higher angles around the velocity **v** of the moving frame than in the rest frame (see e.g. Figure 9.1, where results using Equation 9.22 are plotted).

Interchanging primed and unprimed variables in Equation 9.21 and switching the sign of v, we can exchange the frames of reference. The angle θ of propagation

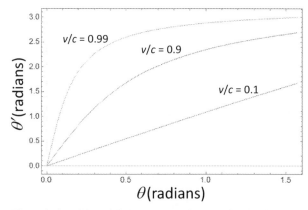

Figure 9.1   The relationship of light propagation angles in the "laboratory" (θ) and as observed in the moving (θ') frames of reference for different relative velocity amplitudes of the frames as labelled. Angles of light propagation θ and θ' are measured relative to the frame velocity **v**.

in the laboratory frame of reference is related to the angle θ' of propagation in the frame of the medium by

$$\tan \theta = \frac{\sin \theta'}{\gamma (\cos \theta' + u'v/c^2)}. \tag{9.23}$$

When the phase velocity u' is equal to the vacuum phase velocity c, Equation 9.23 is identical to the "aberration formula" applicable for light propagation in vacuum (see Equation 2.41). In a medium, the phase velocity u' is determined by the refractive index η. Substituting the phase velocity u' in the frame of the medium by c/η, where η is the refractive index of the medium gives a useful expression relating the angle θ of measured light propagation in the rest frame to the angle θ' of light propagation in the moving frame. We have

$$\tan \theta = \frac{\sin \theta'}{\gamma (\cos \theta' + v/(\eta c))}. \tag{9.24}$$

We can also evaluate expressions for the sine and cosine of θ. From Equation 9.24:

$$\left(\frac{\sin^2 \theta'}{\gamma^2} + \left(\cos \theta' + \left(\frac{v}{\eta c} \right) \right)^2 \right)^{1/2}$$

$$= \left(1 - \left(\frac{v}{c} \right)^2 \left(1 - \frac{1}{n^2} \right) + \left(\frac{v}{c} \right)^2 \cos^2 \theta' + \frac{2v}{\eta c} \cos \theta' \right)^{1/2}$$

upon substituting $1/\gamma^2 = 1 - (v/c)^2$. Expressions for $\sin \theta$ and $\cos \theta$ follow:

$$\sin \theta = \frac{\sin \theta'/\gamma}{\left(1 - (v/c)^2 \left(1 - 1/n^2 \right) + (v/c)^2 \cos^2 \theta' + (2v/\eta c) \cos \theta' \right)^{1/2}}, \tag{9.25}$$

$$\cos \theta = \frac{\cos \theta' + v/(\eta c)}{\left(1 - (v/c)^2 \left(1 - 1/n^2 \right) + (v/c)^2 \cos^2 \theta' + (2v/\eta c) \cos \theta' \right)^{1/2}} \tag{9.26}$$

When $\eta = 1$ these expressions for $\sin \theta$ and $\cos \theta$ reduce to the vacuum propagation expressions (Equations 2.42 and 2.43) as the denominators reduce to $1 + (v/c) \cos \theta'$.

In Section 4.4 we showed that pulses of light propagate at a group velocity $u_g = d\omega/dk$. Equation 4.57 gives a general relationship between the phase velocity u of light and the group velocity u_g. The direction of propagation of a pulse of light is the same as the direction of the **k**-vector of the light, so Equation 4.57 can be modified to have the phase and group velocity in a vector form. Considering the frame of reference of a moving medium of refractive index η, we have that

$$\mathbf{u}'_g = \frac{\mathbf{u}'}{1 + (\omega'/\eta)(d\eta/d\omega')}. \tag{9.27}$$

Equation 9.27 shows that the group velocity $\mathbf{u}_g'$ undergoes Lorentz transformation from the moving medium frame into the laboratory frame in the same way as the phase velocity $\mathbf{u}'$, except for a Doppler modification of the frequency ω'. Equation 9.6 gives the Doppler shift in frequency from the laboratory frame to the frame of the moving medium. Using $k_{||} = k \cos\theta = \omega \cos\theta / c$, we have

$$\omega' = \gamma\omega \left(1 - \frac{v \cos\theta}{c}\right). \tag{9.28}$$

The value of ω' and the derivative of the refractive index $d\eta/d\omega'$ used in Equation 9.27 need to be evaluated at the Doppler-shifted frequency given by Equation 9.28.

We now consider the interaction of a beam of light propagating at an angle θ to the normal of a thick sheet of a uniform material of refractive index η where the medium is moving with a velocity $\mathbf{v}$. We can assume that the velocity $\mathbf{v}$ is directed normal to the planar sheet so that θ is the angle between the light wavevector and the velocity $\mathbf{v}$. If the medium is uniform along directions perpendicular to the normal to the sheet, any velocity component perpendicular to the normal does not affect the propagation of light.

We want to find the angle of the light propagation direction as it passes through and after passing through the moving sheet of material. We need to Lorentz transform θ from the laboratory frame to the medium frame (obtaining θ') and then back to the laboratory frame to obtain the angle θ_T of propagation after transmission. In the frame of the planar medium, the angle of incidence θ' is equal to the angle to the medium normal as light exits the medium. Snell's law will result in a change of angle to θ_1' inside the medium, where $\sin\theta' = \eta \sin\theta_1'$. However, on exiting the medium, the angle to the medium normal (and velocity $\mathbf{v}$) refracts back to θ' in the frame of the medium. A schematic of the transmission of light through a slab medium in different frames of reference is shown in Figures 9.2.

For the Lorentz transform from the laboratory to the medium frame of reference, we need Equations 2.42 and 2.43 modified to transform from stationary to moving frame (i.e. the unprimed to primed frames of references). We have for the angle θ_1' of propagation in the moving medium of refractive index η that

$$\sin\theta_1' = \frac{1}{\eta} \sin\theta' = \frac{1}{\eta} \left(\frac{\sin\theta/\gamma}{1 - (v/c)\cos\theta}\right). \tag{9.29}$$

The transform to the moving frame before light enters the dielectric medium of refractive index η has a cosine value for the angle to the velocity $\mathbf{v}$ given by

$$\cos\theta' = \frac{\cos\theta - v/c}{1 - (v/c)\cos\theta}. \tag{9.30}$$

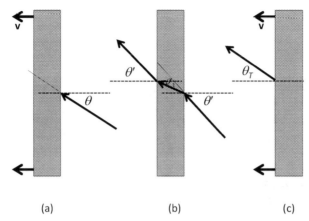

Figure 9.2 The transmission of light through a slab of dielectric medium moving at a relativistic velocity **v** in (a) the laboratory frame with angle of incidence θ, (b) in the frame of the medium, and (c) in the laboratory frame after transmission through the medium.

On exiting the medium, in the frame of the medium the angle θ' of the light path is undeviated from the angle before entering the medium (see Figure 9.2).

For the transform back to the laboratory frame for the light exiting the medium in the frame of the medium, we need Equation 9.24 with a vacuum refractive index ($\eta = 1$):

$$\tan \theta_T = \frac{\sin \theta'/\gamma}{\cos \theta' + v/c} = \frac{(\sin \theta/\gamma^2)/(1 - (v/c) \cos \theta)}{\cos \theta - (v/c) + (v/c)(1 - (v/c) \cos \theta)}$$

$$= \frac{\sin \theta/\gamma^2}{(1 - v^2/c^2) \cos \theta} = \tan \theta \tag{9.31}$$

upon substituting Equation 9.29 for $\sin \theta'$ and Equation 9.30 for $\cos \theta'$. We see that a beam of light is undeviated ($\theta_T = \theta$) in the laboratory frame after passing through a uniform slab of dielectric material moving at a relativistic velocity.

9.3 Angles of Reflection and Doppler Shifts from a Moving Mirror

We now consider the reflection of light from a rapidly moving mirror. The calculation of the frequency and pulse duration changes of light reflected from moving mirrors is important in order to understand the reflection of laser light in plasmas formed by short duration (<100 fs) laser pulse irradiation of targets. With short duration laser pulses, a steep increase of density near a solid target acts as a highly reflecting and rapidly moving mirror.

In order to quantitatively study reflections from a moving mirror, we will determine the Lorentz transformation of the light frequency, the angle of propagation, and the duration of a pulse of light from the laboratory frame into the frame of reference of the mirror and then calculate the Lorentz transformation back to a laboratory frame of reference. This process enables the effect of the reflection in the laboratory frame to be quantified. The angles of reflection in the laboratory frame from a relativistically moving mirror can be examined using the treatments developed for the Lorentz transformation of velocities (see Section 2.3). Changes in the frequency of the light can be found from the Lorentz transformation of frequencies (usually referred to as the Doppler effect, see Section 2.6).

An observer moving with a speeding mirror observes the laws of optics identically to a nominally stationary observer observing a stationary mirror in the laboratory frame of reference. For example, in the frame of reference of the mirror, the angle of incidence of light is equal to the angle of reflection. The reflectivity of the mirror in the frame of the mirror can be found using standard expressions from optics.

In the frame of the mirror moving at velocity $\mathbf{v}$ at an angle α to the normal of the mirror and angle θ' to the incoming light wavevector, the angle of incidence of the light on the mirror can be expressed as $\theta' - \alpha$. The angle of reflection in the mirror frame is then $\theta'_R = \theta' - 2\alpha$. For ease of mathematical manipulation, we will initially assume that the mirror material has a macroscopic velocity $+\mathbf{v}$ normal to its surface opposite to the direction of the light before reflection and is spatially uniform in directions along the planar surface. The angle θ' of incidence of light is then equal to the angle θ'_R of reflection in the mirror frame: $\theta' = \theta'_R$.

In other sections of this book (see Figure 1.2), we assume that the relative velocity between frames is positive when the "moving" frame (denoted by primes) recedes from the stationary frame (unprimed). Assuming here that the mirror is moving towards the light source means that initially the velocity of the moving mirror frame in the laboratory frame is negative.

The Doppler frequency shift between two frames of reference is given by Equation 2.46. When the velocity is towards the light source we have an angle between the $\mathbf{k}$-vector and frame velocity $\mathbf{v}$ of $\pi - \theta$. From Equation 2.46, we then have the frequency ω' in the frame of the mirror:

$$\omega' = \omega\gamma \left(1 - \frac{v}{c}\cos(\pi - \theta)\right) \tag{9.32}$$

To get the Doppler shift of frequencies after reflection in the laboratory frame, we need to Lorentz transform from the laboratory frame to the mirror frame using Equation 9.32 and then transform from the mirror frame back to the laboratory frame. In the transform from the mirror back to the laboratory frame, we have that

$$\omega' = \omega_R \gamma \left(1 - \frac{v}{c} \cos \theta_R \right),$$

where ω_R is the frequency of the light in the laboratory frame after reflection. The initial laboratory frequency ω after reflection in the mirror has a frequency ω_R in the laboratory frame given by

$$\omega_R = \omega \frac{1 + (v/c) \cos \theta}{1 - (v/c) \cos \theta_R}. \qquad (9.33)$$

Positive values of the velocity v mean here that the light is initially propagating in the opposite direction to the mirror velocity which gives an upshift in frequency upon reflection, while negative values of the velocity v imply that the light is initially propagating in the same direction to the mirror velocity which gives an down-shift in frequency upon reflection.

Equation 2.43 relates the cosine of an angle θ in the laboratory frame to the angle θ' in a moving frame of reference. For a positive velocity v towards the light source, we have

$$\cos \theta = \frac{\cos \theta' - v/c}{1 - (v/c) \cos \theta'}. \qquad (9.34)$$

To get the reflected angle θ_R for an angle of incidence θ in the laboratory frame, we need to transform from the laboratory frame to the mirror frame using Equation 9.34 and then transform from the mirror frame back to the laboratory frame (as schematically illustrated in Figure 9.3). In the frame of the mirror, the initial angle between the **k**-vector of the light and the mirror velocity **v** is θ'. When the mirror

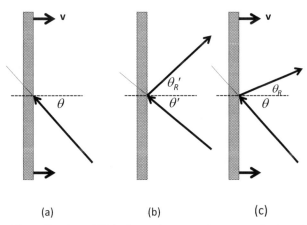

(a)   (b)   (c)

Figure 9.3   The reflection of light in a mirror moving at a velocity **v** is illustrated (a) in the laboratory frame before reflection, (b) in the frame of the moving mirror where the angle of incidence θ' and angle of reflection θ'_R are equal, and (c) in the laboratory frame after reflection.

velocity **v** is parallel to the mirror normal, in the mirror frame the angle of incidence is θ' which equals the angle of reflection θ'_R. The angle θ'_R for reflection in the mirror frame is given by $\theta'_R = \theta'$ with[2]

$$\cos\theta' = \frac{\cos\theta + v/c}{1 + (v/c)\cos\theta}.$$ (9.35)

The angle θ_R of reflection in the laboratory frame is then[3]

$$\cos\theta_R = \frac{\cos\theta'_R + v/c}{1 + (v/c)\cos\theta'_R} = \frac{\left(\frac{\cos\theta + v/c}{1 + (v/c)\cos\theta}\right) + v/c}{1 + (v/c)\left(\frac{\cos\theta + v/c}{1 + (v/c)\cos\theta}\right)},$$

which simplifies to

$$\cos\theta_R = \frac{2(v/c) + (1 + (v/c)^2)\cos\theta}{1 + (v/c)^2 + 2(v/c))\cos\theta}.$$ (9.36)

If the mirror is moving at an angle α to the mirror normal, the angle of reflection θ'_R in the mirror frame is such that $\theta'_R = \theta' - 2\alpha$ with θ' given by Equation 9.35. Substituting $\theta'_R = \theta' - 2\alpha$ into an equation of form

$$\cos\theta_R = \frac{\cos\theta'_R + v/c}{1 + (v/c)\cos\theta'_R}$$ (9.37)

then enables a calculation of the reflected angle θ_R in the laboratory frame. A complete expression for the reflected angle when the mirror velocity **v** is at an angle α to the mirror normal, however, is a little cumbersome and best left to a numerical evaluation.

For positive mirror velocities v approaching the speed of light c, Equation 9.36 shows that θ_R approaches zero. A highly relativistic mirror moving towards a light source reflects the light back along the mirror normal irrespective of the angle of incidence. For v close to c and $\theta_R \approx 0$, Equation 9.33 then reduces as follows:

$$\omega_R = \omega\frac{1 + (v/c)\cos\theta}{1 - (v/c)\cos\theta_R} \approx \omega\frac{1 + (v/c)\cos\theta}{1 - (v/c)} = \omega\frac{(1 + (v/c)\cos\theta)(1 + v/c)}{(1 - v/c)(1 + v/c)}$$

$$= \omega\left(\left(1 + \frac{v\cos\theta}{c}\right)\left(1 + \frac{v}{c}\right)\right)\gamma^2$$

$$= \omega\left(1 + 2\frac{v}{c}\cos^2(\theta/2) - \frac{v}{c}\right)\left(1 + \frac{v}{c}\right)\gamma^2 \approx 4\gamma^2\cos^2(\theta/2)\ \omega$$ (9.38)

The last approximation requires v to be positive (i.e. the velocity is in a direction towards the light source). An increase of reflected frequency by a factor $4\gamma^2$ can occur from a highly relativistic mirror moving towards the light source. Examples

---

[2] The Lorentz transform is now from the laboratory frame (unprimed variables) to the mirror frame (primed variables). Swapping the primed and unprimed frames changes the sign of the velocity v.

[3] The reflected light propagates in the direction of **v** so the sign of v is opposite to that of Equation 9.34.

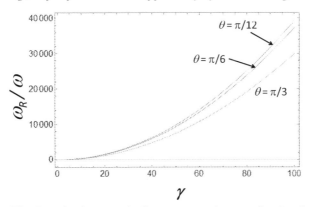

Figure 9.4   The Doppler increase in frequency ω_R/ω on reflection from a plane mirror moving towards the light source as a function of the Lorentz γ of the mirror. Results are shown for angles of incidence θ as labeled.

of the increase in frequency of light reflected from a relativistic mirror calculated using Equations 9.32 and 9.36 are presented in Figure 9.4.

In the laboratory frame, pulses of light reflecting from an approaching relativistic mirror experience a compression of the pulse duration on reflection. To examine the pulse compression, we determine the pulse length in different frames using the Lorentz transform for time (see Equation 2.18). We need to Lorentz transform the temporal duration of a light pulse propagating at an angle θ in the laboratory frame to the mirror frame of reference and then Lorentz transform the reflected pulse from the mirror frame to the laboratory frame. For a pulse of light of duration Δt in the laboratory frame before reflecting in the moving mirror, the spatial length along the velocity axis in the laboratory frame is given by $x = c\Delta t \cos\theta$. The pulse duration $\Delta t'$ in the frame of the mirror is then given by

$$\Delta t' = \gamma(\Delta t - v\Delta t \cos\theta/c).$$

In this section we are defining a positive velocity v for the mirror motion when the mirror moves towards the light source. We now Lorentz transform from the reflected pulse length Δt_R in the laboratory frame back to the mirror frame of reference:

$$\Delta t' = \gamma(\Delta t_R + v\Delta t_R \cos\theta_R/c)$$

Combining these equations gives an expression for the reflected pulse duration Δt_R:

$$\Delta t_R = \frac{1 - (v/c)\cos\theta}{1 + (v/c)\cos\theta_R}\Delta t \tag{9.39}$$

For a perfect mirror, the energy of the pulse is conserved upon reflection in a relativistic mirror. The pulse duration is shortened by a factor ω/ω_R (as calculated for

Equation 9.33) which compensates exactly for the increase in the beam intensity associated with the frequency increase. For a highly relativistic mirror moving towards the light source, we can use the approximations of Equation 9.38 to show that

$$\Delta t_R \approx \frac{1}{4\gamma^2 \cos^2(\theta/2)} \Delta t. \tag{9.40}$$

Einstein considered the reflection from a mirror moving at high speed as a thought experiment [35]. Such mirrors can now be achieved in laser-produced plasmas [60], [61], [90]. Short pulse lasers (pulse duration < 100 fs) interacting with solid targets can accelerate the target surface to relativistic velocities. The target surface becomes a good mirror to the laser pulse as the target quickly ionizes to form a moving plasma with an electron density above the critical electron density.

9.4 The Lorentz Transformation of the Polarization of Light

It is important to see if the polarization of emission from a rapidly moving body is altered when detected in the laboratory frame of reference. Any changes in the polarization of light between frames of reference could affect the reflection from or transmission through a medium as the polarization of light in the medium frame can alter the reflection efficiency and the propagation properties in the medium. This section addresses the issue of polarization change under a Lorentz transformation. We introduce the Stokes parameters which characterizes light polarization and show that the normalized Stokes parameters are invariant under a Lorentz transformation. Lorentz invariance of the Stokes parameters means that the polarization of light does not change between moving frames of reference for any angle of light propagation and any polarization.

For monochromatic light propagating in the X-direction in a Cartesian coordinate system, the electric field components in the Y- and Z- directions can be represented by

$$E_Y(k, t) = E_X \cos(kX - \omega t + \delta_X),$$
$$E_Z(k, t) = E_Y \cos(kX - \omega t + \delta_Y). \tag{9.41}$$

The factors δ_Y and δ_Z allow for phase differences in the Y- and Z- electric fields. For example, circular polarization discussed in Section 4.2 is obtained with $\delta_Y - \delta_Z = \pi/2$. With linear polarization $\delta_Y = \delta_Z$.

A complete knowledge of the polarization of monochromatic electromagnetic waves can be characterized by Stokes parameters I_S, Q_S, U_S, and V_S defined as follows:

$$I_S = E_Y^2 + E_Z^2,$$
$$Q_S = E_Y^2 - E_Z^2,$$
$$U_S = 2E_Y E_Z \cos(\delta_Y - \delta_Z),$$
$$V_S = 2E_Y E_Z \sin(\delta_Y - \delta_Z) \tag{9.42}$$

I_S represents the total intensity of the light, Q_S represents the difference in the Y- and Z-direction linear polarization, U_S represents the components of linear polarization at angles of $\pi/4$ to the Y- and Z-directions and V_S gives a measure of the degree of circular polarization. The degree and nature of the polarization of monochromatic light is determined by Stokes parameters normalized to the I_S value: that is Q_S/I_S, U_S/I_S, and V_S/I_S. For example, a parameter $V_S/I_S = 0$ with $Q_S/I_S \neq 0$ (unless $U_S/I_S = 1$) indicates linear polarization, while $U_S/I_S = 0$ indicates elliptical polarization. For circular polarization, $Q_S/I_S = 0$ and $U_S/I_S = 0$.

Consider Cartesian coordinate systems (x, y, z) for a laboratory frame of reference and a frame of reference (x', y', z') moving with a velocity **v** directed along the x-axis. This is the standard arrangement for frames of reference utilized in this book (see e.g. Figure 1.2). With positive v, we assume here that the moving frame velocity is directed away from the light source. Without loss of generality, we can set the y- and Y-axes as identical. The E_Y component of the electric field is then perpendicular to the velocity **v**, while the corresponding B_Z magnetic field has a component $B_Z \cos \theta$ along the z-axis perpendicular to the velocity **v** (see Figure 9.5). Using Equation 3.24 for the transformation of an electric field normal to **v** from our treatment of the Lorentz transformation of electric fields (Section 3.2), we have for the light electric field value in the moving frame of reference

$$E_Y' = \gamma(E_Y - B_Z v \cos \theta).$$

For light propagating in vacuum, $B_Z = E_Y/c$ so that the electric field Lorentz transformation becomes

$$E_Y' = \gamma E_Y \left(1 - \frac{v \cos \theta}{c}\right). \tag{9.43}$$

The component of electric field E_Z parallel to **v** has a value of $E_Z \sin \theta$. Electric field components parallel to **v** are Lorentz invariant (see Section 3.2). We have for the electric field component parallel to the velocity **v** in the moving frame that

$$E_Z' \sin \theta' = E_Z \sin \theta.$$

The angles that the propagating light has to the velocity **v** are related by Equation 2.42. The Lorentz transform of $\sin \theta$ from the stationary to moving frame is given by

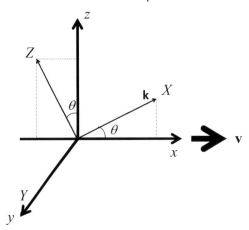

Figure 9.5   The stationary frame Cartesian axes (X, Y, Z) for light propagation in the X-direction superimposed on the axes (x, y, z) for a Lorentz transformation with the moving frame velocity $\mathbf{v}$ parallel to the x-axis. The angle between the $\mathbf{k}$ vector of the light and the frame velocity $\mathbf{v}$ is θ.

$$\sin \theta' = \frac{\sin \theta / \gamma}{1 - (v/c) \cos \theta}.$$

Utilizing the Lorentz invariance of the E_Z component of electric field parallel to the velocity $\mathbf{v}$ means that

$$E_Z' = E_Z \frac{\sin \theta}{\sin \theta'} = \gamma E_Z \left(1 - \frac{v \cos \theta}{c} \right). \qquad (9.44)$$

A similar treatment for the magnetic fields as outlined above for the electric fields shows B_Z and B_Y Lorentz transforming identically to the electric fields E_Z and E_Y (Equations 9.44 and 9.43; see Exercise 9.6).

In Section 3.1 we showed that the phase of an electromagnetic wave is Lorentz invariant. In the transformation to the moving frame, the terms δ_Y and δ_Z used in the definition of the Stokes parameters are therefore constant:

$$\delta_Y' = \delta_Y,$$
$$\delta_Z' = \delta_Z$$

The Stokes parameters characterizing polarization are all proportional to the square of electric field. Given the Lorentz invariance of δ_Y and δ_Z, Equations 9.44 and 9.43 indicate that the Lorentz transformation of the Stokes parameters proceeds such that

$$I_S' = \gamma^2 (1 - v \cos \theta / c)^2 I_S$$
$$Q_S' = \gamma^2 (1 - v \cos \theta / c)^2 Q_S$$
$$U_S' = \gamma^2 (1 - v \cos \theta / c)^2 U_S$$
$$V_S' = \gamma^2 (1 - v \cos \theta / c)^2 V_S, \qquad (9.45)$$

The orientation and nature of the polarization does not change during a Lorentz transformation as all the Stokes parameters are uniformly multiplied by $\gamma^2(1 - v\cos\theta/c)^2$. The normalized Stokes parameters Q_S/I_S, U_S/I_S, and V_S/I_S, defining polarization are Lorentz invariant.

The Lorentz invariance of the polarization of light is important for astrophysical observations of polarized light from, for example, plasma surrounding black holes, pulsars, gamma ray bursts and the cosmic microwave background [19], [23], [70]. Polarization invariance means that observations on earth in relative motion to the sources of the light measure the same polarization as the light source. It also means that the polarization of light does not change between emission and detection in laboratory light sources such as synchrotrons.

The Doppler shift in frequency from ω to ω' in the Lorentz transformation from stationary to moving frame of reference was considered in Section 2.6. Using Equation 2.46, we have that

$$\omega' = \omega\gamma\left(1 - \frac{v}{c}\cos\theta\right). \tag{9.46}$$

The Stokes parameters can therefore be written as

$$I'_S = \left(\frac{\omega'}{\omega}\right)^2 I_S$$

$$Q'_S = \left(\frac{\omega'}{\omega}\right)^2 Q_S$$

$$U'_S = \left(\frac{\omega'}{\omega}\right)^2 U_S$$

$$V'_S = \left(\frac{\omega'}{\omega}\right)^2 V_S, \tag{9.47}$$

The un-normalized Stokes parameters divided by the square of the frequency of the light are Lorentz invariant.

In Section 1.1 we first examined the propagation of light in vacuum and obtained an expression for the intensity (power per unit area) of a beam of light in terms of the electric field strength. The time averaged electric and magnetic energies of a light beam are equal and together have a total energy density $(1/2)\epsilon_0 E_0^2$. Imagine a unit cross-section volume of unit area and length c along the light direction $\mathbf{k}$ in a proper frame of reference. In unit time, the energy of the electromagnetic beam from the volume of c will pass through the unit area. This energy passing per unit time per unit area normal to the direction of the beam is known as the intensity or irradiance I and is given by

$$I = \frac{1}{2}\epsilon_0 c E_0^2. \tag{9.48}$$

In the Lorentz transformation to a moving frame of reference, rather than length c, we need to consider the energy density of a light beam in a volume of unit cross-section and relativistically contracted length $\gamma(c - v\cos\theta)$ (see Equation 2.18). The intensity of light I' in the moving frame propagating at an angle of θ to the velocity $\mathbf{v}$ in the laboratory frame of reference becomes

$$I' = \gamma\left(1 - \frac{v\cos\theta}{c}\right)\left(\frac{1}{2}\epsilon_0 c E_0'^2\right).$$

Using Equations 9.44 and 9.43, it becomes clear that intensities under a Lorentz transformation satisfy

$$I' = \gamma^3\left(1 - \frac{v\cos\theta}{c}\right)^3 I = \frac{\omega'^3}{\omega^3}I. \tag{9.49}$$

This Lorentz invariance scaling confirms our earlier finding in Section 7.1 where we showed that the intensity of light divided by the cube of the frequency (i.e. I/ω^3) is Lorentz invariant.

Exercises

9.1   Fizeau [42] investigated the velocity of light in water by employing an interferometer which measured the phase difference between light propagating parallel to flowing water and light propagating in the opposite direction to water flowing at the same velocity. If light of wavelength λ propagates the same distance z parallel and antiparallel to the water flow and the water has a refractive index of η not affected by any Doppler shift of light frequency, show that the phase difference $\delta\Phi$ of the detected light varies with water velocity v as

$$\delta\Phi = \frac{4\pi z}{\lambda\eta^2}\left(\frac{v}{c}\right)$$

assuming $v \ll c$.

9.2   Laser-produced plasmas created by nanosecond duration lasers expand into vacuum during the irradiation of a solid target with a velocity at the critical density of typically 3×10^5 m s^{-1}. If the laser has a wavelength of 1.060 μm, determine the wavelength of the light reflected back normally from an expanding critical density surface. [1.058 μm]

9.3   High power laser-produced plasmas created by lasers with pulse durations less that 150 fs reflect harmonic light at much higher frequencies than the incoming laser light. The production of higher frequencies has been shown to occur due to rapid velocity oscillations of the reflecting surface in the plasma [121]. If a short pulse laser irradiates a solid plane target at an angle of incidence of $\pi/4$ radians and a specular emission at the 20th harmonic frequency is detected, estimate the velocity of the reflecting surface in the plasma assuming the frequency shift arises due to a Doppler shift. [2.7×10^8 m s^{-1}]

9.4 Consider a scenario where radio waves propagate to earth at small angles $\Delta\theta'$ to the velocity $\mathbf{v}$ of a spacecraft moving directly towards the earth, where $\Delta\theta'$ is measured in the frame of reference of the spacecraft. Show that on earth, the angle $\Delta\theta$ of the radio wave propagation to the spacecraft velocity is given by

$$\Delta\theta \approx \Delta\theta' \left(\frac{1 - v/c}{1 + v/c}\right)^{1/2}$$

for all v in the range 0 to approaching the speed of light c.

9.5 Consider the Stokes parameters of Equation 9.42. Show that

$$Q_S^2 + U_S^2 + V_S^2 = I_S^2.$$

9.6 Consider the Lorentz transformation of the magnetic field associated with an electromagnetic wave propagating in the rest frame at an angle θ to the velocity $\mathbf{v}$ of another frame of reference. For the geometry illustrated in Figure 9.5, show that the magnetic fields in the moving frame of reference are given by

$$B_Y' = \gamma B_Y \left(1 - \frac{v \cos\theta}{c}\right),$$

$$B_Z' = \gamma B_Z \left(1 - \frac{v \cos\theta}{c}\right).$$

[You need the Lorentz transformation of a magnetic field perpendicular to $\mathbf{v}$ [see e.g. Equation 3.22] and to remember that magnetic fields parallel to $\mathbf{v}$ are Lorentz invariant.]

10

Solutions to Selected Exercises

This chapter provides detailed answers to three exercises presented in earlier chapters. We consider a paradox in the ageing of a twin departing on a spacecraft and returning to Earth and the derivation of the Biot-Savart equation for determining magnetic fields around current-carrying wires.

The "paradox" in the twin paradox exercise arises because the exercise involves a twin traveling in a spacecraft and returning to Earth much older than the twin remaining on Earth. Why isn't the aging the other way around? After all, the twin on Earth retreats from the spacecraft in the cosmological twin's frame of reference. We show that the resolution of the paradox as outlined in the question lies in the change of frame of reference due to acceleration undergone by the space-going twin which does not occur for the Earth-bound twin. We consider the twin paradox exercise as posed in the 1967 edition of the textbook *Classical Electrodynamics* by J. D. Jackson [58]. This involves a twin returning to Earth in the year 2020 who finds that the year on Earth is 2336.

The Biot-Savart equation describes the magnetic field generated by a constant electric current. As an example of the equivalence of magnetic and electric fields in different frames of reference, we consider the electrostatic force of moving electrons and stationary ions in a neutral wire on a charge moving at a velocity relative to the wire. The difference in the electrostatic force between moving charge and electrons and moving charge and ions is equivalent to the Lorentz force of the moving charge in a magnetic field. The Biot-Savart expression for the magnetic field is obtained.

10.1 The Twin Paradox

Exercise 3.1: *With an eye to the future, the 1967 edition of the textbook Classical Electrodynamics by J. D. Jackson [58] posed the following question: "Assume that a rocket ship leaves the Earth in the year 2000. One of a set of twins born in 1980*

292

remains on Earth; the other rides in the rocket. The rocket ship is so constructed that it has an acceleration g in its own rest frame (this makes the occupants feel at home). It accelerates in a straight-line path for 5 years (by its own clocks), decelerates at the same rate for 5 more years, accelerates for 5 years, decelerates for 5 years, and lands on Earth. The twin in the rocket is 40 years old. (a) What year is it on Earth? (b) How far away from the Earth did the rocket ship travel?"

The relationship between the acceleration $a'_{||}$ in the frame of reference S' of the rocket ship and acceleration $a_{||}$ in the frame of the Earth S in a direction parallel to the direction of motion is given by Equation 3.33. We have

$$a'_{||} = \gamma^3 a_{||} = \frac{a_{||}}{(1 - v^2/c^2)^{3/2}}. \tag{10.1}$$

The question assumes constant acceleration (or deacceleration) in the frame of the rocket ship, so we can assume throughout that the acceleration $a'_{||}$ in the rocket ship frame is constant. Rearranging, the acceleration in the frame of the Earth can be written as

$$a_{||} = \frac{dv}{dt} = \left(1 - \frac{v^2}{c^2}\right)^{3/2} a'_{||}.$$

This expression in turn can be rearranged to give

$$\frac{dv}{(1 - v^2/c^2)^{3/2}} = a'_{||} dt.$$

Integrating both sides within appropriate limits so that at time $t = 0$, the spaceship velocity is zero:

$$\int_0^v \frac{dv}{(1 - v^2/c^2)^{3/2}} = \int_0^t a'_{||} dt$$

The evaluation of both integrals yields

$$\frac{v}{\sqrt{1 - v^2/c^2}} = a'_{||} t.$$

Squaring this expression and rearranging shows that

$$v^2 = a'^2_{||} t^2 \left(1 - \frac{v^2}{c^2}\right).$$

Gathering the terms in v^2, we have

$$v^2 \left(1 + \frac{a'^2_{||} t^2}{c^2}\right) = a'^2_{||} t^2.$$

Rearranging shows that

$$v = \frac{a'_{||}t}{\sqrt{1 + a'^2_{||}t^2/c^2}}$$

and

$$\frac{1}{\gamma^2} = 1 - \frac{v^2}{c^2} = \frac{1}{1 + a'^2_{||}t^2/c^2}.$$

Using the relationship between dilated time t' in the moving rocket ship and time t on Earth ($t' = t/\gamma$), we can write

$$dt' = \left(1 - \frac{v^2}{c^2}\right)^{1/2} dt = \frac{dt}{\sqrt{1 + a'^2_{||}t^2/c^2}}.$$

We can integrate both sides to relate the total time T' in the rocket ship to the total time T on Earth

$$T' = \int_0^{T'} dt' = \int_0^T \frac{dt}{\sqrt{1 + a'^2_{||}t^2/c^2}}.$$

The result of this integration yields

$$T' = \frac{c}{a'_{||}} \sinh^{-1}(a'_{||}T/c). \tag{10.2}$$

The elapsed time T in the rest frame on Earth is then given by

$$T = \frac{c}{a'_{||}} \sinh(a'_{||}T'/c). \tag{10.3}$$

Equations 10.2 and 10.3 give the relationships between time T in the rest frame and time T' in a body undergoing constant acceleration $a'_{||}$ in the body's frame of reference. As a check, we quickly see that when $a'_{||}T' << c$, the time in the two frames of reference is identical with $T = T'$. When $a'_{||}T'$ becomes significant compared to the speed of light, $T \approx (c/2a'_{||}) \exp(a'_{||}T'/c)$.

The distance traveled by the rocket ship can be found by integrating the rocket ship velocity v in the rest frame time t. We have

$$x = \int_0^T v\, dt = \int_0^T \frac{a'_{||}t}{\sqrt{1 + a'^2_{||}t^2/c^2}} dt = \frac{c^2}{a'_{||}} \left[\left(1 + \left(\frac{a'_{||}T}{c}\right)^2\right)^{1/2} - 1\right]. \tag{10.4}$$

Equation 10.4 is a general expression for the distance traveled by a body in the rest frame time T when the acceleration $a'_{||}$ in the body's frame of reference is constant. When $a'_{||}T \ll c$, expanding $(1 + (a'_{||}T/c)^2)^{1/2} \approx 1 + (1/2)(a'_{||}T/c)^2$ shows that the distance x traveled is given by $x \approx (1/2)a'_{||}T^2$ in agreement with Newtonian mechanics. When $a'_{||}T$ becomes large compared to the speed of light, Equation 10.4 simplifies to $x \approx cT$. To an observer in the rest frame measuring an elapsed time T, the body travels at the speed of light for much of the journey.

The quantitative answer to the question is evaluated using the given acceleration in the rocket ship frame $a'_{||} = g$, where g is the acceleration on earth due to gravity ($g = 9.8$ m s^{-2}). The total journey time is evaluated using Equation 10.3. The total journey time is $4\times$ an evaluation of the time T on earth for a duration $T' = 5$ years on the rocket ship. The distance traveled is twice the distance x evaluated using Equation 10.4 as the distances covered in acceleration and deacceleration are equal.

A year is $365 \times 24 \times 3{,}600 = 3.15 \times 10^7$ s. The time T' of each 5 year acceleration phase on the rocket ship is $T' = 1.577 \times 10^8$ s. Using Equation 10.3, the time on Earth for each 5 year acceleration phase is

$$T = \frac{c}{g} \sinh(gT'/c) \approx \frac{c}{g}(1/2) \exp(gT'/c) = 2.65 \times 10^9 \text{s} = 84 \text{ years.}$$

The total elapsed time on Earth is four times this length of time, namely 336 years. To the twin on the rocket ship, the year of return to Earth is 2020. However, the year of return of the rocket ship to Earth for the human populace remaining on Earth is the year 2336 (the answer to part [a]).

Substituting into Equation 10.4, we can find the distance traveled by the rocket. The two $T = 2.65 \times 10^9$ s acceleration phases on the outward journey each result in a distance traveled from earth given by

$$x = \frac{c^2}{g}\left[\left(1 + \left(\frac{gT}{c}\right)^2\right)^{1/2} - 1\right] \approx cT = 7.95 \times 10^{17} \text{m} \approx 25.8 \text{ parsec.}$$

The total distance traveled is twice this figure at 51.6 parsec (the answer to part [b]).

To give some feeling for the distance of travel (51.6 parsec or 168 light years): The nearest star to earth is Proxima Centauri which is 1.302 parsecs from Earth (4.24 light years). There are a range of nearby stars that the rocket could reach. However, a heavy personal price would be paid by the twin on the rocket ship. As well as spending 20 years (their time) on a rocket ship largely in featureless interstellar space, the space-bound twin would return to a world where anyone they had known (including their twin sibling) had long since died. A consolation perhaps is that the space-bound twin would spend the rest of their life living in what would, at least initially, seem to be "the future."

The twin paradox can be set up in a variety of ways in "thought experiments" used to illustrate special relativity. For example, the acceleration of the spaceship from Earth and for the return journey can be assumed as occurring instantaneously up to a significant fraction of the speed of light. In slightly more realistic scenarios, an outward-going and inward-coming observer can synchronize clocks as they pass each other and as they pass an earth-bound observer. The space-bound observers all age more slowly than the earth-bound observers as their frame of reference changes during the journey. For the exercise as set here, the frame of reference changes continuously due to continuous acceleration. In problems where instantaneous acceleration occurs to a constant velocity, there are typically two changes of the frame of reference: once during the acceleration to start the outward journey and then again to make the return journey. Earth-bound observers remain in the same inertial frame of reference.

Exercise 3.2 *Assume that the rocket ship employed in Exercise 3.1 has a constant rest mass including occupants of 10^3 kg. We can imagine that the rocket is accelerated by some mass-conserving process, say, electron-positron annihilation so that the mass of the spaceship remains approximately constant. (a) Calculate a minimum energy needed for the described spaceflight. (b) Determine the power required to be supplied by the rocket in order to maintain the acceleration g at the end of the two accelerating phases of the flight.*

(a): The minimum energy for the spaceflight is equal to four times the kinetic energy at the end of the two acceleration phases of the flight. The same energy needs to be supplied to produce the two deacceleration phases of the flight. As most of the rocket flight is in interstellar space, we can ignore any acceleration effects arising from gravitational forces, so this minimum energy should be reasonably accurate.

The total kinetic energy W of a mass m_0 is $(\gamma - 1)m_0c^2$. From the answer to Exercise 3.1, we have for the Lorentz factor γ:

$$\gamma = \frac{1}{\sqrt{1 - v^2/c^2}} = \sqrt{1 + \frac{a_{||}^2 t^2}{c^2}}$$

The kinetic energy W of the rocket ship at the end of each of the acceleration phases is given by

$$W = (\gamma - 1)m_0c^2 = \left(\sqrt{1 + \frac{g^2 t^2}{c^2}} - 1\right)m_0c^2. \tag{10.5}$$

The time $t = T$ in the rest frame on Earth for each acceleration phase was found in Exercise 3.1 to be $T = 2.65 \times 10^9$ s. The acceleration $a_{||}' = g$ on the rocket during the acceleration and deacceleration phases ($g = 9.8 \text{ ms}^{-1}$). Quantitatively, for $m_0 = 10^3$ kg, $T = 2.65 \times 10^9$s, and $g = 9.8 \text{ ms}^{-1}$, we get $W = 7.25 \times 10^{21}$ J. The required energy for the complete flight is four times this value (2.9×10^{22} J) as the same energy is needed to deaccelerate the spaceship as is needed for the two acceleration phases.

A convenient measure of large energies is the energy released by a megaton of TNT[1] in an explosion: 1 megaton TNT = 4.184×10^{15} J. The energy content of nuclear bombs is usually quoted in units of equivalent TNT explosive yield. The minimum energy required for the rocket flight in these units is $\approx 7 \times 10^6$ megatons of TNT. This is an incredible amount of energy.

(b): The power dW/dt' to drive the rocket in the frame of the rocket can be found by differentiating with respect to the Earth's time t the expression for the kinetic energy W given in Equation 10.5:

$$\frac{dW}{dt} = \frac{1}{(1 + g^2 t^2/c^2)^{1/2}} m_0 g^2 t$$

The differential with respect to time t' on the rocket is given by

$$\frac{dW}{dt'} = \frac{dW}{dt}\frac{dt}{dt'} = \frac{1}{(1 + g^2 t^2/c^2)^{1/2}} m_0 g^2 t \gamma = m_0 g^2 t \tag{10.6}$$

as $t' = t/\gamma$ is the time on the rocket ship and $\gamma = (1 + g^2 t^2/c^2)^{1/2}$. To maintain a constant acceleration $a'_\parallel = g$, the power supplied at the rocket to propel the rocket needs to increase linearly with the time measured on Earth for most of the journey. The propulsion power in the frame of the rocket ship expressed in Equation 10.6 can also be obtained by simply considering the force $m_0 g$ required to accelerate the rocket multiplied by the velocity gt. At the end of an acceleration phase, the power needed to maintain an acceleration of g is found from Equation 10.6 by setting $t = T = 2.64 \times 10^9$ s $= 84$ years (the time on Earth for each acceleration phase). The power needed is 2.54×10^{14} W. This is equivalent to 60 ktons/second of exploding TNT.

Rocket flights usually accelerate for much less than 84 years. However, due to the large power requirement of all rocket flights, the idea of using nuclear explosions to power rocket flight was seriously considered between 1957 and 1965 (see Dyson [29]). The Partial Test Ban Treaty of 1963 and concerns about nuclear fallout caused the program to stop.

---

[1] TNT is an abbreviation for trinitrotoluene, a pale yellow, solid organic nitrogen compound used as an explosive.

10.2 The Biot-Savart Equation

Exercise 5.1 *Consider a thin wire of short length d**l** with N_e conducting electrons per unit length along which flows a current I_C given by $I_C = N_e e v_e$, where v_e is the velocity of the electrons. A positive test charge q at a distance **r** at an angle of θ to d**l** is electrostatically attracted to the wire by the electrons and repelled by the positively charged ions in the wire. The charges per unit length N_e due to the electrons and ions are equal in magnitude, but opposite in sign so that the overall wire is neutrally charged. (a) If the charged particle is moving at a velocity **v** parallel to d**l** and allowing for relativistic effects on the electric fields produced by the electrons and ions in the wire, show that the respective electrostatic forces of attraction dF_− and repulsion dF_+ between the charge and the electrons and ions respectively are given by*

$$dF_- = -\frac{1}{4\pi\epsilon}\frac{qN_e e dl}{r^2}\left[\left(1 - \frac{(v+v_e)^2}{c^2}\right)^{1/2}\sin\theta + \cos\theta\right],$$

$$dF_+ = \frac{1}{4\pi\epsilon}\frac{qN_e e dl}{r^2}\left[\left(1 - \frac{v^2}{c^2}\right)^{1/2}\sin\theta + \cos\theta\right].$$

(b) Hence, show that the net attractive force $dF = dF_+ + dF_-$ for electron velocities $v_e \ll c$ and a velocity v in the same direction as the current I is given by

$$dF = -\frac{\mu_0}{4\pi}\frac{I_C dl}{r^2}qv\sin\theta.$$

*(c) Assuming the Lorentz equation d**F** = q**v**×d**B** defines a magnetic field d**B** associated with the current carrying element, show that*

$$d\mathbf{B} = \frac{\mu_0}{4\pi}\frac{I_C d\mathbf{l}\times\hat{\mathbf{r}}}{r^2},$$

*where $\hat{\mathbf{r}}$ is a unit vector in the direction of **r**.*

The Biot-Savart equation given in part (c) of this exercise is useful for deducing magnetic fields around electrical current carrying elements. This exercise demonstrates a derivation of the Biot-Savart equation. Rather than starting from Maxwell's equations (specifically Ampere's law), we determine the electrostatic attraction between a test charged particle and the moving electrons and stationary ions in a short length of current carrying element. When the charged particle has a velocity relative to the current carrying element, in the frame of reference of the charge there is a small difference in the Lorentz parameter for the electric fields created by electrons and ions in the current carrying element due to the different relative velocity of the electrons and ions. Assuming the electron and test particle velocities are small compared to the speed of light, the net force the wire exerts on the charge can be equated to the Lorentz force associated with a magnetic field.

The Biot-Savart expression can then be obtained for the magnetic field produced by a short length of current carrying element.

Equation 3.27 summarizes the effect of relativity on electric fields directed perpendicular and parallel to the velocity of a moving frame. Electric fields $\mathbf{E}'_\perp$ directed perpendicular to the velocity $\mathbf{v}$ in the moving frame are related to the perpendicular electric fields $\mathbf{E}_\perp$ in the rest frame by

$$E'_\perp = \gamma E_\perp = \gamma E \sin\theta,$$

while electric fields $\mathbf{E}'_\parallel$ directed parallel to the velocity $\mathbf{v}$ in a moving frame are the same in both moving and rest frame:

$$E'_\parallel = E_\parallel = E \cos\theta$$

Here E is the magnitude of the total electric field and θ is the angle of the total electric field to the velocity $\mathbf{v}$ of the moving frame. For the question, the total electric field $\mathbf{E}$ is directed along the vector $\mathbf{r}$ from $d\mathbf{l}$ to the position of the charge. The charge q is moving parallel to the axis of the wire element at velocity v. In the frame of the charged particle, the electrons have a velocity $v + v_e$ if the current is directed in the same direction as the velocity $\mathbf{v}$. The atoms which are stationary in the laboratory frame have a relative velocity of v to the charge. The magnitude of the electric field from the electrons in the wire experienced by the charge is $E' \sin\theta/\gamma_e + E' \cos\theta$, where the Lorentz factor is given by

$$\gamma_e = \left(\frac{1}{1 - (v + v_e)^2/c^2} \right)^{1/2}.$$

The magnitude of the electric field from the ions in the wire as experienced by the charge is $E' \sin\theta/\gamma_a + E' \cos\theta$. We then have a Lorentz factor

$$\gamma_a = \left(\frac{1}{1 - v^2/c^2} \right)^{1/2}.$$

The electron charge in the element dl of wire is $N_e e dl$ so that in the frame of the electrons in the wire, the electric field at a distance r from the wire due to the electrons has a magnitude given by

$$E' = \frac{1}{4\pi\epsilon} \frac{N_e e dl}{r^2}.$$

The electric field due to the positively charged atoms in the wire in the frame of the atoms has a similar magnitude because the charge on ions in the wire equals the charge on the electrons. However, the electric field due to the ions is opposite in sign. The attractive electrostatic force between the charge and the electrons is then

$$dF_- = -qE'(\sin\theta/\gamma_e + \cos\theta) = -\frac{1}{4\pi\epsilon}\frac{qN_e edl}{r^2}\left[\left(1 - \frac{(v+v_e)^2}{c^2}\right)^{1/2}\sin\theta + \cos\theta\right].$$

The repulsive electrostatic force between the charge and the ions is

$$dF_+ = qE'(\sin\theta/\gamma_a + \cos\theta) = \frac{1}{4\pi\epsilon}\frac{qN_e edl}{r^2}\left[\left(1 - \frac{v^2}{c^2}\right)^{1/2}\sin\theta + \cos\theta\right].$$

This is the answer to part [a].

For small electron velocities $v_e \ll c$ and small particle velocities $v \ll c$:

$$\left(1 - \frac{(v+v_e)^2}{c^2}\right)^{1/2} \approx 1 - \frac{1}{2}\frac{(v+v_e)^2}{c^2} = 1 - \frac{1}{2}\frac{v^2}{c^2} - \frac{1}{2}\frac{v_e^2}{c^2} - \frac{v\,v_e}{c^2} \approx 1 - \frac{1}{2}\frac{v^2}{c^2} - \frac{v\,v_e}{c^2},$$

while

$$\left(1 - \frac{v^2}{c^2}\right)^{1/2} \approx 1 - \frac{1}{2}\frac{v^2}{c^2},$$

so that

$$\left(1 - \frac{(v+v_e)^2}{c^2}\right)^{1/2} - \left(1 - \frac{v^2}{c^2}\right)^{1/2} \approx -\frac{v\,v_e}{c^2}.$$

The net attractive force then becomes

$$dF = dF_+ + dF_- = -\frac{1}{4\pi\epsilon}\frac{N_e edl}{r^2}\frac{v\,v_e}{c^2}q\sin\theta.$$

As $I_C = N_e e v_e$ and $1/c^2 = \epsilon\mu_0$, we obtain the required answer for part [b]:

$$dF = -\frac{\mu_0}{4\pi}\frac{I_C\,dl}{r^2}qv\,\sin\theta$$

For part [c], we now relate the force calculated as above to the Lorentz force $dF = qv \times d\mathbf{B}$. If the magnetic field $d\mathbf{B}$ is perpendicular to the charged particle velocity $\mathbf{v}$ and the current and particle velocity are in the same direction, we have that

$$dF = -qv\,dB.$$

As the force $d\mathbf{F}$ is directed parallel to the vector $\mathbf{r}$ and $d\mathbf{F} = qv \times d\mathbf{B}$, the magnetic field $d\mathbf{B}$ is also normal to $\mathbf{r}$. Comparing $dF = -qv\,B$ to the above (part [b]) expression for dF indicates that

$$dB = \frac{\mu_0}{4\pi}\frac{I_C\,dl}{r^2}\sin\theta$$

This answer for dB is a version of the Biot-Savart equation. As $|d\mathbf{l} \times \hat{\mathbf{r}}| = dl \sin \theta$ and the magnetic field is normal to $\mathbf{r}$, a small element of length $d\mathbf{l}$ carrying a current I_C creates a magnetic field $d\mathbf{B}$ at a position vector $\mathbf{r}$ such that

$$d\mathbf{B} = \frac{\mu_0}{4\pi} \frac{I_C \, d\mathbf{l} \times \hat{\mathbf{r}}}{r^2}, \tag{10.7}$$

where $\hat{\mathbf{r}}$ is a unit vector in the direction of $\mathbf{r}$.

The Biot-Savart equation is useful for determining magnetic fields near current carrying structures where lack of symmetry makes the direct use of Maxwell's equations difficult. A current flowing along any path such as a solid wire can be considered as many infinitesimally small elements creating small increments of magnetic field which can be summed or integrated.

Appendix A

Lorentz Invariant Quantities

Lorentz invariant quantities remain the same in any frame of reference. It is useful to know that a quantity does not change between frames of reference as often a relativity problem can be expressed in terms of a Lorentz invariant quantity. We may also just want to reassure ourselves that a measured quantity is the same, for example, in the laboratory frame of reference as in the frame of reference of a light emitting source. A list of Lorentz invariant quantities is presented in Table A.1.

Table A.1 *Lorentz invariant quantities. The velocity between frames of reference is* **v**. *The section where the invariance is first discussed is listed along with the symbols used.*

Description	Symbols	Section
Absorption coefficient/frequency	K/ν	7.1
Azimuthal angles around **v**	ϕ	3.5
Charge conservation	$\nabla \cdot \mathbf{J} + \partial \rho_c / \partial t = 0$	3.2
Collisional cross-section	σ	8.2
Collisional impact parameter	b	8.2
Distance perpendicular to **v**	y or z	1.2
Einstein coefficient ratios	$A_{21}/B_{12}, A_{21}/B_{21}$	7.2
Electric field parallel to **v**	$E_{\parallel}$	3.2
Emission coefficient/frequency^2	ϵ/ν^2	7.1
Energy^2 – Momentum^2	$W^2 - (pc)^2 = (m_0 c^2)^2$	2.8
Force parallel to **v**	$F_{\parallel}$	3.3
Intensity/frequency^3	I/ν^3	7.1
Light pulse in spacetime	$-c^2 t^2 + x^2 + y^2 + z^2$	2.2
Lineshape function times frequency	$f(\nu)\nu$	7.3
Magnetic field parallel to **v**	$B_{\parallel}$	3.2
Number density photons/energy/volume	$N(E_p)$	7.9
Optical depth	τ	7.1
Phase of light	$\mathbf{k} \cdot \mathbf{r} - \omega t$	3.1
Photon no. emitted/unit volume	$N_2 A_{21}$	7.3
Polarization of light (Stokes parameters)	$Q_s/I_s, U_s/I_s, V_s/I_s$	9.4
Power (energy/unit time)	P	7.9
Rest mass	m_0	2.8
Rest mass energy	$m_0 c^2$	2.8
Source function/frequency^3	S_f/ν^3	7.1
Speed of light in vacuum	c	1.2
Wave equation	$-(1/c^2)\partial^2/\partial t^2 + \nabla^2$	3.2
Wavevector perpendicular to **v**	k_y or k_z	3.1

Appendix B

Modified Bessel Functions of the Second Kind

Bessel functions are solutions of a particular class of second order differential equation. They are useful in some complicated integrations as their values are tabulated and numerical packages for their computational evaluation are available. Bessel functions also have convenient asymptotic expressions for the production of approximate formulas.

The modified Bessel function of the second kind of order α and argument X is conventionally represented by the symbols $K_\alpha(X)$. For α and X real with $\alpha > -1/2$, the modified Bessel function of the second kind is a solution to integrals of form[1]:

$$K_\alpha(X) = \frac{\sqrt{\pi}(2X)^\alpha e^{-X}}{\Gamma(\alpha + 1/2)} \int_{x=0}^{\infty} x^{\alpha-1/2}(x+1)^{\alpha-1/2} \, e^{2Xx} dx, \tag{B.1}$$

where $\Gamma(\alpha + 1/2)$ is the gamma function which for a positive integer α is calculated using

$$\Gamma(\alpha + 1/2) = \frac{(2\alpha)!}{4^\alpha \alpha!}\sqrt{\pi}. \tag{B.2}$$

For large X arguments of the Bessel function $K_\alpha(X)$, the following asymptotic expansion is accurate:

$$K_\alpha(X) \approx \sqrt{\frac{\pi}{2X}} e^{-X} \left(1 + \frac{4\alpha^2 - 1}{8X} + \frac{(4\alpha^2 - 1)(4\alpha^2 - 9)}{2(8X)^2} + \cdots \right) \tag{B.3}$$

The modified Bessel function of the second kind arises in the integration of the three-dimensional Maxwell-Juttner expression for thermalized relativistic particles (Section 6.6), the evaluation of the chemical potential in equation of state determinations (Section 6.7), the calculation of relativistic bremsstrahlung (Section 7.8) and in the determination of relativistic rate coefficients (Section 8.4). We use the modified Bessel function of the second kind and order two evaluated at an inverse

---

[1] See Equation 9.44 of [111], but note the erratum that $(2/X)^\alpha$ should be $(2X)^\alpha$ as written in Equation B.1. Other mathematical relationships used in this appendix can be found at https://dlmf.nist.gov/

temperature value of m_0c^2/k_BT in each of these cases, where m_0c^2 is the rest mass energy of the particles under consideration.

The modified Bessel function of the second kind and order two evaluated at an inverse temperature of m_0c^2/k_BT takes on the form

$$K_2(m_0c^2/k_BT) = \frac{m_0c^2}{k_BT} \int_{\gamma=1}^{\infty} \gamma \left(\gamma^2 - 1\right)^{1/2} \exp\left(-\frac{\gamma m_0c^2}{k_BT}\right) d\gamma. \qquad (B.4)$$

When $k_BT < m_0c^2$, from Equation B.3 we have an approximate value for the Bessel function given by

$$K_2(m_0c^2/k_BT) \approx \frac{\sqrt{\pi}}{2} \left(\frac{2k_BT}{m_0c^2}\right)^{1/2} \exp\left(-\frac{m_0c^2}{k_BT}\right) \left[1 + \frac{15}{8}\frac{k_BT}{m_0c^2}\right]. \qquad (B.5)$$

Figure B.1 shows a comparison between a precise solution of Equation B.4 and the value of the approximation (Equation B.5). We see that the approximation is accurate for $k_BT < m_0c^2$.

A modified Bessel function of the second kind and order one was used to consider relativistic Doppler broadening in Section 7.6. The modified Bessel function of the second kind and order one evaluated at an inverse temperature of m_0c^2/k_BT takes on the form

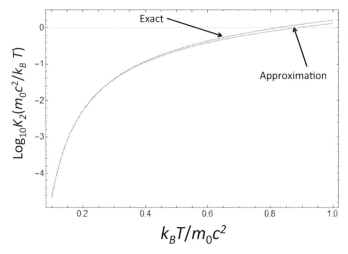

Figure B.1   The modified Bessel function of the second kind and order two with an inverse temperature argument as a function of temperature. The Bessel function is plotted accurately (Equation B.4) along with an asymptotic approximation (Equation B.5). The vertical axis showing the Bessel function is on a logarithmic scale. The horizontal axis is the particle temperature k_BT in units of the particle rest mass energy m_0c^2.

$$K_1(m_0c^2/k_BT) = \int_{\gamma=1}^{\infty} \gamma \left(\gamma^2 - 1\right)^{-1/2} \exp\left(-\frac{\gamma m_0c^2}{k_BT}\right) d\gamma. \qquad (B.6)$$

For large m_0c^2/k_BT, higher order terms in α become small for the modified Bessel function of the second kind and order one (see Equation B.3 with $\alpha = 1$). The modified Bessel function of the second kind and order one for large m_0c^2/k_BT is approximately given by

$$K_1(m_0c^2/k_BT) \approx \frac{\sqrt{\pi}}{2} \left(\frac{2k_BT}{m_0c^2}\right)^{1/2} \exp\left(-\frac{m_0c^2}{k_BT}\right). \qquad (B.7)$$

A comparison of this approximation asymptotic value to an accurate calculation of the modified Bessel function of order one is shown in Figure 7.4.

The modified Bessel function of the second kind and order three is considered in our treatment of the Compton opacity of plasma to a beam of light (Section 7.11). We have

$$K_3(m_0c^2/k_BT) = (1/3) \left(\frac{m_0c^2}{k_BT}\right)^2 \int_{\gamma=1}^{\infty} \gamma(\gamma^2 - 1)^{3/2} \exp\left(-\frac{\gamma m_0c^2}{k_BT}\right) d\gamma. \qquad (B.8)$$

The three modified Bessel functions of the second kind considered in our treatment are accurately plotted in Figure B.2 for different temperatures. At temperatures $k_BT \ll m_0c^2$, the Bessel functions of different order are approximately equal. Factors of approximately three differences in the Bessel function values of different adjacent α orders are apparent for $k_BT \approx m_0c^2$.

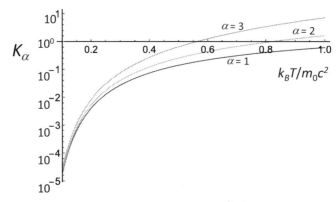

Figure B.2   The modified Bessel function $K_\alpha(m_0c^2/k_BT)$ of the second kind and order $\alpha = 1$, 2, and 3 as a function of temperature. The horizontal axis is the particle temperature k_BT in units of the particle rest mass energy m_0c^2.

Appendix C

The Chemical Potential Variation with Density

An equilibrium particle distribution function giving the number density of particles as a function of velocity, momentum, or energy can be specified in terms of the temperature and the total electron density, or equivalently to the total electron density, by the chemical potential. The chemical potential is defined as the energy required to add another particle to the distribution of particle energies. In the absence of near-full occupancy for free particle quantum states, the chemical potential is large and negative as particles can readily occupy low energy states when added to an ensemble of particles. Quantum states that are fully occupied by fermions according to the Pauli exclusion principle are referred to as being "degenerate." With full occupancy, the number of particles in a quantum state is equal to the degeneracy of the quantum state. For distributions with lower energy degenerate quantum states, an extra particle added to the distribution may need to go into a higher energy quantum state, and then the chemical potential can be positive.

An expression for the electron density n_e as a function of the chemical potential μ for an equilibrium distribution of free electrons with temperature $k_B T$ and chemical potential μ was determined in Section 6.3 when the relativistic Saha-Boltzmann equation was evaluated. Equation 6.30 relates the total electron density n_e and the chemical potential μ when degeneracy and relativistic effects are important:

$$n_e = \frac{8\pi}{h^3} \frac{(m_0 c^2)^3}{c^3} \int_{\gamma=1}^{\infty} \frac{(\gamma^2 - 1)^{1/2} \gamma}{1 + \exp((-\mu + (\gamma - 1)m_0 c^2)/k_B T)} d\gamma \qquad \text{(C.1)}$$

The integration is over the Lorentz parameter γ as the kinetic energy of particles is given by $(\gamma - 1)m_0 c^2$, where $m_0 c^2$ is the rest mass energy (511 keV) for electrons. Equation C.1 can be readily generalized to fermions other than electrons with an appropriate adjustment of the rest mass energy $m_0 c^2$. Values of total electron density n_e as a function of chemical potential μ are plotted using Equation C.1 in Figure C.1. It is apparent that up to extremely high electron densities (e.g. $n_e < 10^{32}$ m^{-3} at temperatures $k_B T > 500$ eV), the chemical potential μ is negative.

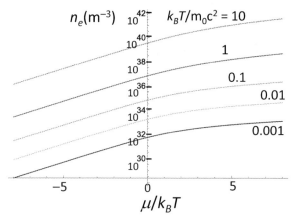

Figure C.1   Electron density n_e as a function of the chemical potential in units of electron temperature (μ/k_BT) for different electron temperatures in units of the electron rest mass energy (k_BT/m_0c^2) with $m_0c^2 = 511$ keV.

References

[1] Alberto, P., Das, S., and Vagenas, E. C. 2018. Relativistic particle in a box: Klein-Gordon versus Dirac equations. *Eur. J. Phys.*, **39**, 025401.

[2] Alexanian, M. 1968. Photon bremsstrahlung from an extreme-relativistic electron gas. *Phys. Rev.*, **165**, 253–257.

[3] Aniculaesei, C., Pathak, V. B., Kim, H. T., et al. 2019. Electron energy increase in a laser wakefield accelerator using up-ramp plasma density profiles. *Sci. Rep.*, **9**, 11249.

[4] Babzien, M., Ben-Zvi, I., Kusche, K., et al. 2006. Observation of the second harmonic in Thomson scattering from relativistic electrons. *Phys. Rev. Lett.*, **96**, 054802.

[5] Beesley, J. J., and Rose, S. J. 2019. Free electron relativistic correction factor to collisional excitation and ionisation rates in a plasma. *HEDP*, **33**, 100716.

[6] Bell, A. R. 1978. The acceleration of cosmic rays in shock fronts. *Mon. Not. R. Ast. Soc.*, **182**, 147–156.

[7] Bell, A. R. 2012. Cosmic ray acceleration. *Astroparticle Phys.*, **43**, 56–70.

[8] Bernstein, J. 2001. *Hitler's Uranium Club: The secret recordings at Farm Hall.* Springer-Verlag, New York.

[9] Bird, D. J., Corbato, S. C., Dai, H. Y., et al. 1995. Detection of a cosmic ray with measured energy well beyond the expected spectral cut off due to cosmic microwave radiation. *Astrophys. J.*, **441**, 144.

[10] Blackburn, T. G., Ridgers, C. P., Kirk, J. G., and Bell, A. R. 2014. Quantum radiation reaction in laser-electron beam collisions. *Phys. Rev. Lett.*, **112**, 015001.

[11] Blumenthal, G. R., and Gould, R. J. 1970. Bremsstrahlung, synchrotron radiation and Compton scattering of high energy electrons traversing dilute gases. *Rev. Mod. Phys.*, **42**, 237–270.

[12] Brian, D. 1996. *Einstein: A life.* Wiley, New York.

[13] Capdessus, R., King, M., Sorbo, D. Del, et al. 2018. Relativistic Doppler-boosted γ-rays in high fields. *Nature Sci. Rep.*, **8**, 9115.

[14] Carlstrom, J. E., Holder, G. P., and Reese, E. D. 2002. Cosmology with the Sunyaev-Zel'dovich effect. *Ann. Rev. Astron. Astrophys.*, **40**, 643–680.

[15] Chalmers, M. 2011. Freeze framing the atomic world. *Phys. World Supplement: Big Science*, **Oct.**, 19–20.

[16] Chen, S., Maksimchuk, A., and Umstadter, D. 1998. Experimental observation of relativistic nonlinear Thomson scattering. *Nature*, **17**, 653–655.

[17] Clemmow, P. C., and Dougherty, J. P. 1990. *Electrodynamics of particles and plasmas.* Addison-Wesley Inc., Redwood City, CA.

[18] Clery, D. 2011. Magnet challenges for ITER. *Phys. World Supplement: Big Science*, **Oct.**, 9–10.

[19] Cocke, W. J., and Holm, D. A. 1972. Lorentz transformation properties of the Stokes parameters. *Nature*, **240**, 161–162.

[20] Cole, J. M., Behm, K. T., Gerstmayr, E., et al. 2018. Experimental evidence of radiation reaction in the collision of a high-intensity laser pulse with laser-wakefield accelerated electron beam. *Phys. Rev. X*, **8**, 011020.

[21] Colgan, J., Kilcrease, D. P., Magee, N. H., et al. 2016. A new generation of Los Alamos opacity tables. *Astrophys. J.*, **817:116**, 1–10.

[22] Collaboration, HESS. 2016. Acceleration of petaelectronvolt protons in the galactic centre. *Nature*, **531**, 476–479.

[23] Collaboration, The Event Horizon Telescope, et al. 2021. First M87 event Horizon telescope results. VII polarization of the ring. *Astrophys. J. Lett.*, **910**, L12.

[24] Cui, Y. Q., Wang, W. M., Sheng, Z. M., Li, Y. T., and Zhang, J. 2013. Laser absorption and hot electron temperature scalings in laser–plasma interactions. *Plasma Phys. Cont. Fus.*, **55**, 085008.

[25] Dirac, P. A. M. 1926. On the theory of quantum mechanics. *Proc. R. Soc. A*, **112**, 661–677.

[26] Dirac, P. A. M. 1978. *Directions in physics: Lectures delivered during a visit to Australia and New Zealand August/September 1975*. Wiley, New York.

[27] Dunkel, J., Hanggi, P., and Hilbert, S. 2009. Non-local observables and lightcone-averaging in relativistic thermodynamics. *Nature Phys.*, **5**, 741–747.

[28] Dyson, F. W., Eddington, A. S., and Davidson, C. 1920. IX. A determination of the deflection of light by the Sun's gravitational field from observation made at the total eclipse of May 29, 1919. *Phil. Trans. R. Soc. Lond.*, **220**, 291–335.

[29] Dyson, G. 2003. *Project Orion: The atomic spaceship*. Penguin, London.

[30] Easwar, N., and MacIntire, D. A. 1991. Study of the effect of relativistic time dilation on cosmic ray muon flux: An undergraduate modern physics experiment. *Am. J. Phys.*, **59**, 589–592.

[31] Eglert, B. G., Scully, M. O., and Walther, H. 1994. The duality in matter and light. *Scientific Am.*, 88–92.

[32] Einstein, A. 1905a. Ist die Trägheit eines Körpers von seinem Energiegehalt abhängig? (Does the inertia of a body depend upon its energy content?). *Ann. Physik*, **18**, 639– 641.

[33] Einstein, A. 1905b. Über die von der molekularkinetischen Theorie der Wärme geforderte Bewegung von in ruhenden Flüssigkeiten suspendierten Teilchen (Investigations on the theory of Brownian movement). *Ann. Physik*, **322**, 549–560.

[34] Einstein, A. 1905c. Über einen die Erzeugung und Verwandlung des Lichtes betreffenden heuristischen Gesichtspunkt (On a Heuristic Point of View about the Creation and Conversion of Light). *Ann. Physik*, **17**, 132– 148.

[35] Einstein, A. 1905d. Zur Elektrodynamik bewegter Körper (On the electrodynamics of moving bodies). *Ann. Phys.*, **17**, 891–921.

[36] Einstein, A. 1916. Strahlungs-Emission und -Absorption nach der Quantentheorie (Light emission and absorption after quantum theory). *Verhandlungen der Deutschen Physikalischen Gesellschaft*, **18**, 318–323.

[37] Einstein, A. 1917. Zur Quantentheorie der Strahlung (On the quantum theory of radiation). *Physikalishche Zeitschrift*, **18**, 121–128.

[38] Fermi, E. 1926. Sulla quantizzazione del gas perfetto monoatomico (Quantization of the monoatomic ideal gas). *Rendiconti Lincei*, **3**, 145–149.

[39] Fermi, E. 1949. On the origin of cosmic radiation. *Phys. Rev*, **75**, 1169–1174.

[40] Feynman, Richard P., Leighton, R. B., and Sands., M. L. 1963. *The Feynman lectures on physics*. Addison-Wesley Pub. Co., Reading, Mass.

[41] Fixsen, D. J. 2009. The temperature of the cosmic microwave background. *Astrophys. J.*, **707**, 916–920.

[42] Fizeau, H. 1851. The hypotheses relating to the luminous aether, and an experiment which appears to demonstrate that the motion of bodies alters the velocity with which light propagates itself in their interior. *Phil. Mag.*, **2**, 568–573.

[43] Fruhling, C., Wang, J., Umstadter, D., Schulzke, C., Romero, M., Ware, M., and Peatross, J. 2021. Experimental observation of polarization-resolved nonlinear Thomson scattering of elliptically polarized light. *Phys. Rev. A*, **104**, 053519.

[44] Gaisser, T. K., Engel, R., and Resconi, E. 2016. *Cosmic rays and particle physics (2nd edition)*. Cambridge University Press, Cambridge.

[45] Gelfand, I. M., and Fomin, S. V. 2000. *Calculus of variations*. Dover Publications Inc., Mineola.

[46] Gerbal, D., and Prud'Homme, M. 1974. Some remarks about relativistic line profiles. *J. Quant. Spect. Rad. Trans.*, **14**, 351–356.

[47] Gibbon, P. 2005. *Short pulse laser interaction with matter: An introduction*. Imperial College Press, London.

[48] Greisen, K. 1966. End to the cosmic ray spectrum? *Phys. Rev. Lett.*, **16**, 748–750.

[49] Henke, B. L., Gullison, E. M., and Davis, J. C. 1993. X-ray interaction: Photoabsorption, scattering, transmission and reflection at $E = 50 - 30000$ eV, $Z = 1–92$. *At. Data Nucl. Data Tables*, **54**, 181–342.

[50] Hestenes, D. 1990. The zitterbewegung interpretation of quantum mechanics. *Foundations of Physics*, **20**, 1213–1232.

[51] Hooker, S. M. 2013. Developments in laser-driven plasma accelerators. *Nature photonics*, **7**, 775–787.

[52] Huang, Y. S. 2020. The classical Doppler-broadened absorption line profile due to thermal effect. *Eur. Phys. Lett.*, **131**, 53002.

[53] Huang, Y. S, Chiue, J. H., Huang, Y. C., and Hsiung, T. C. 2010. Relativistic formulation for the Doppler-broadened line profile. *Phys. Rev. A*, **82**, 010102.

[54] Hughes, T. P. 1962. A new method for the determination of plasma electron temperature and density from Thomson scattering of an optical maser beam. *Nature*, **194**, 268–269.

[55] Hutchinson, I. H. 2002. *Principles of plasma diagnostics*. Cambridge University Press, Cambridge.

[56] Iglesias, C. 1996. Comment on "relativistic corrections to inverse bremsstrahlung in the solar interior" by Tsytovich et al. *Phys. Lett.*, **213**, 313–315.

[57] Iglesias, C. A., and Rose, S. J. 1996. Corrections to bremsstrahlung and Thomson scattering at the solar center. *Astrophys. J.*, **466**, L115–L118.

[58] Jackson, J. D. 1962. *Classical dynamics*. Wiley, New York.

[59] Joshi, C., Mori, W. B., Katsouleas, T., Dawson, J. M., Kindel, J. M., and Forslund, D. W. 1984. Ultrahigh gradient particle acceleration by intense laser-driven plasma density waves. *Nature*, **311**, 525–529.

[60] Kando, M., Pirozhkov, A. S., Kawase, K., et al. 2009. Enhancement of photon number reflected by the relativistic flying mirror. *Phys. Rev. Lett.*, **103**, 235003.

[61] Kando, M., Esirkepov, T. Z., Koga, J. K., Pirozhkov, A. S., and Bulanov, S. V. 2018. Coherent short pulse X-ray generation via relativistic flying mirrors. *Quant. Beam Sci.*, **2, 9**, 1–11.

[62] Klein, O., and Nishina, Y. 1929. Über die Streuung von Strahlung durch freie Elektronen nach der neuen relativistischen Quantendynamik von Dirac (On scattering of

radiation by a free electron using the new quantum electrodynamics of Dirac). *Z. Physik*, **52**, 853–868.

[63] Landau, L. D., and Lifshitz, E. M. 1971. *The classical theory of fields*. Pergamon, Oxford.

[64] Leemans, W. P., Gonsalves, A. J, Nakamura, K., Mao, H. S., et al. 2014. Multi-GeV electron beams from capillary-discharge-guided subpetatwatt laser pulses in the self-trapping regime. *Phys. Rev. Lett.*, **113**, 245002.

[65] Levington, F. M., Fonck, R. J., Gammel, G. M., Kaita, R., Kugel, H. W., Powell, E. T., and Roberts, D. W. 1989. Magnetic field pitch-angle measurements in the PBX-M tokamak using the motional Stark effect. *Phys. Rev. Lett.*, **63**, 2060–2063.

[66] Lewis, G. N. 1926. The conservation of photons. *Nature*, **118**, 874–875.

[67] Lindl, J. 1995. Development of the indirect-drive approach to inertial confinement fusion and the target physics basis for ignition and gain. *Phys. Plasmas*, **2**, 3933–4024.

[68] Lindl, J. D., Amendt, P., Berger, R. L., Glendinning, S. G., Glenzer, S. H., Haan, S. W., Kauffman, R. L., Landen, O. L., and Suter, L. J. 2004. The physics basis for ignition using indirect-drive targets on the National Ignition Facility. *Phys. Plasmas*, **11**, 339.

[69] Lotz, W. 1967. An empirical formula for the electron-impact ionization cross-section. *Z. Physik*, **206**, 205.

[70] Lyutikov, M., Pariev, V. I., and Blandford, R. D. 2003. Polarization of prompt gamma-ray burst emission: Evidence for electromagnetically dominated outflow. *Astrophys. J.*, **597**, 998–1009.

[71] Macchietto, C. D., Benware, B. R., and Rocca, J. J. 1999. Generation of millijoule-level soft-X-ray laser pulses at a 4-Hz repetition rate in a highly saturated tabletop capillary discharge amplifier. *Optics Lett.*, **24**, 1115–1117.

[72] Macintyre, Ben. 2021 (Nov. 6). "Führers of physics" shame the Nobel prize. *The Times* London.

[73] Mangles, S. P. D., Murphy, C. D., Najmudin, Z., Thomas, A. G. R., et al. 2004. Monoenergetic beams of relativistic electrons from intense laser-plasma interactions. *Nature*, **431**, 535–538.

[74] Mangles, S. P. D., Thomas, A. G. R., Kaluza, M. C., Lundh, O., Lindau, F., Persson, A., Tsung, F. S., Najmudin, Z., Mori, W. B., Wahlstrom, C.-G., and Krushelnick, K. 2006. Laser wakefield acceleration of monoenergetic electron beams in the first plasma-wave period. *Phys. Rev. Lett.*, **96**, 215001.

[75] Maxon, S. 1972. Bremsstrahlung rate and spectra from a hot gas ($Z = 1$). *Phys. Rev. A*, **5**, 1630–1633.

[76] Maxwell, James Clerk. 1865. A dynamical theory of the electromagnetic field. *Phil. Trans. Roy. Soc.*, **155**, 459–512.

[77] Meade, D. 2009. 50 years of fusion research. *Nucl. Fus*, **50**, 014004.

[78] Meinecke, J., Doyle, H. W., Miniati, F., Bingham, R., et al. 2014. Turbulent amplification of magnetic field in laboratory laser-produced shock waves. *Nature Phys.*, **7**, 520–524.

[79] Moore, W. J. 2015. *Schrödinger: Life and thought*. Cambridge University Press, Cambridge.

[80] Nuckolls, J., Wood, L., Thiessen, A., and Zimmerman, G. 1972. Laser compression of matter to super-high densities: Thermonuclear (CTR) applications. *Nature*, **239**, 139–142.

[81] Oerter, R. 2006. *The theory of almost everything: The Standard model, the unsung triumph of modern physics*. Penguin, New York.

[82] Olson, R. E., Rochau, G. A., Landen, O. L, and Leeper, R. J. 2011. X-ray ablation rates in inertial confinement fusion capsule materials. *Phys. Plasmas*, **18**, 032706.

[83] Oxenius, J. 1986. *Kinetic theory of particles and photons: Theoretical foundations of non-LTE plasma spectroscopy*. Springer-Verlag, Berlin.

[84] Pais, A. 1997. *A tale of two continents: A physicist's life in a turbulent world*. Princeton University Press, Princeton.

[85] Passoni, M., Bertagna, L., and Zani, A. 2010. Target normal sheath acceleration: Theory, comparison with experiments and future objectives. *New J. Phys.*, **12**, 045012.

[86] Paterniti, M. 2000. *Driving Mr. Albert: A trip across America with Einstein's brain*. Dial Press (Random House), New York.

[87] Pellegrini, C., Marinelli, A., and Reiche, S. 2016. The physics of X-ray free-electron lasers. *Nature Sci. Rep.*, **88**, 015006.

[88] Pert, G. J. 2013. *Introductory fluid mechanics for physicists and mathematicians*. Wiley, Oxford.

[89] Pert, G. J. 2021. *Foundations of Plasma Physics for Physicists and Mathematicians*. Wiley, Oxford.

[90] Pirozhkov, A. S., Bulanov, S. V., Esirkepov, T. Z., Mori, M., Sagisaki, A., and Daido, H. 2006. Attosecond pulse generation in the relativistic regime of the laser-foil interaction: The sliding mirror model. *Phys. Plasmas*, **13**, 013107.

[91] Pitaevskii, L., and Stringari, S. 2003. *Bose-Einstein condensation*. Clarendon, Oxford.

[92] Planck, Collaboration. 2020. Planck 2018 results 1. Overview and the cosmological legacy of Planck. *Astron. Astrophys*, **641**, 916–920.

[93] Pukhov, A. 2002. Strong field interaction of laser radiation. *Rep. Prog. Phys.*, **66**, 47–101.

[94] Purcell, E. M. 1985. *Electricity and magnetism*. McGraw-Hill, New York.

[95] Quesnel, B., and Mora, P. 1999. Simulation of the interaction of ultraintense laser pulses with electrons in vacuum. *Phys. Rev. E*, **58**, 3719–3732.

[96] Quigg, C. 1968. Relativistic correction to plasma bremsstrahlung. *Phys. Fluids*, **11**, 461–462.

[97] Repahei, Y. 1995. Comptonization of the cosmic microwave background: The Sunyaev-Zel'dovich effect. *Ann. Rev. Astron. Astrophys.*, **33**, 541–579.

[98] Robinson, A. 2019. *Einstein on the run: How Britain saved the world's greatest scientist*. Yale University Press, New Haven.

[99] Rose, S. J., and Hatfield, P. W. 2021. Astronomy domine: Advancing science with a burning plasma. *Contemp. Phys.*, doi:10.1080/00107514.2021.1959097.

[100] Rothman, T. 2015. Did Einstein really invent $E = mc^2$? *Sci. American*, **313**, 3.

[101] Rybicki, G. B., and Lightman, A. P. 1979. *Radiative processes in astrophysics*. Wiley, New York.

[102] Sampson, D. H., and Zhang, H. L. 1992. Use of the Van Regemorter formula for collision strengths or cross sections. *Phys. Rev. A*, **45**, 1556–1561.

[103] De Sanctis, E. De, Monti, S., and Ripani, M. 2016. *Energy from nuclear fission: An introduction*. Springer, Basel.

[104] Schekochihin, A. A., and Cowley, S. C. 2007. Turbulence and magnetic fields in astrophysical plasmas. In: Magnetohydrodynamics: *Fluid mechanics and its applications*. Springer, Dordrecht.

[105] Sheffield, J., Froula, D., Glenzer, S. H., and Luhmann, N. C. 2011. *Plasma scattering of electromagnetic radiation: Theory and measurement techniques*. Academic Press, Amsterdam.

[106] S. Kichenassamy, Krikorian, R., and Nikoghossian, A. 1982. The relativistic Doppler broadening of the line absorption profile. *J. Quant. Spect. Rad. Trans.*, **27**, 635–655.

[107] Spitzer, L. 1962. *Physics of fully ionized gases*. Wiley, New York.

[108] Sunyaev, R. A., and Zel'dovich, Y. B. 1980. Microwave background radiation as a probe of the contemporary structure and history of the universe. *Ann. Rev. Astron. Astrophys.*, **18**, 537–560.

[109] Tabak, M., Clark, D. S., Hatchett, S. P., Key, M. H., et al. 2005. Review of progress on fast ignition. *Phys. Plasmas*, **12**, 057305.

[110] Tallents, G. J. 2018. *An introduction to the atomic and radiation physics of plasmas*. Cambridge University Press, Cambridge.

[111] Temme, N. M. 1996. *Special functions: An introduction to the classical functions of mathematical physics*. Wiley-Interscience, New York.

[112] Umstadter, D. 2003. Relativistic laser-plasma interactions. *J. Phys. D.*, **36**, R151–R165.

[113] van Regemorter, H. 1962. Rate of collisional excitation in stellar atmospheres. *Astrophys. J.*, **A132**, 906.

[114] Wilks, S. C., Kruer, W. L., Tabak, M., and Langdon, A. B. 1992. Absorption of ultra-intense laser pulses. *Phys. Rev. Lett.*, **69**, 1383–1386.

[115] Willmott, P. 2019. *An introduction to synchrotron radiation*. Wiley, Hoboken, USA. Hoboken.

[116] Winick, H., and Doniach, S. 1980. *Synchrotron radiation research*. Plenum, New York.

[117] Wolf, E. C., Bock, A., Ford, O. P., Reimer, R., Burckhart, A., Dinklage, A., Hobirk, J., Howard, J., Reich, M., and Stober, J. 2015. Motional Stark effect measurements of the local magnetic field in high temperature fusion plasmas. *JINST*, **10**, P10008.

[118] Yin, L., Albright, B. J., Hegelich, B. M., and Fernandez, J. C. 2006. GeV laser ion acceleration from ultrathin targets: The laser break-out afterburner. *Laser Particle Beams*, **24**, 291–298.

[119] Yin, L., Albright, B. J., Bowers, D., Jung, D., Fernandez, J. C., and Hegelich, B. M. 2011. Three-dimensional dynamics of breakout afterburner ion acceleration using high-contrast short-pulse laser and nanoscale targets. *Phys. Rev. Lett.*, **107**, 045003.

[120] Zekovic, V., Arbutina, B., Dobardzic, A., and Pavlovic, M. 2013. Relativistic nonthermal bremsstrahlung radiation. *Int. J. Mod. Phys.*, **29**, 1350141.

[121] Zepf, M., Tsakisi, G. D., Pretzler, G., Watts, I., et al. 1998. Role of the plasma scale length in the harmonic generation from solid targets. *Phys. Rev. E*, **58**, R5253–R5256.

Index

315

Printed in the United States
by Baker & Taylor Publisher Services